ALLE ZEIT WACH
SJ
1842

W. Weidlich G. Haag (Eds.)

Interregional Migration

Dynamic Theory and Comparative Analysis

With Contributions by
Å. E. Andersson G. Haag I. Holmberg
J. Ledent M. Munz D. Pumain G. Rabino R. Reiner
N. Sarafoglou M. Sonis W. Weidlich

With 129 Figures and 64 Tables

Springer-Verlag
Berlin Heidelberg New York
London Paris Tokyo

Prof. Dr. Dr. h.c. Wolfgang Weidlich
Dr. Günter Haag

Institut für Theoretische Physik der Universität Stuttgart
Pfaffenwaldring 57/III, 7000 Stuttgart 80, FRG

ISBN-13: 978-3-642-73051-1 e-ISBN-13: 978-3-642-73049-8
DOI: 10.1007/978-3-642-73049-8

Library of Congress Cataloging-in-Publication Data
Interregional migration : dynamic theory and comparative analysis /
Wolfgang Weidlich, Günter Haag, eds. ; with contributions by Å. E. Andersson ... [et al.].
p. cm. Bibliography: p.

1. Migration, Interregional–Mathematical models. 2. Migration, Interregional–Case studies.
I. Weidlich, Wolfgang, 1931-. II. Haag, G. (Günter), 1948-. III. Andersson, Åke E., 1936-.
HB1952.I57 1988 304.8–dc 19 88-2002

Softcover reprint of the hardcover 1st edition 1988

Typesetting: Macmillan India Ltd.

Bookbinding: J. Schäffer GmbH & Co. KG., Grünstadt
2142/3140-543210

Preface

This book investigates a sector of dynamic processes in the society, namely interregional migration, in quantitative detail. It has some unusual characteristics, which should be explained to the reader at the outset.

Firstly, its content is the result of a close *international cooperation*. The editors are very pleased, that at the fortunate occasion of the Nato Advanced Studies Institute "Transformations Through Space and Time" in Hanstholm, Denmark (1985) the already existing friendly relations between scientists of six countries, namely Canada, Federal Republic of Germany, France, Israel, Italy and Sweden, could be transformed into an intense cooperation resulting in this book.

Secondly, the scope of the content is *interdisciplinary* in personal and objective respect as well: The authors are regional scientists, demographs and economists on the one side, and physicists on the other side. And correspondingly, the content comprizes theoretical and practical aspects ranging from the methods originating in Theoretical Physics to the socio-economics of population dynamics being central aspects of Regional Science.
Three main aspects have been treated:

a) The development of a dynamic migratory model connecting the microlevel of individual decisions with the macrolevel of the migratory process and making use of the Master Equation method well known in Statistical Physics.
b) The quantitative description and evaluation of interregional migration in different countries in terms of dynamic regional utilities and mobilities. (Here, economists should be informed, that our concept of dynamic utilities, although similar in spirit, must be distinguished from the concept of utility functions in economics!)

And finally

c) The interpretative and quantitative correlation of interregional migration with the socio-economic situation in these countries.

Having available the expert knowledge of the authors about the economic, demographic and migratory situation in their countries, and on the other hand the newly developed model of interregional migration, it was possible to approach the main purpose of our cooperation: The evaluation of migratory processes in several countries on the basis of the *same* migratory model with the intention of assuring full comparability of the results.

In order to achieve this aim, the book had to be organized in the style of a textbook with systematically ordered parts and chapters, and *not* in the style of Proceedings with only loosely connected contributions. Thus, the theoretical parts of the book, namely part I, General Theory, and part IV, Mathematical Methods, are directly and systematically applied to the concrete case studies of part II, which are structurized accordingly. The comparative point of view is taken up in part III, with one chapter comparing migratory models and one chapter comparing the results of the evaluations on the basis of *one* model for different countries.

As all insiders know, true interdisciplinary cooperation is a rare event, often prohibited by too rigid regulations on the side of sponsors. The more we are grateful to the Volkswagen Foundation, because without its help the project could not have been realized. The Volkswagen Foundation not only supported financially two of the co-authors, but also unbureaucratically provided the means for a workshop of the authors in Wüstenrot-Blindenmannshäusle, which proved to be indispensible for the preparation of the manuscript.

Particular thanks are also due to our secretary Mrs. Eva Effenberg, who typed the different versions of the main part of the complicated manuscript with tireless care and high precision. Our thanks also include the secretaries of all authors.

Furthermore we wish to thank Prof. Herwig Birg for his highly valuable comments and advices during the preparation of the chapters 4 and 11.

In view of the special way of its preparation and in view of the composition of the group of its authors we expect, that the audience of this book might consist of two groups: firstly and mainly the group of regional scientists and economists interested in the correlation between migration and the spatially differentiated socio-economic situation, and secondly the group of natural scientists interested in Synergetics and its interdisciplinary applications. Indeed, the development of new ideas has always profited from the discussions over the years about Synergetics and Social Science with Hermann Haken, the friend of one of us (W. W.), to whom go special thanks.

For these reasons, the book appears as a monography belonging to the field of Quantitative Methods and Models in Social Science, but it should be seen on the other hand, to be related to the field of Synergetics.

Last but not least we wish to thank Dr. Werner A. Müller of the Springer-Verlag for undertaking the publication of this book in the approved well-known quality.

Spring 1988 — Wolfgang Weidlich and Günter Haag

List of Authors

Prof. Dr. Åke E. Andersson
Department of Economics, University of Umeå
S-90187 Umeå, Sweden

Dr. Günter Haag
Institut für Theoretische Physik, Universität Stuttgart
Pfaffenwaldring 57/III, D-7000 Stuttgart 80, FRG

Prof. Dr. Ingvar Holmberg
Research Institute of Demography, University of Gothenburg
S-41125 Gothenburg, Sweden

Prof. Dr. Jacques Ledent
INRS-Urbanisation, Université du Quebec
3465 Durocher, Montreal, Quebec H2X2C6, Canada

Dr. Martin Munz
Institut für Theoretische Physik, Universität Stuttgart
Pfaffenwaldring 57/III, D-7000 Stuttgart 80, FRG

Prof. Dr. Denise Pumain
Institut National d'Etudes Demographiques
27, Rue du Commandeur, F-76575 Paris Cedex 14, France

Dr. Giovanni Rabino
Istituto Ricerche Economico-Sociali des Piemonte
Via Bogino 21, I-10123 Torino, Italy

Dipl.-Phys. Rolf Reiner
Institut für Theoretische Physik, Universität Stuttgart
Pfaffenwaldring 57/III, D-7000 Stuttgart 80, FRG

Dipl.-Oec. Nicias Sarafoglou
Department of Economics, University of Umeå
S-90187 Umeå, Sweden

Prof. Dr. Michael Sonis
Department of Geography, Bar-Ilan University
52100 Ramat Gan, Israel

Prof. Dr. Dr. h.c. Wolfgang Weidlich
Institut für Theoretische Physik, Universität Stuttgart
Pfaffenwaldring 57/III, D-7000 Stuttgart 80, FRG

Contents

Introduction

Migration processes in all their theoretical and empiric aspects are a constitutive part of regional science and demography: No demographic theory and no reliable prognosis of the regional evolution can be made without taking into consideration the migration of populations.

Beyond that particular importance to special branches of science, however, migration processes are also embedded into a complex network of social, economic and political developments. The migration behaviour reacts to and interacts with all these sectors and cannot be seen as an isolated event. Vice versa, migration theory reflects this interaction and therefore has an impact on all social sciences.

Furthermore, the field of migration processes is a challenge to develop concepts for its fully dynamic theory, which even could be of interdisciplinary relevance: Since the underlying motivations of individuals to migrate are relatively well defined and specific, they are in principle amenable to inquiry. These motivations finally must result in clear individual decisions to maintain or to change the location in a given interval of time. Their effect, namely the number of relocations per unit of time between regions, can easily be counted. And in terms of such yearly numbers of migrants between regions a rich quantitative empiric material is available in many countries for a considerably long series of years. It provides a good empiric basis for the quantitative treatment of the migratory dynamics.

On the other hand there exists a long tradition in the theoretical quantitative description of dynamic processes in the mathematical and natural sciences. It seems promising to transfer some of the concepts developed there to the theory of migration processes.

It is the intention of this book to make a systematic contribution to all three just mentioned aspects of the migratory system: To the concrete description and prognosis of interregional migration, to its embedding into the socio-economic evolution, and to the development of a systematic fully dynamic theory in terms of equations of motion.

To begin with the most general aspect, the theoretical framework, we do not think of a formal direct and "physicalistic" transfer of, say, models of statistical physics to the treatment of migration processes. Instead, we have to construct a bridge between genuine and typical concepts of social science, as for instance the decision behaviour of individuals under given socio-economic circumstances, and "quasi-physical" equations of motion for the regional redistribution of populations.

The general justification for such a procedure lies in the following fundamental insight of synergetics, the science of cooperative processes in multicomponent systems: Whereas on the microlevel of interacting units each system (including physico-chemical systems as well as the migratory system) has genuine, incommensurable properties of its own, there emerges a universal structure of the macrodynamic behaviour of all systems on the level of aggregate-macrovariables in spite of their different microscopic constitution!

In this sense we can expect – and shall show in detail in section 1 – that the individual random decisions to migrate finally will lead to equations of motion on the macro-level of the migratory process, having a universal mathematical structure similar to that developed for quite different dynamical systems. This holds in particular for the master-equation, which has already thousands of applications in the natural sciences and which turns out to be constitutive for the migratory process, too!

The aspect of embedding migration into its socio-economic environment is also implied in our procedure: The "driving forces" of the migratory dynamics, namely the individual decision probabilities to change location, are governed by "regional utilities" which in their turn depend on socio-economic key-factors. Hence the whole dynamics is driven at least partially by the socio-economic situation. Vice versa, the observed migratory dynamics reflects that influence and allows of drawing conclusions, on which exogenous or endogenous key-factors it may depend.

Finally, the most direct purpose of any migratory theory, the forecasting of the regional distribution of the population can also be answered in the frame of our theory under fortunate circumstances: If the time dependence of the exogenous socio-economic situation is known (for instance, if it remains constant), the equations of motion can be solved to yield the future evolution of the population!

Another intention of this book parallel to the above mentioned aspects is our attempt to combine two lines of interest, which not always go easily together, namely:

1. to make available the knowledge and interpretational expertise about the regional migratory systems in the countries under investigation by contributions of native authors, and

2. to use the same theory, the same method of evaluation and, as far as possible, comparable data for all countries. Here, however, difficulties may arise from administrational restrictions for collecting data and from a non-optimal subdivision of a country into states or provinces from the point of view of migratory evaluations.

Nevertheless it is our hope that the reliability and commensurability of interpretations and conclusions is increased by applying them to equally defined migratory quantities of different countries evaluated on the basis of the same theory.

Guided by these purposes, the content is organized in *four parts*:

Part I contains a concise formulation of the general theory, consisting of the introduction of the conceptional frame (Chapter 1), the derivation of the equations of motion on the stochastic level (master equation) and the deterministic level (meanvalue equations) (see Chapter 2), and the regression analysis for the estimation of parameters (Chapter 3).

Part II is devoted to the analysis of different countries: Country by country is treated in a unified and comparable manner on the basis of the theory of Part I. On the other hand, ample space is left for interpretations specific to the individual countries. These countries are: The Federal Republic of Germany (Chapter 4), Sweden (Chapter 5), France (Chapter 6), Italy (Chapter 7), Canada (Chapter 8) and Israel (Chapter 9).

Part III contains a comparative analysis (Chapters 10 and 11) with the purpose to find out which aspects of migration are universal, and which of them are specific. Furthermore the comparative study will give an insight into alternative approaches and their reach.

Part IV treats mathematical concepts and details of the evaluation procedure in order to keep the main stream of concepts developed in Part I in readable form.

introduction of the conceptional frame (Chapter 1), the derivation of the equations of motion on the stochastic level (master equation) and the deterministic level (meanvalue equations) (see Chapter 2), and the regression [illegible] (Chapter 3).

[illegible]

[illegible] (Chapters 10 and 11) [illegible] which aspects of migration are universal and which are [illegible] the comparative study will give [illegible] their results.

[illegible] in order to keep the main [illegible] developed in Part I [illegible]

I. General Theory

Synopsis of Part I

Part I contains the general theory of interregional migration based on the dynamic model consistently used throughout this book. The purpose of this model is twofold. Firstly it is designed to link the microlevel of individuals deciding to move from one to another region with the macrolevel of aggregate variables like interregional population flows and the regional population configuration. Thereupon equations of motion containing individual migratory trends can be set up for the latter configuration. Secondly the model allows for the interpretation of these individual migratory trends in terms of socio-economic key-factors.

Chapter 1 begins in Section 1.1 with a short discussion of the empirical macrolevel, the population configuration $\{n_i^{(e)}(t)\}$ and the migration matrix $w_{ji}^{(e)}$.

Section 1.2 turns to the microlevel of individuals. Their decisions to migrate are described probabilistically, with the individual conditional probability to move from region i to j in the time interval τ being the basic quantity. On the macrolevel there corresponds the configurational conditional probability; furthermore the probability transition rates are introduced.

In Section 1.3 it is explained that individual decisions to migrate depend on the comparative estimation of the "utility" of the origin and destination region. Therefore the probability transition rates are assumed to depend on certain "dynamic regional utilities" and on mobility factors as well. Thus regional utilities and mobilities express the individual migratory trends.

In Chapter 2 equations of motion are derived for the regional population numbers, that means for the essential quantities on the macrolevel. These equations involve the probabilistic transition rates introduced in Chapter 1.

The exact and fully probabilistic equation of Section 2.1 is the master equation for the configurational probability. The latter not only yields the mean population configuration, but also the probabilities of deviations of this configuration from its meanvalue.

For purposes of comparison with empirical data it is however more appropriate to go over to selfcontained equations for the meanvalues of the population configuration. They are derived in Section 2.2. Only these meanvalue equations are the basis of all empiric case studies of Part II! As shown in Section 2.2.2, the pure migratory process can be separated off from birth-death-processes if the rate of natural growth per person does not depend on the region. Chapter 2 ends with Section 2.2.3, where the stationary solution of the meanvalue equations is found and the concept of "migratory stress" is introduced.

Chapter 3 is devoted to evaluation problems in the comparison of the model with empirical data. This evaluation proceeds in two steps.

In a first step, treated in Section 3.1, the trendparameters directly relating to the migratory process – namely the utilities and mobilities introduced in Chapter 1 and reappearing in the equations of motion of Chapter 2 – are estimated by comparison with empirical migration data using the method of least squares. It turns out that the log-linear estimation of trendparameters in Section 3.1.1 leads to simple analytical expressions for utilities and mobilities in terms of empiric quantities, whereas their nonlinear estimation in Section 3.1.2 leads to a slightly improved, but nonanalytical optimization procedure. The results of the log-linear and nonlinear fitting procedures for different versions of the model (with different forms of the mobility factors) are compared for a concrete case.

In a second step, treated in Section 3.2, the dependence of trendparameters on socio-economic key-factors is evaluated by a "Ranking Regression Analysis" yielding a stepwise selection and simultaneously a ranking of relevance of the socio-economic key-factors representing the utilities and mobilities.

1 Concepts of the Dynamic Migration Model

Günter Haag and Wolfgang Weidlich

Every theory of interregional migration has one central objective: to understand the dynamics of the migration of a population of, say, N members between L regions. In this book we treat regions which are extended areas like the states or provinces of a federally organized country. The intended understanding also includes, how the migration is correlated to other social or economic events and developments in that country.

Like other disciplines of natural science or social science, also the migration theory consists of a microlevel and a macrolevel of consideration, and it is decisive to understand how these levels are linked together [1.3]. In our case the microlevel consists of single individuals moving from an origin to a destination region at certain times, and the macrolevel is defined by a few aggregate variables: regional population numbers and interregional population flows.

Our theoretical concepts will have to include some simplifying, however plausible assumptions, since on the microlevel not all details are known or can be traced to their origin, and on the macrolevel only few macrovariables are empirically known and at our disposal for comparison with the theory.

1.1 Population Configuration and Migration Matrix

We begin with the consideration of the macrolevel. It is plausible to assume that the total population in general consists of subpopulations $\mathscr{P}_\alpha$, $\alpha = 1, 2, \ldots, A$ which are internally homogeneous but differ from each other with respect to their migratory behaviour. (As an example we mention the native population and the guest workers in Germany; the migration behaviour of both groups may differ considerably.) Let n_i^α members of subpopulation $\mathscr{P}_\alpha$ live in region i ($i = 1, 2, \ldots, L$). Then the *population configuration*

$$n = \{n_1^1, \ldots n_1^A, n_2^1, \ldots, n_2^A, \ldots n_L^1, \ldots n_L^A\} \tag{1.1}$$

can be considered as the migratory "state" of the population at a given time.

The assumed number of subpopulations – each with its own migratory dynamics – however depends on the attainable refinement of theoretical description and empiric observation. A model which takes into account subpopulations is presented in [1.4, 5]. We do not consider several distinct subpopulations; instead it seems appropriate to make the coarse-graining assumption of having *one homogeneous population* only. That means that the theory simplified in this way can only take into account an *average* migratory behaviour of possible existing subpopulations. The population configuration then reduces to

$$\boldsymbol{n}=\{n_1, n_2, \ldots, n_L\}\ , \tag{1.2}$$

where n_i is the number of individuals in region i.

Individual migration processes from region i to j or birth/death processes in any region are reflected by changes with time of $\boldsymbol{n}$: A birth (death) process in region i leads to the change:

$$\boldsymbol{n}=\{n_1, \ldots n_i, \ldots n_L\}\rightarrow\boldsymbol{n}^{(i\pm)}=\{n_1, \ldots(n_i\pm 1), \ldots, n_L\} \tag{1.3}$$

and the migration of one individual from region i to region j leads to

$$\begin{aligned}&\boldsymbol{n}=\{n_1, \ldots n_i, \ldots n_j, \ldots n_L\}\\ &\rightarrow\boldsymbol{n}^{(ji)}=\{n_1, \ldots(n_i-1), \ldots(n_j+1), \ldots, n_L\}\end{aligned} \tag{1.4}$$

with obvious notation.

In reality birth/death processes and migration processes take place simultaneously. We shall see however, that mathematically the birth/death processes can be formally separated off by going over to appropriate scaled variables (see Section 2.2). Hence it makes sense to treat the pure migration processes first and to show later how simultaneous birth/death processes can be included. During pure migration the total population number

$$N=\sum_{i=1}^{L} n_i \tag{1.5}$$

of course remains constant. If births and deaths are included, this need not be the case.

The population configuration is not only of theoretical interest, but is also empirically known, at least in principle, by the repeated registration at times $t=1, 2\ldots, T$ of the number of people in each region. We denote the empiric population configuration at time t by the index "e":

$$\boldsymbol{n}^{(e)}(t)=\{n_1^{(e)}(t), n_2^{(e)}(t), \ldots, n_L^{(e)}(t)\}\ , \quad t=1, 2\ldots, T\ . \tag{1.6}$$

The other set of empirically known quantities are the components of the so called *migration matrix* $w_{ji}^{(e)}(t)$; $i, j=1, 2, \ldots, L$ for $j\neq i$. This implies that we do not consider migration, i.e. change of location, *within* the regions. In the dynamic equations introduced later the terms $w_{ii}(t)$, $i=1, 2, \ldots, L$ representing such moves will cancel out.

By definition the number of people migrating from region i to region j in the time interval Δt is $\Delta t\ w_{ji}^{(e)}(t)$. Thus, $w_{ji}^{(e)}(t)$, $i, j=1, \ldots, L$, $j\neq i$, is the number of people migrating from i to j per unit of time. It is convenient to use a notation with the index i of the origin region right of the index j of the destination region. This notation is consistently used throughout the book.

The time interval Δt is given by the registration procedure. It should be short enough that practically nobody migrates twice in Δt.

The difference $n_i^{(e)}(t+\Delta t)-n_i^{(e)}(t)$ is now easily obtained by all people migrating into i minus all people migrating out of i in the time interval between t and $t+\Delta t$, that is by

$$\frac{n_i^{(e)}(t+\Delta t)-n_i^{(e)}(t)}{\Delta t}=\sum_{j=1}^{L}{}' w_{ij}^{(e)}(t)-\sum_{j=1}^{L}{}' w_{ji}^{(e)}(t) . \tag{1.7}$$

The accent ′ at the sum indicates, that the summation index $j=i$ must be excluded.

The relation (1.7) holds if no birth and death processes occur. To be realistic we must include the latter processes, too, and generalize (1.7) to

$$\frac{n_i^{(e)}(t+\Delta t)-n_i^{(e)}(t)}{\Delta t}=\sum_{j=1}^{L}{}' w_{ij}^{(e)}(t)-\sum_{j=1}^{L}{}' w_{ji}^{(e)}(t)$$

$$+w_i^{\beta(e)}(t)-w_i^{\mu(e)}(t) \tag{1.8}$$

where $\Delta t\ w_i^{\beta(e)}(t)$ are the number of births and $\Delta t w_i^{\mu(e)}(t)$ the number of deaths in region i in the time interval between t and $t+\Delta t$.

1.2 The Decision Process

Let us now go over to the microlevel of migration. It consists in the decisions of individuals to migrate from an origin to a destination region and to bring this decision into action in a certain interval of time.

A completely *deterministic* description of individual decisions however is neither feasible nor desirable. On one side the details of individual motivations to change location are not available and on the other hand they would overload the theory with too many details. The way out of this difficulty is the transition to a *probabilistic* description.

Since it is our main purpose to build up a *dynamic* theory, we cannot use, however, the concept of the conventional *static choice theory* between alternatives, because such a model of choice does *not* include the time factor at all!

Instead we choose as the fundamental concept the *conditional probability*, which proves to be constitutive for the dynamic theory of stochastic processes, including the master equation to be derived later (see Section 2.1).

Everyone of the members of the population living in region i at time t has to decide between L alternatives with respect to his region of residence at future times $t+\tau$. Either he decides to stay in region i until time $t+\tau$, which includes moves within the region, or he decides and implements to move to oe of the other $(L-1)$ *regions* $j\neq i$ in the time interval τ.

The conditional probability for individuals

$$p^{\alpha}(j, t+\tau | i, t, \text{past history}) \tag{1.9a}$$

expresses the expected relative frequency of these decisions in an ensemble of people in region i at time t; in other words: the expression (1.9a) is the probability to find the individual in region j at time $(t+\tau)$, given that it was in region i at time t.

In general this conditional probability may depend on the specific subpopulation $\mathscr{P}_\alpha$ to which the individual belongs, and also on the past history of that individual (for instance on his places of residence *before* the time t).

We make however some simplifying assumptions which decisively structurize the further calculations:

a) *Homogeneity Assumption.* As already indicated in Section 1.1, we assume that the population is homogeneous with respect to the migration behaviour. That means, the conditional probability should be the same for all members of the total population irrespective of being a member of certain subpopulations $\mathscr{P}_\alpha$.

b) *Markov Assumption.* Furthermore, the conditional probability should not depend on the migratory states of the individual *before* the time t. Hence it is assumed, that the decision maker is unprejudiced. He only takes into account his present location and a comparative valuation of all regions at time t for his migration decisions. But he is not influenced by his former "migratory biography".

Assumptions a) and b) mean, that the conditional probability should neither depend on α nor on the past history. Under these conditions the individual probability (1.9a) assumes the form

$$p(j,\, t+\tau \,|\, i,\, t)\,. \tag{1.9b}$$

c) *Assumption of Statistical Independence.* It is also presumed, that the decisions of the individuals are statistically independent. Hence, probabilistic correlations between the decision of two or more individuals are excluded.

(The simultaneous and correlated migration of families or small groups can be taken into account by considering them as new migratory units. We shall however not do this here, because the content of the meanvalue equations is not changed by this extension.)

d) *Dependence on Endogenous and Exogenous Variables.* The conditional probability, and in particular the transition probability introduced below, however, may depend on endogenous variables like the population configuration and on exogenous socio-economic variables like the regional distribution of income, rents, employment etc. The explicit form of this dependence is discussed in Section 1.3 and 3.

The "ideal" assumptions a), b) and c) may not always be fulfilled in reality. Nevertheless, the effects of possible deviations from these assumptions will cancel out to a high degree on the macrolevel.

The individual conditional probability (1.9b) obeys the probability normalization condition

$$\sum_{j=1}^{L} p(j, t+\tau \,|\, i, t) = 1 \tag{1.10}$$

since the migrant must be in someone of the L regions at any time $t+\tau$ and it satisfies the initial conditions:

$$p(j, t\,|\,i, t)=\delta_{ji} \tag{1.11}$$

since the migrant cannot have left his origin i in the zero time interval $\tau=0$.

Here,

$$\delta_{ji}=\begin{cases}1 & \text{for } j=i\\ 0 & \text{for } j\neq i\end{cases} \tag{1.12}$$

is the Kronecker δ-symbol.

Let us now consider the structure of the individual conditional probability for very small time intervals τ. Expanding $p(j, t'\,|\,i, t)$ with respect to the variable $t'=t+\tau$ in a Taylor series around t one obtains

$$\begin{aligned}p(j, t+\tau\,|\,i, t)&=p(j, t\,|\,i, t)+\tau p_{ji}(t)+O(\tau^2)\\ &=\delta_{ji}+\tau p_{ji}(t)+O(\tau^2)\end{aligned} \tag{1.13}$$

where

$$p_{ji}(t)=\left.\frac{\partial p(j,t'\,|\,i,t)}{\partial t'}\right|_{t'=t}. \tag{1.14}$$

For sufficiently small τ the higher order terms like $O(\tau^2)$ can be neglected in (1.13). Inserting (1.11) in (1.13) and making use of (1.10) one thus obtains for very small intervals τ in the case $j\neq i$

$$p(j,t+\tau\,|\,i,t)=\tau\,p_{ji}(t) \tag{1.15a}$$

and in the case $j=i$

$$\begin{aligned}p(i,t+\tau\,|\,i,t)&=1+\tau\,p_{ii}(t)\\ &=1-\tau\sum_{j=1}^{L}{}' p_{ji}(t)\,.\end{aligned} \tag{1.15b}$$

The new quantities $p_{ji}(t)$ are denoted as "individual probability transition rates" or "individual transition probabilities", although the latter notation is not fully consistent, since by definition the $p_{ji}(t)$ are *not* probabilities, but changes of probability per unit of time with the dimension 1/(time). The transition rates $p_{ji}(t)$ with $j\neq i$ must be positive, since $p(j,t+\tau\,|\,i,t)$ is positive by definition, and $\tau\,p_{ji}(t)$ describes the amount of probability transferred from region i to region j in the time interval τ, given that region i had probability $=1$ at time t. Similarly, $p(i,t+\tau\,|\,i,t)$ is the probability of staying in region i during the time interval τ. Therefore

$$\tau\,p_{ii}(t)=-\tau\sum_{j=1}^{L}{}' p_{ji}(t) \tag{1.16}$$

is negative and $\tau p_{ii}(t)$ describes the *decrease* of the probability to be in region i from 1 to $\left[1-\tau\sum_{j=1}^{L}{}' p_{ji}(t)\right]$ during the small time interval τ.

The individual probability transition rates will turn out to be the central quantities of the dynamic migration theory. If they are known in terms of endogenous and exogenous variables the whole dynamics of the migratory system can be derived (see Sections 2 and 3).

Before this can be accomplished, we must however go over from individual migration processes to the corresponding changes of the population configuration on the macrolevel. Such a change has the general form

$$\boldsymbol{n}=\{n_1, n_2, \ldots n_L\}\rightarrow\boldsymbol{n}'=\{n'_1, n'_2, \ldots, n'_L\}$$
$$\text{with} \quad \sum_{i=1}^{L} n_i=N \quad \text{and} \quad \sum_{i=1}^{L} n'_i=N \tag{1.17}$$

where $\boldsymbol{n}$ and $\boldsymbol{n}'$ are the initial and final distribution of the total population among the L regions.

Because the individual migration acts are probabilistic processes, the changes of configurations (1.17) are probabilistic, too. And corresponding to the individual conditional probabilities we can introduce the *configurational conditional probability*

$$P(\boldsymbol{n}',t+\tau|\boldsymbol{n},t)\equiv P(n'_1 \ldots n'_L, t+\tau|n_1 \ldots n_L, t)\,. \tag{1.18}$$

This quantity describes the probability to find the configuration $\boldsymbol{n}'$ at time $t+\tau$, given the configuration $\boldsymbol{n}$ at time t.

The initial condition for $\tau=0$ corresponds to (1.11) and reads

$$P(n'_1 \ldots n'_L, t|n_1 \ldots n_L, t)=\delta_{n'_1 n_1}\delta_{n'_2 n_2}\ldots\delta_{n'_L n_L}\equiv\delta_{\boldsymbol{n}'\boldsymbol{n}} \tag{1.19}$$

and the probability normalization condition

$$\sum_{\boldsymbol{n}'} P(\boldsymbol{n}',t+\tau|\boldsymbol{n},t) = 1 \tag{1.20}$$

has to be fulfilled for any τ in analogy to (1.10). Furthermore, the short time behaviour is obtained by expanding $P(\boldsymbol{n}',t'|\boldsymbol{n},t)$ into a Taylor series in $t' = t+\tau$ around t:

$$\begin{aligned} P(\boldsymbol{n}',t+\tau|\boldsymbol{n},t) &= P(\boldsymbol{n}',t|\boldsymbol{n},t)+\tau w_{\boldsymbol{n}'\boldsymbol{n}}(t)+O(\tau^2) \\ &=\delta_{\boldsymbol{n}'\boldsymbol{n}}+\tau\, w_{\boldsymbol{n}'\boldsymbol{n}}(t)+O(\tau^2) \end{aligned} \tag{1.21}$$

with

$$w_{\boldsymbol{n}'\boldsymbol{n}}(t)=\left.\frac{\partial P(\boldsymbol{n}',t'|\boldsymbol{n},t)}{\partial t'}\right|_{t'=t}\,. \tag{1.22}$$

For very small τ again the higher order terms $O(\tau^2)$ can be neglected. The normalization condition (1.20) yields

$$w_{\boldsymbol{n}\boldsymbol{n}}(t)=-\sum_{\boldsymbol{n}'(\neq\boldsymbol{n})} w_{\boldsymbol{n}'\boldsymbol{n}}(t)\,. \tag{1.23}$$

The $w_{\boldsymbol{n}'\boldsymbol{n}}(t)$ are denoted as "*configurational probability transition rates*" or as "*configurational transition probabilities*". Again, the $w_{\boldsymbol{n}'\boldsymbol{n}}(t)$ are *not* probabilities but probability changes per unit of time with the dimension 1/(time).

The important question now arises, how the *configurational* conditional probability (1.18) can be expressed by the *individual* conditional probability (1.9b). The derivation of this relation is straight-forward, but somewhat tedious; therefore it is deferred to Part IV (Section 12). Nevertheless a simple and also intuitively plausible result is obtained for the short time conditional probabilities or, equivalently, for the configurational probability transition rates. The result reads:

$$\left.\begin{array}{l} w_{\boldsymbol{n}^{ji},\boldsymbol{n}}(t)\equiv w_{ji}(\boldsymbol{n},t)=n_i p_{ji}(t) \\ \text{for the final configuration} \\ \boldsymbol{n}'=\boldsymbol{n}^{(ji)}=\{n_1,\ldots(n_i-1),\ldots(n_j+1),\ldots n_L\} \end{array}\right\} \tag{1.24a}$$

and

$$\left.\begin{array}{l} w_{\boldsymbol{n}',\boldsymbol{n}}(t)=0 \\ \text{for all other final configurations } \boldsymbol{n}'\neq\boldsymbol{n} \text{ and } \boldsymbol{n}'\neq\boldsymbol{n}^{(ji)}\,. \end{array}\right\} \tag{1.24b}$$

The interpretation of this result is very simple: The probability transition rate $w_{ji}(\boldsymbol{n},t)$ from the configuration $\boldsymbol{n}$ to the neighbouring configuration $\boldsymbol{n}^{(ji)}$ is n_i times the individual probability transition rate $p_{ji}(t)$, since anyone of the n_i persons in region i can independently migrate into region j with individual probability transition rate $p_{ji}(t)$. It will be shown in Part IV, that configurational transition rates to other end configurations $\boldsymbol{n}'$ can be excluded because these transitions occur in higher order in τ only.

Inserting (1.24) into (1.21) and making use of (1.23), the short time configurational conditional probability assumes the following form:

$$\left.\begin{array}{l} P(\boldsymbol{n}^{(ji)},t+\tau|\boldsymbol{n},t)=\tau w_{ji}(\boldsymbol{n},t)+O(\tau^2) \\ \text{for } \quad i,j=1,2,\ldots,L;\quad j\neq i \end{array}\right\} \tag{1.25a}$$

and

$$\left.\begin{array}{l} P(\boldsymbol{n}',t+\tau|\boldsymbol{n},t)=O(\tau^2) \\ \text{for all other final configurations } \boldsymbol{n}'\neq\boldsymbol{n} \quad\text{and}\quad \boldsymbol{n}'\neq\boldsymbol{n}^{(ji)}\,. \end{array}\right\} \tag{1.25b}$$

The normalization (1.20) yields:

$$P(\boldsymbol{n},t+\tau|\boldsymbol{n},t)=1-\sum_{i,j}^{L}{}' \tau w_{ji}(\boldsymbol{n},t)+O(\tau^2)\,. \tag{1.25c}$$

Henceforth we use the notation, that the accent $'$ at the sum excludes the summation over the indices $i=j$.

The result (1.25) will be the starting point for the derivation of the master equation for migratory systems in Section 2.

1.3 Transition Probabilities as Functions of Dynamic Utilities and Mobilities

We have now seen, that the crucial quantity of the migratory process, the configurational conditional probability $P(\boldsymbol{n}^{(ji)}, t+\tau|\boldsymbol{n}, t)$ for very short intervals τ can be traced back to the individual probability transition rates $p_{ji}(\ldots)$ defined by (1.14).

In this section we discuss the important question, how the migratory trends in the decisions of individuals can optimally be reflected in the functional form of the transition rates. (The points in $p_{ji}(\ldots)$ indicate, that we have to find this form now where a dependence on time t is implied.)

Two general remarks are in place before we begin this discussion:

a) We shall introduce new "factors" like "*dynamic utilities*" and "*mobilities*", on which the transition rates p_{ji} will functionally depend. Because we need *numerical values* of the p_{ji} to describe an explicit dynamics, the concept of "utility" to be used here cannot be an *ordinal* one (only taking into account the *order* of preferences), but must be a *cardinal* one in terms of explicit numerical values of utilities!

b) There may exist relations of our concept of "utility" and "mobility" to other conceptual frameworks (like the static random choice utility theory), and we shall even discuss some of these relations. On the other hand it must be stated that our concept of *dynamic utilities* is a fully selfcontained and selfexplanatory one in the following sense: The relevant dynamic quantities, namely the $w_{ji}(\boldsymbol{n}, t)$ will be uniquely represented by the dynamic utilities and mobilities. This implies that their functional form – and only this – *implicitly defines* the meaning and interpretation of "utility" and "mobility". Vice versa, it will be shown in Section 3 how the *numerical values* of utility and mobility factors can be extracted from empirical data by a regression analysis.

The explicit construction of the individual transition rates $p_{ji}(\ldots)$ with $j \neq i$ has the purpose to attribute the information contained in the migration flows to a few parameters only. It begins with the observation, that the decision trend to migrate from i to j – reflected in the magnitude of $p_{ji}(\ldots)$ – is proportional to two effects – the mobility effect and the push/pull effect [6] – which have a rather different and independent interpretation. Thus we make a decomposition of the individual probability transition rate into a *symmetric mobility factor* $v_{ji}(t)$ and a *push/pull factor* $G_{ji}(t)$

$$p_{ji}(t) = v_{ji}(t) \cdot G_{ji}(t) . \tag{1.26}$$

The *mobility factor* $v_{ji}(t)$ should include all effects which will either facilitate or impede a transition from i to j *independently* of any gain of utility in such a transition. This mobility will in particular depend on the "*effective distance*" between the regions i and j. The general concept of effective distance includes

- *geographic effects* in terms of the length of routes between the centres of regions i and j,

- *economic effects* in terms of transport costs and transport facilities between i and j and
- *social effects* in terms of a decrease of information about far-distant regions leading to a reduction of mobility.

All these effects are primarily symmetric with respect to regions i and j. More generally we incorporate in $v_{ji}(t)$ *all* effects symmetrical in i and j by putting

$$v_{ij}(t)=v_{ji}(t) \quad \text{for} \quad i, j=1, 2, \ldots, L; \quad i \neq j \tag{1.27}$$

whereas all influences which are asymmetric are taken into account by the push/pull factor $G_{ji}(t)$.

Furthermore it is not only plausible but also validated in the regression analysis of Section 3, that the mobility factor has the reduced form

$$\left.\begin{array}{l} v_{ji}^{(r)}(t)=v_0(t) f_{ji} \\ \text{with time independent } f_{ji}=f_{ij} \end{array}\right\}. \tag{1.28}$$

Here, $v_0(t)$ is a global time dependent mobility, which reflects global effects of the economic evolution and the "deterrence factor" contains the (approximately) time independent interregional distance effects. Defining $v_0(t)$ as the meanvalue of all interregional mobilities,

$$v_0(t)=\frac{1}{L(L-1)} \sum_{i,j}{}' v_{ji}^{(r)}(t) \tag{1.29}$$

it follows from (1.28) and (1.29) that the f_{ji} must fulfill the normalization condition

$$\frac{1}{L(L-1)} \sum_{i,j}^{L}{}' f_{ji}=1 . \tag{1.30}$$

An "effective distance" $D_{ji}=D_{ij}$ comprising all geographic, economic and social distance effects may now be introduced by writing f_{ji} in the form

$$\frac{f_{ji}}{L(L-1)}=\exp(-D_{ji}) . \tag{1.31}$$

Since

$$0<\frac{f_{ji}}{L(L-1)}<1 \tag{1.32}$$

must hold for each f_{ji}, there follows

$$D_{ji}=D_{ij}>0 \tag{1.33}$$

for all effective distances. Only if it turns out that the effective distance is more or less proportional to the geographical distance d_{ji}, that means, if

$$D_{ji}=\beta d_{ji} \tag{1.34}$$

is fulfilled, the mobility factors assume the form

$$\frac{f_{ji}}{L(L-1)}=\exp(-\beta d_{ji}) . \tag{1.35}$$

In this case the mobility reads:

$$v_{ji}^{(R)}(t) = L(L-1)\,v_0(t)\exp(-\beta d_{ji})\,. \tag{1.36}$$

It will be checked explicitly in Section 3 whether the form (1.36) of the mobility is acceptable or not.

The *push/pull factor* $G_{ji}(t)$ on the other hand has to take into account the specific usefulness which a migration from i to j might have for an individual and how the relative preferences are between different choices of alternatives. In general, this usefulness will be *asymmetric*, that means that $G_{ji}(t) \neq G_{ij}(t)$ must be expected. The factor G_{ji} thus must be a quantitative expression of the fact, that the potential migrant compares the "utility" of several regions, at least of the origin region and the prospective destination region, with respect to his own demands and wishes before he decides to move from i to j.

Therefore it is indicated to introduce beforehand a quantitative measure $u_i(t)$ for the attractiveness or utility of each region i to every member of the population in view of his migration decisions. The $u_i(t)$ are denoted as "*regional dynamic utilities*".

We have introduced the word "dynamic utilities" in order to distinguish our approach from other procedures where the concept of utility functions is also used. In most of these conventional procedures the state of the system is found by "maximizing the utility" of the participants under certain constraints. This is however a *static* approach which does not contain the *dynamics* of the process. In contrast, our dynamic utilities are not maximized but serve to define dynamic quantities, namely the transition probabilities which finally lead to equations of motion. The relation of our utilities to those used in the static random choice utility theory is discussed at the end of Section 2.

The push/pull factor G_{ji}, hence the probability transition rate p_{ji}, is now assumed to depend on the (dynamic) utility u_j of the destination region j as well as on the utility u_i of the origin. That means, we put

$$G_{ji} = g(u_j, u_i) \tag{1.37}$$

where the function $g(u_j, u_i)$ has to express the migration preferences of the members of the population due to different attractiveness of the regions. Thus, $g(u_j, u_i)$ must fulfill the following minimal requirements:

1. $g(u_j, u_i)$ must be positive definite for arbitrary real values of u_j and u_i, since G_{ji} and p_{ji} with $j \neq i$ are positive definite.

2. If $u_j > u_i$ holds, $g(u_j, u_i) > g(u_i, u_j)$ must be fulfilled, as the individual transition rate p_{ji} from i to j will exceed the rate p_{ij} for the inverse transition from j to i if j is more attractive than i.

3. The function $g(u_j, u_i)$ must be a monotonously increasing function of the difference $(u_j - u_i)$, since an increasing difference between utilities of regions j and i will induce a higher transition rate from i to j.

The most reasonable form of $g(u_j, u_i)$ satisfying the conditions 1. to 3. is an exponential function

$$g(u_j, u_i) = \exp(u_j - u_i) = \exp(u_j) \cdot \exp(-u_i) . \tag{1.38}$$

In (1.38) the first factor $\exp(u_j)$ can be interpreted as the *pulling influence* of the destination region j growing with *increasing* utility u_j, and the second factor $\exp(-u_i)$ can be seen as the *pushing effect* of the origin region i growing with *decreasing* utility u_i.

Composing the parts of the individual probability transition rate (1.26) as defined in (1.27) through (1.38) one obtains the formulas

$$\left.\begin{array}{ll} & p_{ji}(t) = \nu_{ji}(t) \exp[u_j(t) - u_i(t)] \\ \text{with} & \nu_{ji}(t) = \nu_{ij}(t) \end{array}\right\} \tag{1.39a}$$

or using the reduced mobilities

$$\left.\begin{array}{ll} & p_{ji}(t) = \nu_0(t) f_{ji} \exp[u_j(t) - u_i(t)] \\ \text{with} & f_{ji} = f_{ij} . \end{array}\right\} . \tag{1.39b}$$

As a consequence of (1.24) and (1.39) the configurational probability transition rates read:

$$w_{ji}(\boldsymbol{n}, t) = n_i \nu_{ji}(t) \exp[u_j(t) - u_i(t)] \tag{1.40a}$$

or, taking the reduced mobilities

$$w_{ji}(\boldsymbol{n}, t) = n_i \nu_0(t) f_{ji} \exp[u_j(t) - u_i(t)] . \tag{1.40b}$$

Although the proposed form (1.39a, b) and (1.40a, b) of the transition rates is highly plausible, it must and can be subject to empiric verification. This will be done in detail in Section 3.

Also modifications of (1.39) and (1.40) can be analysed. We mention one feasible modification which was tested in the regression analysis: The form of $p_{ji}(t)$ could for instance be generalized to read:

$$p_{ji}(t) = \nu_{ji}(t) \exp[u_j(t) - \mu u_i(t)] . \tag{1.41}$$

Here, μ is a "*sensitivity factor*" which changes the scaling of the "origin utilities" against the "destination utilities". The introduction of this factor corresponds to a frequently used argument, that another weight is given to the advantages or disadvantages of the *origin region* as compared to *destination regions*. The optimization of the factor μ in an empiric regression analysis for five countries (Canada, France, Germany, Italy, Sweden) however revealed the interesting result that in all five cases the variation of μ around 1 was very small with an average of 0.99. Therefore we shall omit it in the further discussion.

Since in formulas (1.39) and (1.40) there only appear the *differences* of the utilities of the origin and destination region, all $u_i(t)$ are only defined except for an arbitrary common additive constant. This constant can be adjusted such that the utilities always fulfil the condition

$$\sum_{i=1}^{L} u_i(t) = 0 . \tag{1.42}$$

For further use in Part 2 we may also introduce the regional variance of the utilities,

$$\sigma_u^2(t) = \frac{1}{(L-1)} \sum_{i=1}^{L} u_i^2(t) \tag{1.43}$$

which is a measure for the inhomogeneity of regions with respect to their migratory attractivity.

The setting up of the transition rates in terms of mobilities and utilities is, though highly important, only the first part of the socio-economic analysis of the migration process. Even more important is the second part, to be implemented in Section 3, namely the *correlation of the mobilities and utilities to socio-economic variables*.

It is our point of view to perform both parts of the analysis separately and sequentially because of the following reasons: The mobilities and utilities do already completely determine the dynamics of the migratory system whatever their composition of socio-economic variables might be. Therefore these dynamically relevant factors should be determined first by a regression analysis of empiric data. In a second and independent step it can then be decided, which – or which alternative sets of – key-factors correlate best to mobilities and utilities and hence are expected to be influential on the migratory process.

The type and the ranking of socio-economic key-factors will be discussed in detail in Section 3. Here we only mention that there exist two types of key-f actors: The *endogenous* ones built up from the population configuration itself, and the *exogenous* ones describing climatic, economic and social regional influences.

Anticipating the result of the analysis of concrete regional systems, the utilities will appear to be composed as follows

$$u_i(t) = \kappa \frac{[n_i^{(e)}(t) - \bar{n}(t)]}{\bar{n}(t)} + \sigma \frac{[n_i^{(e)2}(t) - \overline{n^2}(t)]}{\bar{n}^2(t)} + \delta_i(t) \,. \tag{1.44}$$

The first two terms are endogenous as they depend on the (empiric) population number $n_i^{(e)}(t)$ and its square. The first term describes an agglomeration effect, if κ is positive, and the second term a saturation effect, if σ is negative. Hence, both terms are regional size effects. The last term $\delta_i(t)$ is denoted as *regional preference*, because it describes that part of the regional attractiveness which is exogenous and independent of the population size. Because of (1.42) the preferences fulfill the condition

$$\sum_{i=1}^{L} \delta_i(t) = 0 \,. \tag{1.45}$$

The regional variance of the preferences

$$\sigma_\delta^2(t) = \frac{1}{(L-1)} \sum_{i=1}^{L} \delta_i^2(t) \tag{1.46}$$

can in analogy to (1.43) be introduced as a measure of their inhomogeneity.

Since endogenous variables may appear in mobilities and utilities, we have the interesting situation, that the microlevel of individual transition rates is coupled to the macrolevel of the population configuration. The dynamic consequences of this structure are discussed in the next section.

2 The Migratory Equations of Motion

Günter Haag and Wolfgang Weidlich

We shall now derive equations of motion for the dynamics of the population configuration. The description of the dynamics takes place on *two levels*: The *stochastic* and the *quasi-deterministic* level. Only the stochastic or probabilistic level is the fully consistent one for simple reasons: Since the individual decision process is described in probabilistic terms, the evolution on the macrolevel can only be a probabilistic one, too. Therefore only the fully probabilistic treatment gives the insight how the decisions on the microlevel of individuals induce probabilistic fluctuations on the macrolevel. The mean square deviations etc. on the macrolevel can however be very small because of mutual cancellations of fluctuations. The equation describing the full probabilistic evolution including the probability of deviations from the mean path is the *master equation.*

A short but selfcontained derivation of this master equation is given in Section 2.1. Because of the fundamental importance of the master equation approach which goes far beyond migration theory, we give a more detailed derivation of this equation in Section 12. The latter derivation will be embedded into the general concepts of Markov-processes.

Although the treatment by means of the master equation is conceptually satisfactory and consistent, it has the disadvantage that the master equation must be solved numerically in most cases and that its huge amount of information can only be compared with a relatively poor empiric material.

Hence, it is appropriate to go over to a simple treatment by only considering the dynamics of the *meanvalues* of the population configuration, omitting all probabilistic deviations from the mean path. This is denoted as the *quasideterministic approach,* because the approximate equations of motion of the meanvalues are a selfcontained set of coupled deterministic differential equations. These will be derived from the master equation in Section 2.2.

On both the statistical and the quasideterministic level it will be easy to generalize the equations so that they include migration as well as birth/death processes. Thereupon it will be shown how birth/death processes can be formally separated off in the meanvalue equations.

2.1 The Master Equation for the Population Configuration

The underlying process of the evolution of the population configuration consists in the random decisions of individuals to migrate. Their probabilistic description was given in Section 1. Therefore also on the macrolevel in the strict sense only probabilistic statements are possible. The fundamental quantity for such statements is the probability to find at a given time t a certain population configuration $\boldsymbol{n}=\{n_1, n_2, \ldots, n_L\}$.

This *configurational probability* is denoted as

$$P(\boldsymbol{n}, t) \equiv P(n_1, n_2, \ldots, n_L, t) . \tag{2.1}$$

Of course, $P(\boldsymbol{n}, t)$ must satisfy at all times t the probability normalization condition

$$\sum_{\boldsymbol{n}}^{\text{P.C.}} P(\boldsymbol{n}, t) = 1 \tag{2.2}$$

where the sum extends over all possible population configurations P.C. The equation of motion to be derived for $P(\boldsymbol{n}, t)$ is the "*master equation*".

A very special case of a configurational probability was already introduced in Section 1.2, namely the *conditional configurational probability* (see (1.18)) $P(\boldsymbol{n}, t+\tau \mid \boldsymbol{n}', t)$. It is nothing else than a configurational probability with a special initial condition, because by definition $P(\boldsymbol{n}, t+\tau \mid \boldsymbol{n}', t)$ is the probability to find a certain configuration $\boldsymbol{n}=\{n_1, n_2, \ldots, n_L\}$ at the time $t+\tau$ under the condition that another configuration $\boldsymbol{n}'$ was realized at the initial time t.

Being only a special case in this sense the conditional configurational probability turns out to be fundamental on the other hand, because the evolution with time from, say, t until $t+\tau$ of *every* configurational probability (2.1) can be expressed in terms of $P(\boldsymbol{n}, t+\tau \mid \boldsymbol{n}', t)$.

Indeed, a simple consideration of the meaning of definitions leads to the conclusion that the fundamental relation must hold:

$$P(\boldsymbol{n}, t+\tau) = \sum_{\boldsymbol{n}'}^{\text{P.C.}} P(\boldsymbol{n}, t+\tau \mid \boldsymbol{n}', t) P(\boldsymbol{n}', t) . \tag{2.3a}$$

This relation says: The probability to find configuration $\boldsymbol{n}$ at time $t+\tau$ is composed of a sum of the probabilities of mutually exclusive cases: Each of these cases consists of the event α) to find at time t one of the configurations $\boldsymbol{n}'$ and the event β) to proceed to state $\boldsymbol{n}$ at time $(t+\tau)$ starting from state $\boldsymbol{n}'$ at time t. Since the events α) and β) are statistically independent, the joint event $\alpha) \cap \beta)$ has the probability $P(\boldsymbol{n}, t+\tau \mid \boldsymbol{n}', t)\, P(\boldsymbol{n}', t)$ and all mutually exclusive joint events have the probability (2.3a) where the sum extends over all population configurations P.C. The right hand side of (2.3a) can be split into terms $\boldsymbol{n}'=\boldsymbol{n}$ and $\boldsymbol{n}' \neq \boldsymbol{n}$ so that one obtains

$$\left. \begin{aligned} P(\boldsymbol{n}, t+\tau) = {} & P(\boldsymbol{n}, t+\tau \mid \boldsymbol{n}, t)\, P(\boldsymbol{n}, t) \\ & + \sum_{\boldsymbol{n}'(\neq \boldsymbol{n})}^{\text{P.C.}} P(\boldsymbol{n}, t+\tau \mid \boldsymbol{n}', t) P(\boldsymbol{n}', t) . \end{aligned} \right\} \tag{2.3b}$$

The master equation now follows directly from (2.3b) if one considers infinitesimally small time intervals τ for which the conditional probabilities assume the form (1.25a, b, c), yielding

$$\left.\begin{aligned} P(\boldsymbol{n}, t+\tau) = {} & P(\boldsymbol{n}, t) - \sum_{i,j}^{L}{}' \tau w_{ji}(\boldsymbol{n}, t) P(\boldsymbol{n}, t) \\ & + \sum_{i,j}^{L}{}' \tau w_{ji}(\boldsymbol{n}^{(ij)}, t) P(\boldsymbol{n}^{(ij)}, t) + O(\tau^2) \end{aligned}\right\} \tag{2.4}$$

or, after rearranging terms:

$$\left.\begin{aligned} & \frac{P(\boldsymbol{n}, t+\tau) - P(\boldsymbol{n}, t)}{\tau} \\ & = \sum_{i,j}^{L}{}' w_{ji}(\boldsymbol{n}^{(ij)}, t) P(\boldsymbol{n}^{(ij)}, t) - \sum_{i,j}^{L}{}' w_{ji}(\boldsymbol{n}, t) P(\boldsymbol{n}, t) + \frac{1}{\tau} O(\tau^2) . \end{aligned}\right\} \tag{2.5}$$

In the limit $\tau \to 0$ there follows from (2.5) the master equation:

$$\frac{dP(\boldsymbol{n}, t)}{dt} = \sum_{i,j}^{L}{}' w_{ji}(\boldsymbol{n}^{(ij)}, t) P(\boldsymbol{n}^{(ij)}, t) - \sum_{i,j}^{L}{}' w_{ji}(\boldsymbol{n}, t) P(\boldsymbol{n}, t) . \tag{2.6}$$

This equation has a very direct and intuitively appealing interpretation: The l.h.s. of (2.6) is the change per time of the probability of a given configuration $\boldsymbol{n}$. The r.h.s. of (2.6) tells us the composition of this change out of partial processes: The first sum describes the probability flow from neighbouring configurations $\boldsymbol{n}^{(ij)}$ to the given configuration $\boldsymbol{n}$ and the second sum describes the probability flow from configuration $\boldsymbol{n}$ to all neighbouring configurations $\boldsymbol{n}^{(ji)}$.

In more detail: The probability flow (=probability transfer per unit of time) from configuration $\boldsymbol{n}^{(ij)}$ to $\boldsymbol{n}$ consists of the probability that $\boldsymbol{n}^{(ij)}$ is realized times the probability transition rate from $\boldsymbol{n}^{(ij)}$ to $\boldsymbol{n}$; that means this partial probability flow is $w_{ji}(\boldsymbol{n}^{(ij)}, t) P(\boldsymbol{n}^{(ij)}, t)$. The total probability flow into $\boldsymbol{n}$ is the sum of partial probability flows from all those neighbouring configurations $\boldsymbol{n}^{(ij)}$ for which nonvanishing rates $w_{ji}(\boldsymbol{n}^{(ij)}, t)$ to configuration $\boldsymbol{n}$ exist (see (1.24a)). The analogous consideration holds for the composition of the total probability flow out of $\boldsymbol{n}$.

Summarizing we can state that the master equation is nothing but a probability rate equation describing the change of the probability of every configuration in terms of the net probability flow between $\boldsymbol{n}$ and all neighbouring configurations $\boldsymbol{n}^{(ij)}$.

Finally we generalize the master equation (2.6) to include birth/death processes. These processes lead to the following transitions between population configurations:

$$\begin{aligned} & \boldsymbol{n} = \{n_1, \ldots n_i, \ldots n_L\} \\ & \to \boldsymbol{n}^{(i\pm)} = \{n_1, n_2, \ldots (n_i \pm 1), \ldots n_L\} . \end{aligned} \tag{2.7}$$

The corresponding short-time configurational conditional probabilities can be expanded in the same manner as in (1.25)

$$P(\boldsymbol{n}^{(i+)}, t+\tau|\boldsymbol{n}, t) = \tau w_i^\beta(\boldsymbol{n}, t) + O(\tau^2) \tag{2.8a}$$

$$P(\boldsymbol{n}^{(i-)}, t+\tau|\boldsymbol{n}, t) = \tau w_i^\mu(\boldsymbol{n}, t) + O(\tau^2) \tag{2.8b}$$

where

$$w_i^\beta(\boldsymbol{n}, t) \equiv w_{\boldsymbol{n}^{i+}, \boldsymbol{n}}(t) = \beta_i(t) n_i \tag{2.9a}$$

and

$$w_i^\mu(\boldsymbol{n}, t) \equiv w_{\boldsymbol{n}^{i-}, \boldsymbol{n}}(t) = \mu_i(t) n_i \tag{2.9b}$$

are configurational probability transition rates for the births and deaths in region i, respectively. These rates are assumed to be proportional to the number n_i of individuals in i; $\beta_i(t)$ is the average birth rate per person and $\mu_i(t)$ the average death rate per person in region i. The normalization condition (1.20) leads to a modification of (1.25c) which now reads:

$$\left.\begin{aligned} P(\boldsymbol{n}, t+\tau|\boldsymbol{n}, t) = 1 &- \sum_{i,j}^{L}{}' \tau w_{ji}(\boldsymbol{n}, t) \\ &- \sum_{i=1}^{L} \tau w_i^\beta(\boldsymbol{n}, t) - \sum_{i=1}^{L} \tau w_i^\mu(\boldsymbol{n}, t) . \end{aligned}\right\} \tag{2.10}$$

The generalized master equation can now be derived from (2.3) as before making use of the configurational probabilities (1.25a), (2.8a, b) and (2.10). The resulting master equation including migration and birth/death processes reads:

$$\left.\begin{aligned} \frac{dP(\boldsymbol{n}, t)}{dt} = &\sum_{i,j}^{L}{}' w_{ji}(\boldsymbol{n}^{(ij)}, t) P(\boldsymbol{n}^{(ij)}, t) - \sum_{i,j}^{L}{}' w_{ji}(\boldsymbol{n}, t) P(\boldsymbol{n}, t) \\ &+ \sum_{i=1}^{L} w_i^\beta(\boldsymbol{n}^{(i-)}, t) P(\boldsymbol{n}^{(i-)}, t) - \sum_{i=1}^{L} w_i^\beta(\boldsymbol{n}, t) P(\boldsymbol{n}, t) \\ &+ \sum_{i=1}^{L} w_i^\mu(\boldsymbol{n}^{(i+)}, t) P(\boldsymbol{n}^{(i+)}, t) - \sum_{i=1}^{L} w_i^\mu(\boldsymbol{n}, t) P(\boldsymbol{n}, t) . \end{aligned}\right\} \tag{2.11}$$

The first, second and third line on the r.h.s. of (2.11) describe the net probability flows by migration, birth and death processes, respectively.

In order to make some forthcoming derivations more elegant we shall now introduce some formal concepts. At first we define translation operators $E_i^{\pm 1}$ acting on every function $F(n_1, \ldots, n_L)$ as follows:

$$\left.\begin{aligned} &E_i^{\pm 1} F(n_1, \ldots n_i, \ldots n_L) = F(n_1, \ldots (n_i \pm 1), \ldots n_L) \\ &\text{or } E_i^{\pm 1} F(\boldsymbol{n}) = F(\boldsymbol{n}^{(i\pm)}) . \end{aligned}\right\} \tag{2.12}$$

This definition implies the relations:

$$\left.\begin{aligned} &n_k E_i^{\pm 1} F(\boldsymbol{n}) = E_i^{\pm 1}(n_k \mp \delta_{ik}) F(\boldsymbol{n}) \\ &n_k E_i^{+1} E_j^{-1} F(\boldsymbol{n}) = E_i^{+1} E_j^{-1}(n_k + \delta_{jk} - \delta_{ik}) F(\boldsymbol{n}) \end{aligned}\right\} \tag{2.13}$$

and

$$E_i^{+1}E_j^{-1}F(\boldsymbol{n}) = F(\boldsymbol{n}^{(ij)}) . \tag{2.14}$$

Making use of the translation operators, the master equation (2.11) can also be written in the equivalent form

$$\left.\begin{aligned} \frac{dP(\boldsymbol{n},t)}{dt} &= \sum_{i,j}^{L}{}' (E_i^{+1}E_j^{-1}-1)w_{ji}(\boldsymbol{n},t)P(\boldsymbol{n},t) \\ &+ \sum_{i=1}^{L}(E_i^{-1}-1)w_i^{\beta}(\boldsymbol{n},t)P(\boldsymbol{n},t) \\ &+ \sum_{i=1}^{L}(E_i^{+1}-1)w_i^{\mu}(\boldsymbol{n},t)P(\boldsymbol{n},t) . \end{aligned}\right\} \tag{2.15}$$

Furthermore we introduce general configurations (G.C.)

$$\boldsymbol{n} = \{n_1, n_2, \ldots, n_L\}; \quad n_i \gtreqless 0 \tag{2.16}$$

whose components n_i can be positive or negative integers or zero. The true population configurations (P.C.) consist of components n_i which are all positive integers or zero, whereas "virtual" configurations (V.C.) contain at least one negative component $n_k < 0$.

The functions $P(\boldsymbol{n},t)$, $w_{ji}(\boldsymbol{n},t)$, $w_i^{\beta}(\boldsymbol{n},t)$, $w_i^{\mu}(\boldsymbol{n},t)$ introduced so far are only defined for true population configurations. We can however easily extend their definition to all general configurations (*G.C.*) (comprising population configurations P.C. and virtual configurations V.C.) as follows:

$$F(\boldsymbol{n}) = \begin{cases} F(\boldsymbol{n}) & \text{for} \quad \boldsymbol{n} \in P.C. \\ 0 & \text{for} \quad \boldsymbol{n} \in V.C. . \end{cases} \tag{2.17}$$

In all forthcoming calculations we make use of this formal extension. It implies that sums over $F(\boldsymbol{n})$ with $\boldsymbol{n}$ running through all true population configurations can formally be extended to sums over all general configurations (G.C.):

$$\sum_{\boldsymbol{n}}^{P.C.} F(\boldsymbol{n}) = \sum_{\boldsymbol{n}}^{G.C.} F(\boldsymbol{n}) \tag{2.18}$$

and it is evident, that the relations hold

$$\sum_{\boldsymbol{n}}^{G.C.} E_i^{\pm 1}F(\boldsymbol{n}) = \sum_{\boldsymbol{n}}^{G.C.} E_i^{+1}E_j^{-1}F(\boldsymbol{n}) = \sum_{\boldsymbol{n}}^{G.C.} F(\boldsymbol{n}) \tag{2.19}$$

because the sum $\sum_{\boldsymbol{n}}^{G.C.}$ includes summation over all $-\infty < n_i,\ n_j < +\infty$. This formalism can now be used to show, that the master equation (2.15) is compatible with the probability normalization condition (2.2). Taking the sum $\sum_{\boldsymbol{n}}^{P.C.}$ on both sides of (2.15), which can be formally extended to $\sum_{\boldsymbol{n}}^{G.C.}$ because of (2.18), one obtains:

$$\left.\begin{aligned}\frac{d}{dt}\sum_{\boldsymbol{n}}^{G.C.} P(\boldsymbol{n}, t) &= \sum_{i,j}^{L}{}' \sum_{\boldsymbol{n}}^{G.C.} (E_i^{+1} E_j^{-1} - 1) w_{ji}(\boldsymbol{n}, t) P(\boldsymbol{n}, t) \\ &+ \sum_{i=1}^{L} \sum_{\boldsymbol{n}}^{G.C.} (E_i^{-1} - 1) w_i^{\beta}(\boldsymbol{n}, t) P(\boldsymbol{n}, t) \\ &+ \sum_{i=1}^{L} \sum_{\boldsymbol{n}}^{G.C.} (E_i^{+1} - 1) w_i^{\mu}(\boldsymbol{n}, t) P(\boldsymbol{n}, t) \\ &= 0 \, .\end{aligned}\right\} \tag{2.20}$$

On the r.h.s. of (2.20) repeated use has been made of the relation (2.19). The result (2.20) shows, that the configurational sum over all probabilities $P(\boldsymbol{n}, t)$ remains constant with time and hence can be normalized to 1. In the next section we shall use the formal operations to give a concise derivation of the meanvalue equations.

2.2 The Meanvalue Equations

We have already stated that the probability distribution $P(\boldsymbol{n}, t)$ contains too much information in comparison with the available empiric data. Whereas only one population configuration is realized, the distribution informs us about the probability at time t of all possible configurations $\boldsymbol{n}$. Normally, however, $P(\boldsymbol{n}, t)$ will be a sharply peaked unimodal distribution and it is to be expected, that the *meanvalue* $\bar{\boldsymbol{n}}(t)$ of $\boldsymbol{n}$ – that is the average of all possible $\boldsymbol{n}$ weighted by the probability of their appearance – practically coincides with the *realized* configuration. Therefore it is highly desirable to derive selfcontained equations of motion for the meanvalues $\bar{n}_i(t)$ of the components of the population configuration.

2.2.1 Derivation of their General Form

We begin with the definition of the meanvalue $\overline{f(\boldsymbol{n})}$ of an arbitrary function $f(\boldsymbol{n})$ of $\boldsymbol{n}$[1]:

$$\overline{f(\boldsymbol{n})} = \sum_{\boldsymbol{n}}^{P.C.} f(\boldsymbol{n}) P(\boldsymbol{n}, t) = \sum_{\boldsymbol{n}}^{G.C.} f(\boldsymbol{n}) P(\boldsymbol{n}, t) \, . \tag{2.21}$$

In particular the mean regional population numbers are given by

$$\bar{n}_k(t) = \sum_{\boldsymbol{n}}^{G.C.} n_k P(\boldsymbol{n}, t) \tag{2.22}$$

[1] This relation holds even if $f(\boldsymbol{n}) \neq 0$ for virtual configurations, because $P(\boldsymbol{n}, t)$ is zero for the latter.

The equation of motion for $\bar{n}_k(t)$ follows by taking the time derivative on both sides of (2.22) and by inserting the master equation (2.15)

$$\left.\begin{aligned}\frac{d\bar{n}_k(t)}{dt} &= \sum_{\boldsymbol{n}}^{G.C.} n_k \frac{dP(\boldsymbol{n},t)}{dt}\\ &= \sum_{i,j}^{L}{}' \sum_{\boldsymbol{n}}^{G.C.} n_k (E_i^{+1}E_j^{-1}-1) w_{ji}(\boldsymbol{n},t)P(\boldsymbol{n},t)\\ &+ \sum_{i=1}^{L} \sum_{\boldsymbol{n}}^{G.C.} n_k (E_i^{-1}-1) w_i^{\beta}(\boldsymbol{n},t)P(\boldsymbol{n},t)\\ &+ \sum_{i=1}^{L} \sum_{\boldsymbol{n}}^{G.C.} n_k (E_i^{+1}-1) w_i^{\mu}(\boldsymbol{n},t)P(\boldsymbol{n},t)\,.\end{aligned}\right\} \tag{2.23}$$

Making use of the relations (2.13) and (2.19) on the r.h.s. of (2.23) one obtains

$$\left.\begin{aligned}\frac{d\bar{n}_k(t)}{dt} &= \sum_{i,j}^{L}{}' \sum_{\boldsymbol{n}}^{G.C.} E_i^{+1}E_j^{-1}(\delta_{jk}-\delta_{ik}) w_{ji}(\boldsymbol{n},t)\,P(\boldsymbol{n},t)\\ &+ \sum_{i=1}^{L} \sum_{\boldsymbol{n}}^{G.C.} E_i^{-1}\delta_{ik} w_i^{\beta}(\boldsymbol{n},t)P(\boldsymbol{n},t)\\ &- \sum_{i=1}^{L} \sum_{\boldsymbol{n}}^{G.C.} E_i^{+1}\delta_{ik} w_i^{\mu}(\boldsymbol{n},t)P(\boldsymbol{n},t)\\ &= \sum_{i=1}^{L}{}' \sum_{\boldsymbol{n}}^{G.C.} w_{ki}(\boldsymbol{n},t)P(\boldsymbol{n},t) - \sum_{j=1}^{L}{}' \sum_{\boldsymbol{n}}^{G.C.} w_{jk}(\boldsymbol{n},t)P(\boldsymbol{n},t)\\ &+ \sum_{\boldsymbol{n}}^{G.C.} w_k^{\beta}(\boldsymbol{n},t)P(\boldsymbol{n},t)\\ &- \sum_{\boldsymbol{n}}^{G.C.} w_k^{\mu}(\boldsymbol{n},t)P(\boldsymbol{n},t)\,.\end{aligned}\right\} \tag{2.24}$$

The right hand side of (2.24) consists of certain meanvalues only, so that we obtain the exact meanvalue equations:

$$\begin{aligned}\frac{d\bar{n}_k(t)}{dt} &= \sum_{i=1}^{L}{}' \overline{w_{ki}(\boldsymbol{n},t)} - \sum_{j=1}^{L}{}' \overline{w_{jk}(\boldsymbol{n},t)}\\ &+ \overline{w_k^{\beta}(\boldsymbol{n},t)} - \overline{w_k^{\mu}(\boldsymbol{n},t)}\,.\end{aligned} \tag{2.25}$$

The intuitive interpretation of these equations is clear: The change of the meanvalue $\bar{n}_k(t)$ with time is due to the mean immigrant rate from all regions $i\ (\neq k)$ to k minus the mean outmigrant rate from k to all regions $j\ (\neq k)$ and to the mean birth rate in k minus the mean death rate in k. In this case the total population N is no longer constant.

Comparing (2.25) with the empiric relation (1.8) one can see, that the following identifications hold between theoretical and empiric quantities:

$$\frac{n_k^{(e)}(t+\Delta t)-n_k^{(e)}(t)}{\Delta t} \simeq \frac{d\bar{n}_k(t)}{dt} \tag{2.26a}$$

$$\begin{aligned} w_{ki}^{(e)}(t) &= \text{empiric migration matrix element from } i \text{ to } k \\ = \overline{w_{ki}(\boldsymbol{n},\, t)} &= \text{meanvalue of the configurational transition rate from } i \text{ to } k \end{aligned} \tag{2.26b}$$

$$\begin{aligned} w_k^{\beta(e)}(t) &= \text{empiric birth rate in } k \\ = \overline{w_k^{\beta}(\boldsymbol{n},\, t)} &= \text{configurational transition rate for birth process in } k \end{aligned} \tag{2.26c}$$

$$\begin{aligned} w_k^{\mu(e)}(t) &= \text{empiric death rate in } k \\ = \overline{w_k^{\mu}(\boldsymbol{n},\, t)} &= \text{configurational transition rate for death process in } k\,. \end{aligned} \tag{2.26d}$$

The exact meanvalue equations (2.25) still have one disadvantage: They are not selfcontained equations, because one needs the probability distribution $P(\boldsymbol{n}, t)$ in order to calculate the right hand side. If however it can be assumed that $P(\boldsymbol{n}, t)$ is a wellbehaved sharp-peaked unimodal distribution, the approximate relation holds [2.1]

$$\overline{f(\boldsymbol{n},\, t)} \cong f(\bar{\boldsymbol{n}}(t),\, t) \tag{2.27}$$

which says: The meanvalue of a function of $\boldsymbol{n}$ is approximately equal to that function of the meanvalue of $\boldsymbol{n}$ in the case of steep unimodal distributions.

Inserting the approximation (2.27) into (2.25) we obtain a set of selfcontained – in general nonlinear – equations of motion for the meanvalues $\bar{n}_k(t)$:

$$\left.\begin{aligned} \frac{d\bar{n}_k(t)}{dt} = {} & \sum_{i=1}^{L}{}' \, w_{ki}(\bar{\boldsymbol{n}}(t),\, t) - \sum_{j=1}^{L}{}' \, w_{jk}(\bar{\boldsymbol{n}}(t),\, t) \\ & + w_k^{\beta}(\bar{\boldsymbol{n}}(t),\, t) - w_k^{\mu}(\bar{\boldsymbol{n}}(t),\, t)\,. \end{aligned}\right\} \tag{2.28}$$

The equations become fully explicit by using the form (1.40a) and (2.9a, b) of the transition rates w_{ki}, w_{jk}, w_k^{β}, w_k^{μ}:

$$\left.\begin{aligned} \frac{d\bar{n}_k(t)}{dt} = {} & \sum_{i=1}^{L}{}' \, \bar{n}_i \, p_{ki}(\,.\,.) - \sum_{j=1}^{L}{}' \, \bar{n}_k \, p_{jk}(\,.\,.) \\ & + \rho_k(t)\, \bar{n}_k \end{aligned}\right\} \tag{2.29}$$

with (see (1.39a))

$$\left.\begin{aligned} & p_{ki}(t) = \nu_{ki}(t) \exp[u_k(t) - u_i(t)] \\ & \text{for} \quad k \neq i \end{aligned}\right\} \tag{2.30}$$

and

$$\rho_k(t) = [\beta_k(t) - \mu_k(t)] \, . \tag{2.31}$$

The factor $\rho_k(t)$ is the birth rate minus the death rate per person or the rate of natural increase. In general $\rho_k(t)$ may depend on time and on region. Formally the sums $\sum_{i=1}^{L}{}'$ and $\sum_{j=1}^{L}{}'$ in (2.28) and (2.29) can now be extended over all i and j (including k), because the terms w_{kk} are cancelling out anyway.

A further remarkable structure can now be read off from (2.29): If the utilities u_k in the definition of the individual transition rates (1.39) depend on time t and/or *exogenous* socio-economic factors only, we have meanvalue equations which are *linear* in the population numbers $\bar{n}_k$. In this case an exact relationship between solutions of the master equation and solutions of the meanvalue equations can be established (see Part IV, Chapter 13).

If, on the other hand, the utilities u_k also depend on the *endogenous* variables $\bar{n}_k$ – and it will turn out in Section 3 that this assumption is realistic – the meanvalue equations become *nonlinear* differential equations in the variables $\bar{n}_k$ because the individual transition rates to be inserted in (2.29) now read

$$p_{ki}(\bar{\boldsymbol{n}}, t) = v_{ki}(t) \exp[u_k(\bar{\boldsymbol{n}}, t) - u_i(\bar{\boldsymbol{n}}, t)] \, . \tag{2.30a}$$

The consequences, which the nonlinear structure of the meanvalue equations can have, are important. For instance there may occur "migratory phase transitions", that means instabilities of population configurations and sudden generation of new stable configurations, for instance the concentrations of the population in a few regions. Models of this type have been disscussed in [2.1] and [2.2].

Most of the further analyses of this book focussing on the concrete evaluation of interregional migration are based on the meanvalue equations (2.28, 29) with (2.30, 30a).

2.2.2 Separation of the Birth/Death Processes

Because we are mainly interested in migration, whereas migration and birth/death processes always occur simultaneously, it is interesting to note, that one can separate off the birth/death processes in eq. (2.29). We do it for the most important case that the rate of natural increase $\rho_k(t)$ does not depend on the specific region, that is under the assumption

$$\rho_k(t) = \rho(t) \tag{2.32}$$

At first the evolution with time of the total population

$$N(t) = \sum_{k=1}^{L} \bar{n}_k(k) \tag{2.33}$$

is easily obtained from (2.29). Taking the sum over all k in (2.29) yields

$$
\left.\begin{aligned}
\frac{dN(t)}{dt} &= \sum_{k=1}^{L}\left(\sum_{i=1}^{L}\bar{n}_i p_{ki} - \sum_{j=1}^{L}\bar{n}_k p_{jk}\right)\\
&\quad + \sum_{k=1}^{L}\rho(t)\bar{n}_k(t)\\
&= \rho(t)N(t)
\end{aligned}\right\} \tag{2.34}
$$

with the solution

$$
\left.\begin{aligned}
&N(t) = N(0)\exp[R(t)]\\
\text{where}\quad &R(t) = \int_0^t \rho(t')dt' .
\end{aligned}\right\} \tag{2.35}
$$

Let us now introduce relative population numbers $x_k(t)$ by the substitution

$$
\bar{n}_k(t) = N(t)x_k(t) \tag{2.36}
$$

where the $x_k(t)$ because of (2.33) fulfill

$$
\sum_{k=1}^{L} x_k(t) = 1 \tag{2.37}
$$

Inserting (2.36) into (2.29) one obtains

$$
\left.\begin{aligned}
&\frac{dN}{dt}x_k + N\frac{dx_k}{dt}\\
&= \sum_{i=1}^{L} N x_i p_{ki}(N\boldsymbol{x}, t) - \sum_{j=1}^{L} N x_k p_{jk}(N\boldsymbol{x}, t)\\
&+ \rho(t)N x_k .
\end{aligned}\right\} \tag{2.38}
$$

This equation yields the pure migratory equation

$$
\frac{dx_k}{dt} = \sum_{i=1}^{L} x_i p_{ki}(N\boldsymbol{x}, t) - \sum_{j=1}^{L} x_k p_{jk}(N\boldsymbol{x}, t) \tag{2.39}
$$

for the relative population numbers $x_k(t)$ after taking into account the differential equation (2.34) for the total population. (Since $N(t)$ is empirically known, it provides no difficulty to go over to the $x_k(t)$ instead of $\bar{n}_k(t)$ in all functions depending on the population numbers $\bar{n}_k(t)$.)

Inserting the individual transition rate (2.30) into (2.39) we obtain the explicit form

$$
\left.\begin{aligned}
&\frac{dx_k}{dt} = \sum_{i=1}^{L} x_i v_{ki}(t)\exp[u_k(t)-u_i(t)] - \sum_{j=1}^{L} x_k v_{jk}(t)\exp[u_k(t)-u_j(t)]\\
&\text{for}\quad k = 1,2,\ldots,L .
\end{aligned}\right\} \tag{2.40}
$$

2.2.3 The Stationary Solution

In general the solutions $x_k(t)$ of (2.40) will never approach equilibrium values, because the mobilities $\nu_{ki}(t)$ and utilities $u_j(t)$ are functions of time. Let us however consider the theoretical case, that the mobilities and utilities remain – after a certain moment – constant with time. It can easily be read off, that the then emerging equation

$$\frac{dx_k}{dt}=\sum_{i=1}^{L} x_i \nu_{ki}\exp[u_k-u_i]-\sum_{j=1}^{L} x_k \nu_{jk}\exp[u_j-u_k] \tag{2.40a}$$

has the stationary solution

$$\left.\begin{aligned} x_k^{st} &= C\exp(2u_k) \\ \text{with}\quad C &= \left[\sum_{i=1}^{L}\exp(2u_i)\right]^{-1}. \end{aligned}\right\} \tag{2.41}$$

Furthermore it can be proved that all timedependent $x_k(t)$ of (2.40a) finally evolve into this unique stationary solution:

$$x_k(t)\underset{(t\to\infty)}{\longrightarrow} x_k^{st} \tag{2.42}$$

In other words: The values x_k^{st} are "virtual equilibrium values" of the relative population numbers, into which the momentary $x_k(t)$ would evolve, given that from now on the mobilities and utilities would remain constant!

It is therefore of some interest to ask how far apart from their virtual equilibrium values x_k^{st} the real $x_k(t)$ are: The more they are away from this equilibrium, the more "*migratory stress or migratory unrest*" is present in the society in which the real population distribution has not yet adjusted to the regional utilities as seen by the migrants. (It should be clear however, that even at equilibrium the migratory flows between the regions do not vanish; but then the flow from i to j is equal to the flow from j to i for each pair i, j of regions.)

The correlation coefficient

$$\left.\begin{aligned} r(\boldsymbol{x}, \boldsymbol{x}^{st}) &= \frac{\sum_{i=1}^{L}(x_i-\bar{x})(x_i^{st}-\bar{x})}{\sqrt{\sum_{i=1}^{L}(x_i-\bar{x})^2\cdot\sum_{j=1}^{L}(x_j^{st}-\bar{x})^2}} \\ \text{with}\quad \bar{x} &= 1/L \end{aligned}\right\} \tag{2.43}$$

can now be used to define a measure for the migratory stress. Because of

$$-1\leq r(\boldsymbol{x}, \boldsymbol{x}^{st})\leq +1 \tag{2.44}$$

and

$$r(\boldsymbol{x}, \boldsymbol{x}^{st}) = 1, \qquad \text{iff} \quad \boldsymbol{x} = \boldsymbol{x}^{st} \tag{2.45}$$

the migratory stress $s(\boldsymbol{x}, \boldsymbol{x}^{st})$ can be reasonably introduced by

$$\left.\begin{array}{l} s(\boldsymbol{x}, \boldsymbol{x}^{st}) = \frac{1}{2}[1 - r(\boldsymbol{x}, \boldsymbol{x}^{st})] \\ \text{with} \quad 0 \leq s(\boldsymbol{x}, \boldsymbol{x}^{st}) \leq 1 \,. \end{array}\right\} \tag{2.46}$$

Let us finally compare our dynamic theory with the "static random utility theory". The latter does not contain the concept of evolution with time, and comes to the result: If v_i is the "utility" attributed to the alternative i (out of L alternatives $i = 1, 2, \ldots, L$, then the (static) probability $p(k)$ of chosing the alternative k is given by [2.3]

$$p(k) = \frac{\exp(v_k)}{\sum_{i=1}^{L} \exp(v_i)} \,. \tag{2.47}$$

This corresponds to our stationary solution (2.41) if we interpret $x_k^{st} \to p(k)$ as the stationary probability to find an individual in region k, that is in his k-th alternative. In this sense one can identify our dynamic utility u_k with $1/2\, v_k$. It must however be stressed that our theory goes far beyond any static or equilibrium theory because it allows for the description of the full dynamics of the population configuration starting from any non-equilibrium initial state.

3 The Estimation of Parameters

Günter Haag, Martin Munz and Rolf Reiner

In this section we develop the methods necessary for applying the theory of Sections 1 and 2 to concrete migratory systems. As already indicated this regression analysis consists of two parts:

In the *first* part (that is in Section 3.1) all parameters of the theory, which *directly* relate to the migration process, are estimated by comparison with empirical data. The explicit form of the probability transition rates consisting of mobilities and utilities turn out to be functions of the regions involved and of time.

The intention of the *second* part (see Section 3.2) is to provide insights into the dependence of the migratory process on socio-economic key-factors. Of course it is well-known that it is difficult if not impossible to establish a *direct and unique causal* relationship between the socio-economic situation and migratory dynamics. Instead we expect that many influences merge with different intensities in producing the migratory behaviour. Therefore we take the degree of the regional and temporal correlation between the socio-economic variables and the migratory variables as *an indirect measure* of their relevance in generating or influencing migration. In particular the correlation of certain key-factors with the utility function $u_i(t)$ and with the global mobility $v_0(t)$ will be determined and simultaneously a method of *ranking their relevance* will be introduced.

3.1 The Regression Analysis for Trendparameters

In a migratory system consisting of one homogeneous population living in regions the following empiric quantities can be registered: The number of persons, $n_i^{(e)}(t)$ living in region i $(i=1, 2, \ldots, L)$ at time t $(t=1, 2, \ldots, T)$ and the number of persons $w_{ji}^{(e)}(t)$ migrating from i to j $(i \neq j;\ i, j=1, 2, \ldots, L)$ in the time interval $[t, t+1]$. These observed quantities, all with index (e) for "empiric", are listed in Table 3.1.

It was shown in (2.26b) that in comparing the theoretical meanvalue equations with empirical migration, the mean configurational transition rate from i to j must be identified with the empiric migration matrix element. Thus we obtain with (1.40a) and (2.26)

Table 3.1. Observed data

destination region	destination population	number of migrations in time interval $[t,t+1]$ from origin region i to destination region j					
1	$n_1^{(e)}(t)$	-	$w_{12}^{(e)}(t)$	....	$w_{1i}^{(e)}(t)$	....	$w_{1L}^{(e)}(t)$
2	$n_2^{(e)}(t)$	$w_{21}^{(e)}(t)$	-	....	$w_{2i}^{(e)}(t)$	....	$w_{2L}^{(e)}(t)$
⋮	⋮	⋮	⋮		⋮		⋮
j	$n_j^{(e)}(t)$	$w_{j1}^{(e)}(t)$	$w_{j2}^{(e)}(t)$	....	$w_{ji}^{(e)}(t)$	....	$w_{jL}^{(e)}(t)$
⋮	⋮	⋮	⋮		⋮		⋮
L	$n_L^{(e)}(t)$	$w_{L1}^{(e)}(t)$	$w_{L2}^{(e)}(t)$	....	$w_{Li}^{(e)}(t)$	....	-
origin population		$n_1^{(e)}(t)$	$n_2^{(e)}(t)$	....	$n_i^{(e)}(t)$	....	$n_L^{(e)}(t)$
origin region		1	2	....	i	....	L

$$w_{ji}(t)=v_{ji}\exp[u_j(t)-u_i(t)]\,n_i^{(e)}(t)$$
$$\cong w_{ji}^{(e)}(t) \tag{3.1}$$

with

$$v_{ji}(t)=v_{ij}(t)\,. \tag{3.2}$$

The relation (3.1) contains a large number of parameters, namely the utilities $u_i(t)$ and the mobilities $v_{ji}(t)$ which all have to be estimated in the fitting procedure.

The number of observations is given by the following relation: If the observation period lasts T intervals, there are

$$q=T\,(L^2-L) \tag{3.3}$$

observed matrix elements $w_{ji}^{(e)}(t)$, whereas the $T\cdot L$ utilities $u_i(t)$ and the $T(L^2-L)/2$ mobilities $v_{ji}(t)$ add up to

$$k=\tfrac{1}{2}\,T(L^2+L) \tag{3.4}$$

fitting parameters. The number of degrees of freedom, defined as $f=q-k$ thus amounts to

$$f=q-k=\frac{T}{2}(L^2-3L)\,. \tag{3.5}$$

The number of fitting parameters is however reduced, if, according to (1.28) the mobility $v_{ji}(t)$ is assumed to have the "reduced" form

$$v_{ji}^{(r)}(t)=v_0(t)f_{ji}=v_0(t)L(L-1)\exp(-D_{ji}) \tag{3.6}$$

with a *time dependent* global mobility $v_0(t)$ and a *time independent* factor $f_{ji}=L(L-1)\exp(-D_{ji})$ describing the influence of an effective distance between i and j. The number of independent mobility parameters is then reduced to $T+\frac{1}{2}(L^2-L)$ and we obtain in this case:

$$k_r=T(L+1)+\tfrac{1}{2}(L^2-L) \tag{3.7}$$

fitting parameters and

$$f_r=q-k_r=T(L^2-2L-1)-\tfrac{1}{2}L(L-1) \tag{3.8}$$

degrees of freedom.

The number of fitting parameters is further reduced, if we assume the mobility to have the form

$$v_{ji}^{(R)}(t)=L(L-1)\,v_0(t)\exp(-\beta d_{ji}) \tag{3.9}$$

introduced in (1.36). With $T+1$ independent mobility parameters the total number of fitting parameters is now

$$k_R=T(L+1)+1 \tag{3.10}$$

and in this case we have

$$f_R=q-k_R=T(L^2-2L-1)-1 \tag{3.11}$$

degrees of freedom. For illustration let us take the later discussed case Germany with $L=11$ regions and $T=27$ years. In this case the numbers q, k, k_r, k_R and f, f_r, f_R are, according to (3.3) . . . (3.11)

$$\left.\begin{aligned} & k=1782; \quad k_r=379 \quad k_R=325; \\ q=2970; \quad & f=1188; \quad f_r=2591 \quad f_R=2645\,. \end{aligned}\right\} \tag{3.12}$$

It is intuitively clear that the form (3.6) or (3.9) of $v_{ji}(t)$ acquires plausibility if it turns out, that the $w_{ji}(t)$ can be fitted to the $w_{ji}^{(e)}(t)$ under such assumptions with a high accuracy, although the number of parameters has been drastically reduced from k to k_r or to k_R.

The trendparameters – i.e. the utilities and mobilities – are determined by minimizing the sum of the squared deviations between theoretical and empirical expressions. In other words, we apply the *method of least squares.*

It seems to be highly plausible that the fluctuations of the value of the empiric migration matrix $w_{ji}^{(e)}(t)$ around their meanvalue are Poisson-distributed [3.1, 2]. Empirical tests support this assumption. The minimization of the sum of square deviations between $w_{ji}(t)$ and $w_{ji}^{(e)}(t)$ can be expected to provide good fitting results. But also the log-linear estimation, that means the minimization of the sum of square deviations between $\ln[w_{ji}(t)]$ and $\ln[w_{ji}^{(e)}(t)]$ leads to almost equally good results [3.2]. We shall begin with the latter estimation, since this procedure provides simple analytical expressions of all trend parameters in terms of empirical quantities. The further procedure of Section 3.1 is organized according to Table 3.2.

Table 3.2. Organisation of Section 3.1

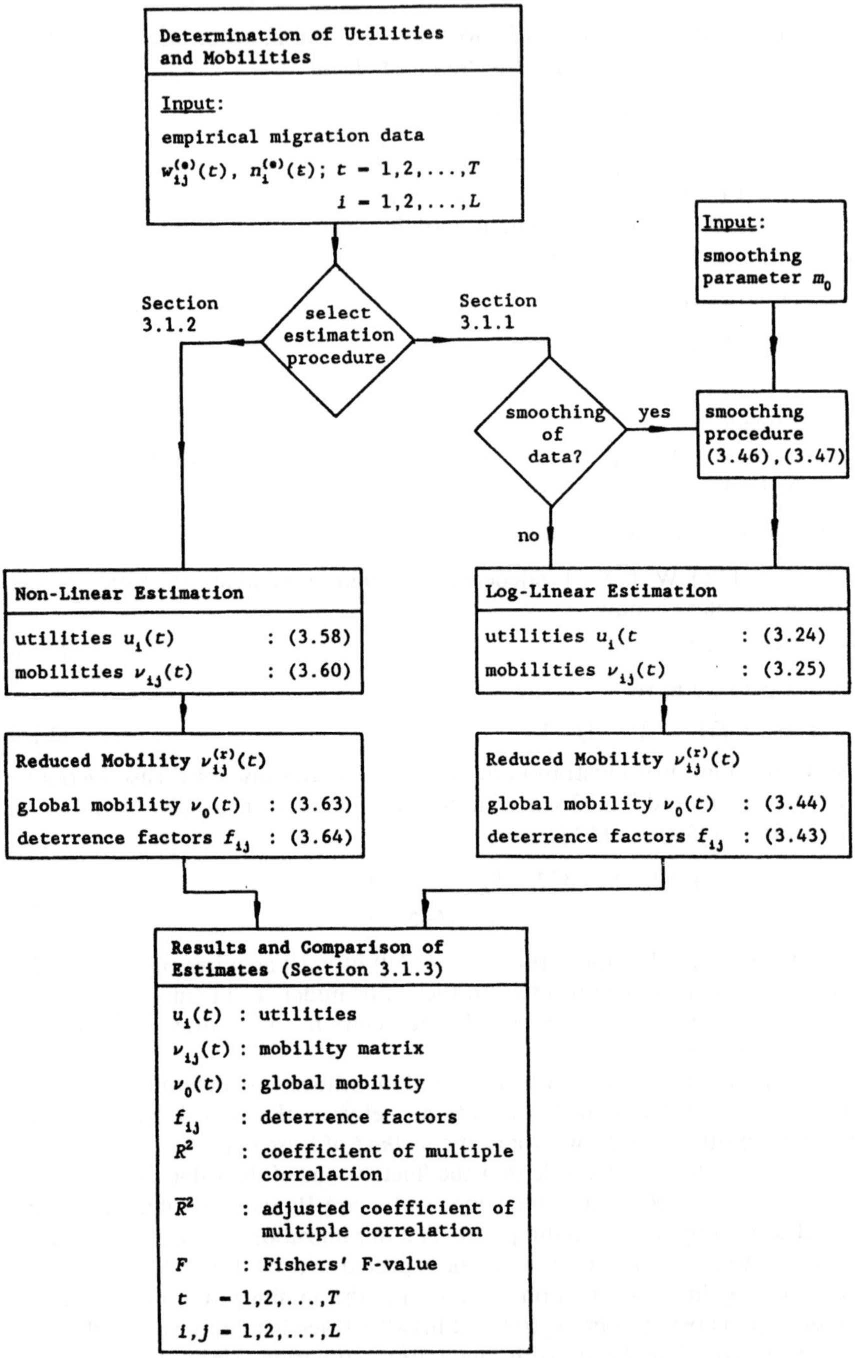

3.1.1 Log-Linear Estimation of Trendparameters

The log-linear estimation procedure consists in minimizing the expression [3.3]

$$F(v, u)=\sum_{t=1}^{T}\sum_{k,l=1}^{L}{}' \{\ln[w_{kl}^{(e)}(t)]-\ln[n_l^{(e)}(t)\,v_{kl}(t)\exp(u_k(t)-u_l(t))]\}^2 \qquad (3.13)$$

by appropriate choice of the parameters $v_{kl}(t)=v_{lk}(t)$ and $u_k(t)$. (The prime at the sum over k and l indicates that $k \neq l$.) Because the probability transition rates depend on the *differences* of the utilities only, we are allowed to impose without restriction of generality the additional constraint

$$\left.\begin{aligned}&\sum_{i=1}^{L} u_i(t)=0\\ &\text{for}\quad t=1, 2, \ldots, T\end{aligned}\right\} \qquad (3.14)$$

in order to obtain uniquely determined values of the $u_i(t)$. The necessary condition for $F[v, u]$ to assume its minimum is, that its variation vanishes:

$$\left.\begin{aligned}\delta F=\sum_{t=1}^{T}\Bigg\{&\sum_{i=1}^{L}\left[\frac{\partial F}{\partial u_i(t)}+\lambda(t)\right]\delta u_i(t)\\ &+\sum_{i,j}^{L}\frac{1}{2}\left[\frac{\partial F}{\partial v_{ij}(t)}+\frac{\partial F}{\partial v_{ji}(t)}\right]\delta v_{ij}(t)\Bigg\}\overset{!}{=}0\end{aligned}\right\} \qquad (3.15)$$

where the constraint (3.14) has been taken into account using Lagrange parameters $\lambda(t)$. The variations $\delta u_i(t)$ and $\delta v_{ij}(t)=\delta v_{ji}(t)$ are here considered as independent of each other. Hence, each of the corresponding variational terms must vanish independently, yielding

$$\frac{\partial F}{\partial u_i(t)}+\lambda(t)=0, \quad \text{for}\quad i=1, 2, \ldots, L \quad \text{and} \quad t=1, 2, \ldots, T \qquad (3.16)$$

and

$$\left.\begin{aligned}&\frac{\partial F}{\partial v_{ji}(t)}+\frac{\partial F}{\partial v_{ij}(t)}=0 \qquad \text{for}\quad i, j=1, 2, \ldots, L,\ i \neq j\\ &\qquad\qquad\qquad\qquad\quad\ \text{and}\quad t=1, 2, \ldots, T.\end{aligned}\right\} \qquad (3.17)$$

Using the explicit form (3.13) one obtains

$$\left.\begin{aligned}\frac{\partial F}{\partial u_i(t)} &= \sum_{k=1}^{L}{}' 2\{\ln[v_{ik}(t)]+[u_i(t)-u_k(t)]-\ln[p_{ik}^{(e)}(t)] \\ &\qquad -\ln[v_{ki}(t)]+[u_i(t)-u_k(t)]+\ln[p_{ki}^{(e)}(t)]\} \\ &= \sum_{k=1}^{L}{}' 4\left\{[u_i(t)-u_k(t)]-\ln\left[\frac{p_{ik}^{(e)}(t)}{p_{ki}^{(e)}(t)}\right]\right\}\end{aligned}\right\} \tag{3.18}$$

and

$$\frac{\partial F}{\partial v_{ji}(t)} = \frac{2}{v_{ji}(t)}\{\ln[v_{ji}(t)]+[u_j(t)-u_i(t)]-\ln[p_{ji}^{(e)}(t)]\} \tag{3.19}$$

where we have introduced the empiric individual transition rate

$$p_{ji}^{(e)}(t) = \frac{w_{ji}^{(e)}(t)}{n_i^{(e)}(t)} \tag{3.20}$$

and have made use of $v_{ik}(t)=v_{ki}(t)$ in (3.18). The conditions (3.16) and (3.17) now take the form

$$\sum_{k=1}^{L}{}' 4\left\{[u_i(t)-u_k(t)]-\frac{1}{2}\ln\left[\frac{p_{ik}^{(e)}(t)}{p_{ki}^{(e)}(t)}\right]\right\}+\lambda(t)=0 \tag{3.21}$$

and, again using $v_{ik}(t)=v_{ki}(t)$,

$$\frac{2}{v_{ji}(t)}\{\ln[v_{ji}^2(t)]-\ln[p_{ji}^{(e)}(t)p_{ij}^{(e)}(t)]\}=0\ . \tag{3.22}$$

Evidently eqs. (3.21) and (3.22) are uncoupled equations for the determination of the $u_i(t)$ and $v_{ij}(t)=v_{ji}(t)$, respectively. From (3.21) there follows by summation over $i=1, 2, \ldots, L$, that the Lagrange parameter $\lambda(t)$ vanishes:

$$\lambda(t) = \frac{1}{L}\sum_{i,k}^{L}{}' \ln\left[\frac{p_{ik}^{(e)}(t)}{p_{ki}^{(e)}(t)}\right]=0\ . \tag{3.23}$$

Thereupon $u_i(t)$ can be determined uniquely if the constraint (3.14) is inserted in (3.21). The result is

$$\left.\begin{aligned}&u_i(t) = \frac{1}{2L}\sum_{k=1}^{L}{}' \ln\left[\frac{p_{ik}^{(e)}(t)}{p_{ki}^{(e)}(t)}\right], \\ &\text{for } i=1, 2, \ldots, L \quad\text{and}\quad t=1, 2, \ldots, T.\end{aligned}\right\} \tag{3.24}$$

Finally (3.22) yields

$$\left.\begin{aligned}&v_{ji}(t)=v_{ij}(t)=[p_{ji}^{(e)}(t)p_{ij}^{(e)}(t)]^{1/2} \\ &\text{for } j, i = 1, 2, \ldots, L \qquad (j \neq i) \\ &\text{and } t = 1, 2, \ldots, T.\end{aligned}\right\} \tag{3.25}$$

Formulas (3.24) and (3.25) express the utilities and mobilities in terms of the empirically known individual transition rates.

*Utilities and Mobilities as Functions of Explanatory Variables**

So far we have treated $u_i(t)$ and $v_{ji}(t)=v_{ij}(t)$ as independent trendparameters. Theoretical considerations may however lead to the conclusion that $u_i(t)$ and $v_{ji}(t)$ depend on a few independent parameters only. For instance let us take the case

$$\left.\begin{array}{ll} u_i(t)=u_i[\Gamma_\beta(t)]\,, & \beta=1,2,\ldots,B \\ v_{ji}(t)=v_{ji}[\Omega_\alpha(t)], & \alpha=1,2,\ldots,A\,. \end{array}\right\} \tag{3.26}$$

where $\Gamma_\beta(t)$ and $\Omega_\alpha(t)$ are the independent parameters to be determined by the least square method. The variation principle must now be modified. Expressing the variations $\delta u_i(t)$ and $\delta v_{ji}(t)$ as

$$\left.\begin{array}{l} \delta u_i(t)=\displaystyle\sum_\beta \frac{\partial u_i(t)}{\partial \Gamma_\beta(t)}\,\delta\Gamma_\beta(t) \\ \delta v_{ji}(t)=\displaystyle\sum_\alpha \frac{\partial v_{ji}(t)}{\partial \Omega_\alpha(t)}\,\delta\Omega_\alpha(t) \end{array}\right\} \tag{3.27}$$

the variational principle now assumes the form

$$\delta F=\sum_{t=1}^{T}\left\{\sum_{\beta=1}^{B}\left[\frac{\partial F}{\partial \Gamma_\beta(t)}+\lambda_\beta(t)\right]\delta\Gamma_\beta(t)+\sum_{\alpha=1}^{A}\frac{\partial F}{\partial \Omega_\alpha(t)}\,\delta\Omega_\alpha(t)\right\} \tag{3.28}$$

with

$$\left.\begin{array}{l} \dfrac{\partial F}{\partial \Gamma_\beta(t)}=\displaystyle\sum_{i=1}^{L}\frac{\partial F}{\partial u_i(t)}\cdot\frac{\partial u_i(t)}{\partial \Gamma_\beta(t)} \\ \dfrac{\partial F}{\partial \Omega_\alpha(t)}=\displaystyle\sum_{i,j}^{L}{}'\,\frac{\partial F}{\partial v_{ji}(t)}\cdot\frac{\partial v_{ji}(t)}{\partial \Omega_\alpha(t)} \\ \lambda_\beta(t)=\displaystyle\sum_{i=1}^{L}\lambda(t)\frac{\partial u_i(t)}{\partial \Gamma_\beta(t)}\,. \end{array}\right\} \tag{3.29}$$

Because $\delta\Gamma_\beta(t)$ and $\delta\Omega_\alpha(t)$ are independent variations by assumption, the variational principle now leads to

$$\left[\frac{\partial F}{\partial \Gamma_\beta(t)}+\lambda_\beta(t)\right]=0;\qquad \beta=1,2,\ldots,B \tag{3.30a}$$

and

$$\frac{\partial F}{\partial \Omega_\alpha(t)}=0;\qquad \alpha=1,2,\ldots,A\,. \tag{3.30b}$$

* This section can be omitted on a first reading

Another case of particular interest is the determination of the mobility under the assumption that it has the reduced form

$$\nu_{ji}^{(r)}(t)=\nu_0(t)f_{ji} \tag{3.31}$$

with time-independent $f_{ji}=f_{ij}$ satisfying the constraint (see (1.30))

$$\sum_{i,j}^{L}{}' f_{ji}=L(L-1)\,. \tag{3.32}$$

(The prime at the sum indicates that $i \neq j$.)

The variation principle with respect to $\nu_{ji}^{(r)}(t)$ reads

$$\left.\begin{aligned}
&\delta\left[F(\nu,u)-\Lambda\sum_{j,i}^{L}{}' f_{ji}\right]\\
&=\sum_{t=1}^{T}\sum_{j,i}^{L}{}'\left[\frac{\partial F}{\partial \nu_{ji}(t)}\delta\nu_{ji}(t)+\lambda(t)\delta f_{ji}\right]\\
&=\sum_{t=1}^{T}\sum_{j,i}^{L}{}'\left\{\left[\frac{\partial F}{\partial \nu_{ji}(t)}f_{ji}\right]\delta\nu_0(t)+\left[\frac{\partial F}{\partial \nu_{ji}(t)}\nu_0(t)+\lambda(t)\right]\delta f_{ji}\right\}\\
&\overset{!}{=}0
\end{aligned}\right\} \tag{3.33}$$

where now

$$\delta\nu_{ji}(t)=\delta\nu_0(t)f_{ji}+\nu_0(t)\delta f_{ji} \tag{3.34}$$

had to be used. Since the variations $\delta\nu_0(t)$, $\delta f_{ji}=\delta f_{ij}$ are independent, eq. (3.33) yields

$$\sum_{k,l}^{L}{}'\frac{1}{2}\left[\frac{\partial F}{\partial \nu_{lk}(t)}+\frac{\partial F}{\partial \nu_{kl}(t)}\right]f_{lk}=0; \qquad t=1,2,\ldots,T \tag{3.35}$$

and

$$\left.\begin{aligned}
&\sum_{t=1}^{T}\frac{1}{2}\left[\frac{\partial F}{\partial \nu_{ji}(t)}+\frac{\partial F}{\partial \nu_{ij}(t)}\right]\nu_0(t)+\Lambda=0\\
&\text{with}\quad \Lambda=\sum_{t=1}^{T}\lambda(t)\qquad\text{and}\qquad i,j=1,2,\ldots,L;\quad i\neq j\,.
\end{aligned}\right\} \tag{3.36}$$

On inserting (3.19) and (3.31) into (3.35) and (3.36) there follow the equations

$$\left.\begin{aligned}
&\frac{1}{\nu_0(t)}\sum_{k,l}^{L}{}'\{\ln[\nu_0^2(t)]+\ln[f_{lk}^2]-\ln[p_{lk}^{(e)}(t)p_{kl}^{(e)}(t)]\}=0\\
&\text{for}\quad t=1,2,\ldots,T
\end{aligned}\right\} \tag{3.37}$$

and

$$\left.\begin{aligned}&\sum_{t=1}^{T}\{\ln[v_0^2(t)]+\ln[f_{ji}^2]-\ln[p_{ji}^{(e)}(t)\,p_{ij}^{(e)}(t)]\}+f_{ji}\Lambda=0\\&\text{for}\quad i,j=1,2,\ldots,L;\quad i\neq j\,.\end{aligned}\right\}\tag{3.38}$$

By taking the sum over j and i (with $j \neq i$) of (3.38) and using (3.37) it follows that the Lagrange parameter Λ must vanish

$$\Lambda=0\tag{3.39}$$

Solving (3.37) for $v_0(t)$ and (3.38) for f_{ji} one obtains the results:

$$v_0(t)=\prod_{k,l}{}' \frac{[p_{lk}^{(e)}(t)\,p_{kl}^{(e)}(t)]^{\frac{1}{2L(L-1)}}}{[f_{lk}]^{\frac{1}{L(L-1)}}}\tag{3.40}$$

where the accent at the product again excludes factors with $l=k$, and

$$f_{ji}=\prod_{t=1}^{T}\frac{[p_{ji}^{(e)}(t)\,p_{ij}^{(e)}(t)]^{\frac{1}{2T}}}{[v_0(t)]^{\frac{1}{T}}}\,.\tag{3.41}$$

Eliminating $v_0(t)$ in (3.41) by inserting (3.40) one is lead to

$$f_{ji}=\frac{\prod\limits_{k,l}^{L}{}' [f_{kl}]^{\frac{1}{L(L-1)}}\cdot\prod\limits_{t=1}^{T}[p_{ji}^{(e)}(t)\,p_{ij}^{(e)}(t)]^{\frac{1}{2T}}}{\prod\limits_{t=1}^{T}\prod\limits_{k,l}^{L}{}' [p_{lk}^{(e)}(t)\,p_{kl}^{(e)}(t)]^{\frac{1}{2TL(L-1)}}}\,.\tag{3.42}$$

Evidently, the f_{ji} are only determined up to a common factor which cancels out in (3.42). Therefore instead of (3.42), f_{ji} can be written in the form

$$f_{ji}=C\cdot\prod_{t=1}^{T}[p_{ji}^{(e)}(t)\,p_{ij}^{(e)}(t)]^{\frac{1}{2T}}\tag{3.43}$$

where C is to be determined by the constraint (3.32). The final form of $v_0(t)$ is now found by inserting (3.43) into (3.40)

$$v_0(t)=\frac{1}{C}\,\frac{\prod\limits_{k,l}^{L}{}' [p_{lk}^{(e)}(t)\,p_{kl}^{(e)}(t)]^{\frac{1}{2L(L-1)}}}{\prod\limits_{k,l}^{L}{}'\prod\limits_{t=1}^{T}[p_{lk}^{(e)}(t)\,p_{kl}^{(e)}(t)]^{\frac{1}{2TL(L-1)}}}\,.\tag{3.44}$$

The results (3.43) and (3.44) are simple and plausible: Apart from normalization factors it turns out, that the *time-independent* factor f_{ji} is given by the *temporal geometric meanvalue* of the symmetrical $(p_{ji}^{(e)}(t)\,p_{ij}^{(e)}(t))^{1/2}$, whereas the *space-independent* factor $v_0(t)$ is proportional to the *regional geometric meanvalue* of the $(p_{ji}^{(e)}(t)\,p_{ij}^{(e)}(t))^{1/2}$. With (3.43) and (3.44) the reduced mobility can easily be composed according to (3.31):

$$v_{ji}^{(r)}(t) = \frac{\prod_{k,l}^{L}{}' \, [p_{lk}^{(e)}(t) p_{kl}^{(e)}(t)]^{\frac{1}{2L(L-1)}} \prod_{t=1}^{T} [p_{ji}^{(e)}(t) p_{ij}^{(e)}(t)]^{\frac{1}{2T}}}{\prod_{k,l}^{L}{}' \prod_{t=1}^{T} [p_{lk}^{(e)}(t) p_{kl}^{(e)}(t)]^{\frac{1}{2TL(L-1)}}} . \tag{3.45}$$

This formula must be compared with (3.25) obtained *without* the assumption that $v_{ji}(t)$ factorizes into a time-dependent and region-dependent term.

*Smoothing of Migration Matrices**

It seems reasonable to use smoothed values of the empiric migration matrices in the expressions just derived because of two reasons: Firstly there will exist ambiguities in the counting of migration acts which can lead to considerable deviations from the true values of the migration matrices, if the latter are small and secondly the value $w_{ji}^{(e)}(t) = 0$ – which could accidently arise for very small regions i or j in some year – should be avoided because it leads to divergencies in the formula (3.21) for the utilities.

A natural assumption for the smoothed migration matrix is

$$w_{ij}^{(s)}(t) = \sum_{t'=1}^{T} A_{ij}(t) g_{ij}(t-t') \, w_{ij}^{(e)}(t') \tag{3.46}$$

where

$$\left. \begin{aligned} & g_{ij}(t-t') = \exp\left\{ -\frac{\overline{w_{ij}^{(e)}}}{m_0} |t-t_0| \right\} \\ \text{with} \quad & \overline{w_{ij}^{(e)}} = \frac{1}{T} \sum_{t=1}^{T} w_{ij}^{(e)}(t) \end{aligned} \right\} \tag{3.47}$$

is a weighting factor for the influence of migration in neighbouring years t' on the smoothed migration matrix $w_{ij}^{(s)}(t)$ and $A_{ij}(t)$ follows from the normalization condition

$$\sum_{t'=1}^{T} A_{ij}(t) g_{ij}(t-t') = 1 . \tag{3.48}$$

The parameter m_0 determines the width of the smoothing function. The two limiting cases

$$\left. \begin{aligned} & w_{ij}^{(s)}(t) = w_{ij}^{(e)}(t), \qquad t = 1, 2, \ldots, T \\ & \quad \text{for} \quad m_0 \to 0 \\ & w_{ij}^{(s)}(t) = \frac{1}{T} \sum_{t'=1}^{T} w_{ij}^{(e)}(t') = \overline{w_{ij}^{(e)}}, \qquad t = 1, 2, \ldots, T \\ & \quad \text{for} \quad m_0 \to \infty \end{aligned} \right\} \tag{3.49}$$

* This section can be omitted on a first reading

are included in the general formula (3.46). For $m_0 \to 0$ no smoothing takes place, whereas for $m_0 \to \infty$ the time dependence of $w_{ij}^{(s)}(t)$ is smeared out completely so that $w_{ij}^{(s)}$ becomes the meanvalue of the migration matrices $w_{ij}^{(e)}(t)$ in the interval $[0, T]$. For realistic regional systems which include small migration matrices like the Canadian system, a value $m_0 \approx 60$ proves to be optimal in the sense that then the quality of the prediction of migration flows reaches its optimum.

3.1.2 Nonlinear Estimation of Trendparameters

Although the log-linear estimation of the mobilities and utilities has lead us to analytical and unique expressions for $u_i(t)$ and $v_{ji}(t)$ we expect even better results of the optimization procedure, if we apply the least square method directly to the residuals

$$\varepsilon_{kl}(t) = w_{kl}^{(e)}(t) - w_{kl}(t) \tag{3.50}$$

The determination of the minimum of the functional

$$\tilde{F}(\boldsymbol{v}, \boldsymbol{u}) = \sum_{t=1}^{T} \sum_{k,l}^{L}{}' \{w_{kl}^{(e)}(t) - n_l^{(e)}(t) v_{kl}(t) \exp[u_k(t) - u_l(t)]\}^2 \tag{3.51}$$

under the constraint (3.14) now leads to the variation principle

$$\delta\tilde{F} = \sum_{t=1}^{T} \left\{ \sum_{i=1}^{L} \left[\frac{\partial \tilde{F}}{\partial u_i(t)} + \tilde{\lambda}(t) \right] \delta u_i(t) + \sum_{i,j}^{L}{}' \frac{1}{2} \left[\frac{\partial \tilde{F}}{\partial v_{ji}(t)} + \frac{\partial \tilde{F}}{\partial v_{ij}(t)} \right] \delta v_{ji}(t) \right\} = 0 \tag{3.52}$$

in complete analogy (3.15). If $u_i(t)$ and $v_{ji}(t) = v_{ij}(t)$ can be varied independently, (3.52) yields the equations

$$\left[\frac{\partial \tilde{F}}{\partial u_i(t)} + \tilde{\lambda}(t) \right] = 0, \quad \text{for} \quad i = 1, 2, \ldots, L \quad \text{and} \quad t = 1, 2, \ldots, T \tag{3.53}$$

and

$$\left[\frac{\partial \tilde{F}}{\partial v_{ji}(t)} + \frac{\partial \tilde{F}}{\partial v_{ij}(t)} \right] = 0, \quad \text{for} \quad i, j = 1, 2, \ldots, L \quad \text{and} \quad t = 1, 2, \ldots, T. \tag{3.54}$$

Inserting the explicit expressions

$$\left.\begin{aligned} &\frac{\partial \tilde{F}}{\partial u_i(t)} = \\ &\sum_{k}^{L}{}' \{2[w_{ki}^{(e)}(t) n_i^{(e)}(t) v_{ki}(t) g(u_k(t), u_i(t)) - n_i^{(e)2}(t) v_{ki}^2(t) g^2(u_k(t), u_i(t))] \\ &\qquad - 2[w_{ik}^{(e)}(t) n_k^{(e)}(t) v_{ik}(t) g(u_i(t), u_k(t)) - n_k^{(e)2}(t) v_{ik}^2(t) g^2(u_i(t), u_k(t))]\} \end{aligned}\right\} \tag{3.55}$$

and

$$\frac{\partial \tilde{F}}{\partial v_{ji}(t)} = 2[n_i^{(e)2}(t) v_{ji}(t) g^2(u_j(t), u_i(t)) - n_i^{(e)}(t) w_{ji}^{(e)}(t) g(u_j(t), u_i(t))] \tag{3.56}$$

with the abbreviation

$$g(u_j(t), u_i(t)) = \exp[u_j(t) - u_i(t)] \tag{3.57}$$

into (3.53) and (3.54), it can be easily seen that (3.53) reduces to

$$\frac{\partial \tilde{F}}{\partial u_i(t)} = 0, \qquad \text{for} \quad i = 1, 2, \ldots, L \quad \text{and} \quad t = 1, 2, \ldots, T \tag{3.58}$$

because it follows by summation of (3.53) over i, that the Lagrange parameter vanishes:

$$\tilde{\lambda}(t) = 0\,. \tag{3.59}$$

Furthermore (3.54) can be solved for $v_{ji}(t) = v_{ij}(t)$:

$$v_{ji}(t) = \frac{n_i^{(e)}(t) w_{ji}^{(e)}(t) g(u_j(t), u_i(t)) + n_j^{(e)}(t) w_{ij}^{(e)}(t) g(u_i(t), u_j(t))}{n_i^{(e)2}(t) g^2(u_j(t), u_i(t)) + n_j^{(e)2}(t) g^2(u_i(t), u_j(t))}\,. \tag{3.60}$$

The differences to the log-linear estimation become now obvious: The equations (3.58) and (3.60) which correspond to (3.21) and (3.22) remain coupled equations for $v_{ji}(t)$ and $u_i(t)$. Furthermore they are highly nonlinear equations for which only numerical methods of solution exist. It is however safe to assume that they will have unique solutions which only slightly differ from the analytical solutions of the log-linear estimation procedure. Inserting (3.60) into (3.58) one obtains a system of L equations for the utilities only. These equations are algebraic in the variables $\exp(u_k(t))$ and can be solved by numerical standard methods.

Finally we can also determine the factors $v_0(t)$ and f_{ji} of the reduced mobility (3.31) in the same manner as in the log-linear estimation. The equations following from the variational principle

$$\sum_{k,l}^{L}{}' \frac{1}{2}\left[\frac{\partial \tilde{F}}{\partial v_{lk}(t)} + \frac{\partial \tilde{F}}{\partial v_{kl}(t)}\right] f_{lk} = 0 \qquad \text{for} \quad t = 1, 2, \ldots, T \tag{3.61}$$

and

$$\left.\begin{aligned} &\sum_{t=1}^{T} \frac{1}{2}\left[\frac{\partial \tilde{F}}{\partial v_{ji}(t)} + \frac{\partial \tilde{F}}{\partial v_{ij}(t)}\right] v_0(t) = 0 \\ &\text{for} \quad i, j = 1, 2, \ldots, L; \quad i \neq j \end{aligned}\right\} \tag{3.62}$$

fully correspond to (3.35) and (3.36) of the log-linear estimation. Inserting of (3.56) into (3.61) and (3.62) yields the results:

$$v_0(t) = \frac{\sum_{i,j}^{L}{}' f_{ji} w_{ji}^{(e)}(t) n_i^{(e)}(t) g(u_j(t), u_i(t))}{\sum_{i,j}^{L}{}' f_{ji}^2 n_i^{(e)2}(t) g^2(u_j(t), u_i(t))} \tag{3.63}$$

and

$$f_{ji}=\frac{\sum_{t=1}^{T} \nu_0(t)[w_{ji}^{(e)}(t)n_i^{(e)}(t)g(u_j(t), u_i(t))+w_{ij}^{(e)}(t)n_j^{(e)}(t)g(u_i(t), u_j(t))]}{\sum_{t=1}^{T} \nu_0^2(t)[n_i^{(e)2}(t)g^2(u_j(t),u_i(t))+n_j^{(e)2}(t)g^2(u_i(t),u_j(t))]}. \tag{3.64}$$

The coupled equations (3.63) and (3.64) for $\nu_0(t)$ and f_{ji} still contain the utilities, which must be determined beforehand in the above described way. Afterwards (3.63) and (3.64) can be solved with a fast converging iteration algorithm: Beginning with the zeroth order of the f_{ji} by choosing $f_{ji}^{(0)}=1$ for all i, j, the first order approximation $\nu_0^{(1)}(t)$ of $\nu_0(t)$ is found by inserting $f_{ji}^{(0)}$ in (3.63) and the first order $f_{ji}^{(1)}$ approximation of f_{ji} follows by inserting $\nu_0^{(1)}(t)$ in (3.64). This procedure is repeated to obtain higher approximants until their values converge satisfactorily.

3.1.3 Comparison of the Quality of Estimates

In the preceding Sections 3.1.1 and 3.1.2 we have worked out different versions of estimates of the theoretical migration matrices $w_{ji}(t)$ which lead to the determination of the utilities and of different versions of the mobilities.

The cases are:

1a) the log-linear estimation of $u_i(t)$ and $\nu_{ji}(t)$
1b) the log-linear estimation of $u_i(t)$ and $\nu_{ji}^{(r)}(t)=\nu_0(t)f_{ji}$
1c) the log-linear estimation of $u_i(t)$ and $\nu_{ji}^{(R)}(t)=\nu_0(t)L(L-1)\exp(-\beta d_{ji})$
2a) the non-linear estimation of $u_i(t)$ and $\nu_{ji}(t)$
2b) the non-linear estimation of $u_i(t)$ and $\nu_{ji}^{(r)}(t)=\nu_0(t)f_{ji}$
2c) the non-linear estimation of $u_i(t)$ and $\nu_{ji}^{(R)}(t)=\nu_0(t)L(L-1)\exp(-\beta d_{ji})$.

We compare these six estimations in the case of Germany taken as an example. Here we have $L=11$ regions, $T=27$ years of registration and $q=2970$ empirical migration matrix elements. Some comparative informations about the six estimations including the significance tests are summarized in Table 3.3.

Furthermore the Figures 3.1a,b,c and 3.2a,b,c give an illustrative account of the validity of the general theoretical form (3.1) of the migration matrices and of the comparative reliability of the estimation procedures.

In these figures the theoretical migration matrices $w_{ji}(t)$ are plotted versus the empirical matrices $w_{ji}^{(e)}(t)$ with $w_{ji}(t)$ as ordinate and $w_{ji}^{(e)}(t)$ as abscissa for the six cases 1a,b,c and 2a,b,c listed in Table 3.2. An ideal fit would require all points to lie on the straight 45°-line. Deviations from this ideal show a decreasing quality of the fit.

As expected the non-linear estimations lead in all cases to slightly better results than the log-linear estimation although the latter is already quite satisfactory. The transition from general mobilities $\nu_{ji}(t)$ (case a) to reduced mobilities $\nu_{ji}^{(r)}(t)$ (case b) of course leads to an increase of deviations from the straight 45°-line. On the other hand the fits b) are still quite good, although the number of fitting parameters is

Table 3.3. Comparison of six versions of estimations in the case of Germany

	1. Log-linear estimation			2. Non-linear estimation		
case	a	b	c	a	b	c
number of parameters	1782	379	325	1782	379	325
form of mobility	$\nu_{ij}(t)$	$\nu_{ij}^{(r)}(t)$	$\nu_{ij}^{(R)}(t)$	$\nu_{ij}(t)$	$\nu_{ij}^{(r)}(t)$	$\nu_{ij}^{(R)}(t)$
R^2	0.9973	0.9798	0.7895	0.9992	0.9812	0.7895
$\bar{R}^2$	0.9933	0.9768	0.7636	0.9981	0.9784	0.7636
F	248.1	331.0	30.5	882.4	355.7	30.5

drastically reduced. The case c) however, working with a mobility $\nu_{ji}^{(R)}(t)$ expressed in terms of the *geometrical distance* between regions gives a much worse fit, although the number of fitting parameters is only slightly reduced against case b). The form $\nu_{ji}^{(R)}(t) = \nu_0(t) L(L-1) \exp(-\beta d_{ji})$ is therefore ruled out in the further discussion.

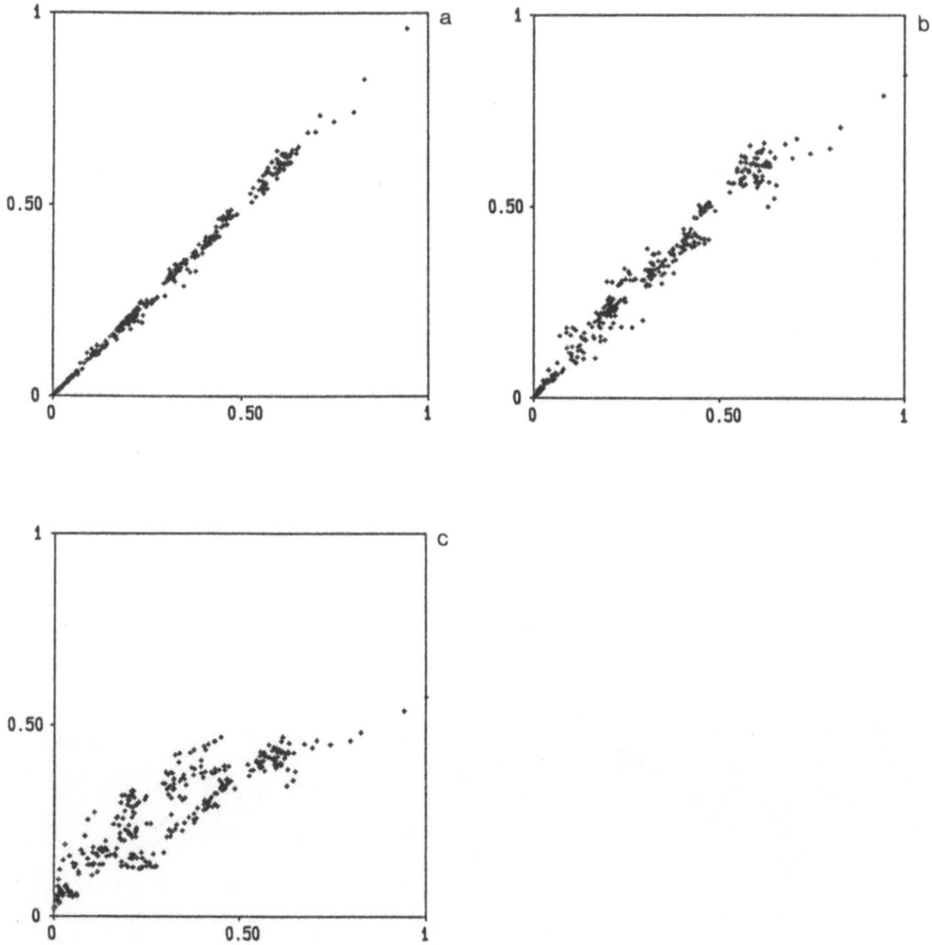

Fig. 3.1a,b,c. Theoretical migration matrices $w_{ji}(t)$ plotted versus empirical matrices $w_{ji}^{(e)}(t)$ for the log-linear estimation and three assumed forms of the mobilities (see Table 3.3)

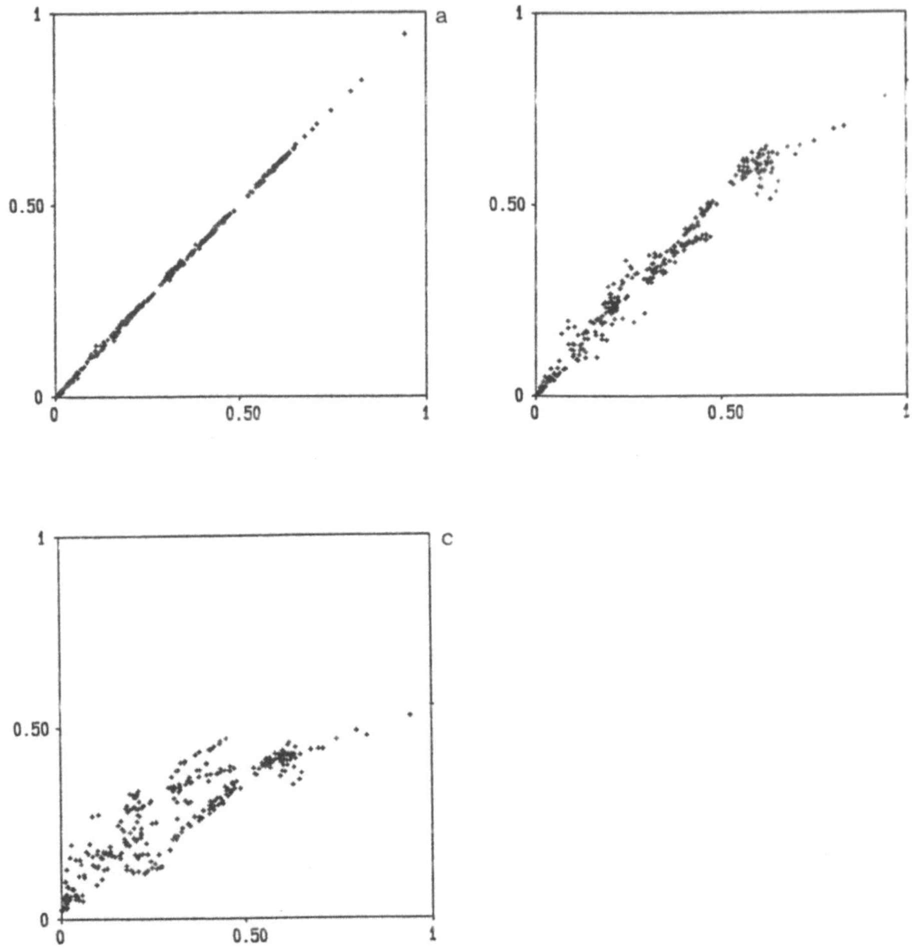

Fig. 3.2a,b,c. Theoretical migration matrices $w_{ji}(t)$ plotted versus empirical matrices $w_{ji}^{(e)}(t)$ for the non-linear estimation and three assumed forms of the mobilities (see Table 3.3)

3.2 The Dependence of Trendparameters on Socio-Economic Key-Factors

In the preceding section we have found by a regression analysis all trendparameters directly determining the dynamics of the migratory system. Now we turn to the dependence of the trendparameters, in particular the utilities, on the socio-economic situation.

Several time and space dependent variables are available to characterize this situation, for example income per capita, the labour force, the rate of unemployment, the housing stock or starts, retail sales, rents, taxes, services and so on. Beyond variables *exogenous* to the migration process also *endogenous* variables like the regional population numbers themselves and their powers may play a role.

Although it cannot be contested, that an influence of such socio-economic variables on the migratory behaviour exists, it is very difficult on the other hand to establish this causal relationship *directly*. Therefore we need an *indirect* indicator for this imputed causal influence. We shall take the *degree* and the *statistical significance* of the spatio-temporal correlation of a socio-economic variable to the regional utilities $u_i(t)$ and/or to the mobilities $v_{ji}(t)$ as a *measure of the (causal) relevance* of this variable for migration. Those endogenous or exogenous socio-economic variables which can be considered as important in this sense are denoted as *key-factors.* Their importance will become explicit by representing the $u_i(t)$ and the $v_{ji}(t)$ as a function of key-factors.

The spirit of this argumentation implies two further consequences:

a) If socio-economic variables have the role of "causes" and migratory trendparameters the role of "effects", time delays – up to a few years – between the effected migration trends and the causing key-factors may arise since the effect may need some "delay time" to adapt to the cause. Therefore we shall take into account the possibility of time lags from the very beginning.

b) The degree of influence of key-factors on migration will vary considerably. Therefore we need an algorithm for the *ranking of relevance* of socio-economic variables. The purpose of the algorithm – to be developed in Section 3.2.3 – is to select in an objective mathematical manner a few key-factors with highest correlation, hence highest explanatory value out of many socio-economic variables. On the other hand this algorithm must not anticipate the interpretation of the key-factors.

We shall now introduce in Section 3.2.1 the vector space of spatial-temporal functions. In Section 3.2.2 the search for an optimal representation of the utility vector in terms of key-factor will simultaneously lead to the ranking of the latter. The organization of Section 3.2.1 until 3.2.3 is graphically represented in Table 3.4.

Table 3.4. Organisation of Section 3.2

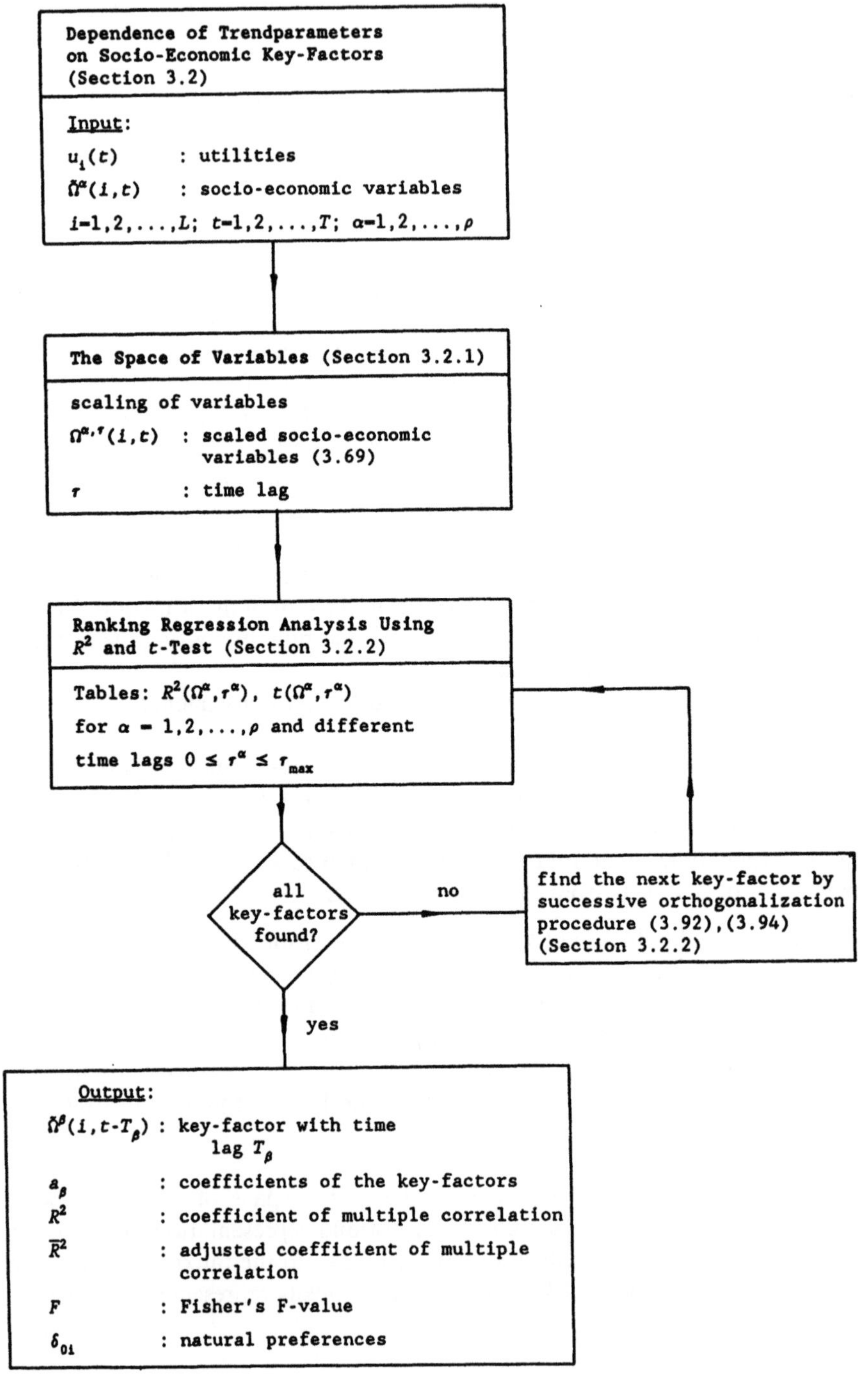

3.2.1 The Space of Variables

In the following we shall treat the dependence of spatial-temporal functions $f(i, t)$ on the region or space index $i=1, 2, \ldots, L$ and the time index $t=1, 2, \ldots, T$ in a formally equivalent way. Therefore it is better now to write the time dependent regional utilities in the form

$$u_i(t) \rightarrow u(i, t) \qquad \text{for} \quad i=1, \ldots, L \quad \text{and} \quad t=1, \ldots, T. \tag{3.65}$$

The ρ socio-economic variables empirically available for a system of L regions and for T time steps are denoted as

$$\left.\tilde{\Omega}^\alpha(i, t); \quad \begin{array}{l} \alpha=1, 2, \ldots, \rho \\ i=1, 2, \ldots, L \\ t=1, 2, \ldots, T \end{array}\right\} \tag{3.66}$$

As yet these variables have different dimensions and scales and it is not yet decided whether they are relevant for migration or not. In order to take into account time lags appropriately we also introduce time-shifted socio-economic variables by[1]

$$\tilde{\Omega}^{\alpha,\tau}(i, t) \equiv \tilde{\Omega}^\alpha(i, t-\tau) \tag{3.67}$$

$$\text{for} \quad i=1, \ldots, L; \quad t=1, \ldots, T; \quad \tau=0, 1, 2, \ldots$$

where $\tau=0$ reproduces the variables (3.66). Making use of the regional meanvalues

$$\bar{\Omega}^{\alpha,\tau}(t) = \frac{1}{L} \sum_{i=1}^{L} \tilde{\Omega}^{\alpha,\tau}(i, t) \tag{3.68}$$

one can now introduce scaled dimensionless variables

$$\Omega^{\alpha,\tau}(i, t) = \frac{\tilde{\Omega}^{\alpha,\tau}(i, t) - \bar{\Omega}^{\alpha,\tau}(t)}{\bar{\Omega}^{\alpha,\tau}(t)} \tag{3.69}$$

with the property

$$\sum_{i=1}^{L} \Omega^{\alpha,\tau}(i, t) = 0 . \tag{3.70}$$

It is now very convenient to consider all spatio-temporal functions $f(i, t)$, in particular $u(i, t)$ and $\Omega^{\alpha,\tau}(i, t)$ as elements of an $L \cdot T$-dimensional linear *vector space* $\mathscr{V}$. The vectors are denoted as

$$\left.\begin{array}{lll} |u\rangle & \text{with components} & u(i, t) \\ |\Omega^{\alpha,\tau}\rangle & \text{with components} & \Omega^{\alpha,\tau}(i, t) \\ \text{and generally} & & \\ |f\rangle & \text{with components} & f(i, t) \end{array}\right\} \tag{3.71}$$

[1] It is implied that $\tilde{\Omega}^\alpha(i, t)$ must be known in the time interval $[1-\tau, T-\tau]$.

Vectors can be multiplied with real numbers and added component-wise by the prescription that

$$\left.\begin{array}{l} |h\rangle = |a\cdot f + b\cdot g\rangle \in \mathscr{V} \\ \quad \text{has components} \quad a f(i,t) + b g(i,t)\,. \end{array}\right\} \tag{3.72}$$

For any two elements $|f\rangle$, $|g\rangle$ of $\mathscr{V}$ a scalarproduct $\langle g|f\rangle$ with the properties

$$\left.\begin{array}{c} \langle g|f\rangle = \langle f|g\rangle \\ \langle g|a_1 f_1 + a_2 f_2\rangle = a_1\langle g|f_1\rangle + a_2\langle g|f_2\rangle \\ \text{for real numbers } a_1,\, a_2 \end{array}\right\} \tag{3.73}$$

can be introduced as follows:

$$\langle g|f\rangle = \sum_{t=1}^{T}\sum_{i=1}^{L} g(i,t) f(i,t)\,. \tag{3.74}$$

The norm of a vector which is a measure for its "length" is given by

$$\|f\| = \sqrt{\langle f|f\rangle}\,. \tag{3.75}$$

If a vector g has the norm $\|g\| = 1$, it is denoted as "normalized" vector. Two special relations between two vectors are of further interest:

$$|f\rangle = c|g\rangle, \quad \text{or} \quad f(i,t) = c g(i,t) \tag{3.76}$$

$$\langle h|k\rangle = \sum_{t=1}^{T}\sum_{i=1}^{L} h(i,t) k(i,t) = 0\,. \tag{3.77}$$

In case (3.76) $|f\rangle$ is said to be parallel to $|g\rangle$ and
in case (3.77) $|h\rangle$ is said to be orthogonal to $|k\rangle$.

Let us now set up a tentative relation between the utility and socio-economic variables in component notation or, alternatively, in vector notation. If the regio-temporal dependence of $u(i, t)$ is to be "explained" by the corresponding dependence on r selected "relevant" socio-economic variables $\Omega^{\alpha,\tau_\alpha}(i, t)$, $\alpha = 1, 2, \ldots, r$, we expect that $u(i, t)$ can be represented essentially as a (linear) superposition of these variables:

$$u(i,t) = \sum_{\alpha=1}^{r} b_\alpha \Omega^{\alpha,\tau_\alpha}(i,t) + \Delta(i,t) \tag{3.78a}$$

or in vector notation

$$|u\rangle = \sum_{\alpha=1}^{r} b_\alpha |\Omega^{\alpha,\tau_\alpha}\rangle + |\Delta\rangle\,. \tag{3.78b}$$

Here, the τ_α are eventual time lags and the b_α describe the amplitude of the influence of $|\Omega^{\alpha,\tau_\alpha}\rangle$ on the utility vector $|u\rangle$. The vector $|\Delta\rangle$ is the residual utility whose spatio-temporal form cannot be explained by the r key-factors $|\Omega^{\alpha,\tau_\alpha}\rangle$. It can be interpreted as a residual "natural" utility vector of the regions i. Because of (3.70) and with

$$\sum_{i=1}^{L} u(i,t)=0 \qquad \text{for} \quad t=1,2,\ldots,T \tag{3.79}$$

there follows that the residual utilities satisfy

$$\sum_{i=1}^{L} \Delta(i,t)=0 \qquad \text{for} \quad t=1,2,\ldots,T. \tag{3.80}$$

3.2.2 The Standard Regression

Proceeding in a direct manner, the *optimal choice* of the b_α follows by the least square requirement

$$\delta\sum_{t=1}^{T}\sum_{i=1}^{L}\left[u(i,t)-\sum_{\alpha=1}^{r} b_\alpha\Omega^{\alpha,\tau_\alpha}(i,t)\right]^2=0 \tag{3.81a}$$

or, equivalently, by

$$\left.\begin{aligned}
&\delta\|\Delta\|^2\\
&=\delta\left\langle u-\sum_{\beta=1}^{r} b_\beta\Omega^{\beta,\tau_\beta}\middle| u-\sum_{\alpha=1}^{r} b_\alpha\Omega^{\alpha,\tau_\alpha}\right\rangle\\
&=-2\sum_{\beta=1}^{r}\delta b_\beta\left\langle \Omega^{\beta,\tau_\beta}\middle| u-\sum_{\alpha=1}^{r} b_\alpha\Omega^{\alpha,\tau_\alpha}\right\rangle\\
&=0\,.
\end{aligned}\right\} \tag{3.81b}$$

Since the variations δb_β, $\beta=1,2,\ldots r$ are independent, (3.81b) yields the equations

$$\left.\begin{aligned}
&\sum_{\alpha=1}^{T} b_\alpha\langle\Omega^{\beta,\tau_\beta}|\Omega^{\alpha,\tau_\alpha}\rangle=\langle\Omega^{\beta,\tau_\beta}|u\rangle\\
&\text{for} \quad \beta=1,2,\ldots,r
\end{aligned}\right\} \tag{3.82}$$

which may serve to determine the b_α, We see that simultaneously the norm or the "length" of the rest utility $|\Delta\rangle$ has been minimized. Furthermore it turns out by taking the scalar product of (3.78b) with $\langle\Omega^{\beta,\tau_\beta}|$ and using (3.82) that

$$\left.\begin{aligned}&\langle\Omega^{\beta,\tau_\beta}|\Delta\rangle\\&=\langle\Omega^{\beta,\tau_\beta}|u\rangle-\sum_{\alpha=1}^{r}b_\alpha\langle\Omega^{\beta,\tau_\beta}|\Omega^{\alpha,\tau_\alpha}\rangle=0\,.\end{aligned}\right\}\tag{3.83}$$

That means, the residual utility $|\Delta\rangle$ is orthogonal to all $|\Omega^{\beta,\tau_\beta}\rangle$ used in representing $|u\rangle$.

The just described standard regression is formally elegant but has some serious disadvantages: All variables $\Omega^{\alpha,\tau_\alpha}$ must simultaneously be taken into account. The full set of variables however contains relevant and irrelevant ones, and some of the variables may be almost equivalent spatio-temporal functions so that one of them could suffice as a proxy. Furthermore no criterion exists which time lag τ_α should be attributed to variable Ω^α (the choice $\tau_\alpha=0$ would be arbitrary). Therefore we prefer a *modified optimization method* to be developed in the next section. It will lead to the successive determination and the ranking of the most relevant key-factors $|\Omega^{\alpha,\tau_\alpha}\rangle$ including the optimal choice of their time lag τ_α: Finally, however, we shall also end up with a representation of $|u\rangle$ of the form (3.78) where now the $|\Omega^{\alpha,\tau_\alpha}\rangle$, $\alpha=1, 2, \ldots, r$ are ordered in the sequence of their relevance.

3.2.3 Ranking of Relevance of Key-Factors and Representation of the Utility Vector

We shall now introduce an alternative regression algorithm denoted as Ranking Regression Analysis which avoids the shortcomings of the standard regression and automatically leads to a sequential selection and ranking of relevance of key-factors in the representation of the utility vector. The procedure of finding successively the sequence of the most relevant key-factors $|\hat{\Omega}^{1,\tau_1}\rangle$, $|\hat{\Omega}^{2,\tau_2}\rangle, \ldots,$ $|\hat{\Omega}^{r,\tau_r}\rangle$ is described in three steps:

At first we describe explicitly the initial step leading to the determination of the most important key-factor $|\hat{\Omega}^{1,\tau_1}\rangle$ and the first order representation of $|u\rangle$.

Secondly we proceed by inductive reasoning: Starting from the assumption that the first γ key-factors $|\hat{\Omega}^1\rangle, \ldots, |\hat{\Omega}^\gamma\rangle$ and the γth order representation of $|u\rangle$ have already been found, we show how the $(\gamma+1)$-st key-factor $|\hat{\Omega}^{(\gamma+1)}\rangle$and the $(\gamma+1)$-st order representation of $|u\rangle$ can be constructed systematically.

Thirdly we show how after ending with the r-th order the final representation of $|u\rangle$ can be transformed into the form (3.78b).

The first step consists in representing (in first order) $|u\rangle$ in the form

$$|u\rangle=a_1|\hat{\Omega}^{1,\tau_1}\rangle+|u_{\perp 1}\rangle\tag{3.84}$$

where $|\hat{\Omega}^{1,\tau_1}\rangle$ is normalized to $\|\hat{\Omega}^{1,\tau_1}\|=1$ and must be found as that variable out of all socio-economic variables $|\Omega^{\alpha,\tau_\alpha}\rangle$ which "fits best" to $|u\rangle$. Afterwards the α's are renamed to give this variable the index 1. The coefficient a_1 is of course optimized to lead to the minimal norm of the residual utility $|u_{\perp 1}\rangle$, with the result

$$a_1 = \langle \hat{\Omega}^{1,\tau_1} | u \rangle \tag{3.85}$$

and it follows from (3.84) and (3.85) that

$$\langle \hat{\Omega}^{1,\tau_1} | u_{\perp 1} \rangle = 0 \tag{3.86}$$

so that the residual utility $|u_{\perp 1}\rangle$ is orthogonal to the key-factor $|\hat{\Omega}^{1,\tau_1}\rangle$.

We have now to find *criteria for the selection* of the best key-factor $|\hat{\Omega}^{1,\tau_1}\rangle$ out of all $|\Omega^{\alpha,\tau}\rangle$.

Intuitively the functional form of $\Omega^{1,\tau_1}(i, t)$ as compared to all other $\Omega^{\alpha,\tau}(i, t)$ should be the most similar to the form of the utility function $u(i, t)$. In other words: The vector $|\Omega^{1,\tau_1}\rangle$ should be "the most parallel" one to $|u\rangle$ among all. Hence we choose among all (normalized) variables $|\hat{\Omega}^{\alpha,\tau}\rangle$ that variable $|\hat{\Omega}^{1,\tau_1}\rangle$ for which the value of the coefficient $|a_1|$ is larger or at least equal to other values $|\langle \hat{\Omega}^{\alpha,\tau} | u \rangle|$.

In order to make the choice of the $|\hat{\Omega}^{1,\tau_1}\rangle$ of highest explanatory value and highest significance unique, the values of the *corrected coefficient of multiple correlation* $\bar{R}^2$ must be compared for all $|\Omega^{\alpha,\tau}\rangle$; furthermore the comparative *t-test* must be made. The details of these statistical significance tests are given in Part IV. Their result is that the most significant variable $|\hat{\Omega}^{1,\tau_1}\rangle$ with the optimal time lag τ_1 can be found *unambiguously*. The same holds for the $|\hat{\Omega}^{2,\tau_2}\rangle$, $|\hat{\Omega}^{3,\tau_3}\rangle, \ldots$ found in the higher orders.

The first step ends with the decomposition of all $|\Omega^{\alpha}\rangle$ into a term proportional to $|\hat{\Omega}^{1,\tau_1}\rangle$ and a residual term $|\Omega^{\alpha}_{\perp 1}\rangle$ orthogonal to $|\hat{\Omega}^{1,\tau_1}\rangle$:

$$\left.\begin{aligned} |\Omega^{1,\tau_1}\rangle &= c_{11} |\hat{\Omega}^{1,\tau_1}\rangle \\ |\Omega^{\alpha}\rangle &= c_{1\alpha} |\hat{\Omega}^{1,\tau_1}\rangle + |\Omega^{\alpha}_{\perp 1}\rangle \\ &\text{for all other } \alpha \neq 1 \,. \end{aligned}\right\} \tag{3.87}$$

This decomposition is unique since the orthogonality condition

$$\langle \hat{\Omega}^{1,\tau_1} | \Omega^{\alpha}_{\perp 1} \rangle = 0 \tag{3.88}$$

leads to the determination of the coefficients:

$$c_{1\alpha} = \langle \hat{\Omega}^{1,\tau_1} | \Omega^{\alpha} \rangle, \qquad \alpha = 1, 2, 3, \ldots \,. \tag{3.89}$$

Evidently, the residual term $|\Omega^{\alpha}_{\perp 1}\rangle$ of $|\Omega^{\alpha}\rangle$ describes that part of the spatio-temporal function $\Omega^{\alpha}(i, t)$, which is not proportional to $\Omega^{1,\tau_1}(i, t)$, but orthogonal to it.

The second step: Let us now make the inductive assumption

a) that the first γ key-factors $|\hat{\Omega}^{\beta,\tau_\beta}\rangle$, $\beta = 1, 2, \ldots, \gamma$ have already been found and that they are an orthonormal system satisfying

$$\langle\hat{\Omega}^{\alpha,\tau_\alpha}|\hat{\Omega}^{\beta,\tau_\beta}\rangle=\delta_{\alpha\beta} \qquad \text{for} \quad \alpha,\beta=1,2,\ldots,\gamma. \tag{3.90}$$

b) that the γ-th order representation of $|u\rangle$ reads

$$\left.\begin{aligned} &|u\rangle=\sum_{\beta=1}^{\gamma} a_\beta|\hat{\Omega}^{\beta,\tau_\beta}\rangle+|u_{\perp\gamma}\rangle \\ &\text{with} \\ &a_\beta=\langle\hat{\Omega}^{\beta,\tau_\beta}|u\rangle \\ &\text{and} \\ &\langle\hat{\Omega}^{\beta,\tau_\beta}|u_{\perp\gamma}\rangle=0 \\ &\text{for} \quad \beta=1,2,\ldots,\gamma \end{aligned}\right\} \tag{3.91}$$

c) and that all socio-economic variables can be decomposed as follows

$$\left.\begin{aligned} &|\Omega^\alpha\rangle=\sum_{\beta=1}^{\alpha} c_{\beta\alpha}|\hat{\Omega}^{\beta,\tau_\beta}\rangle \qquad \text{for} \quad \alpha\leq\gamma \\ &|\Omega^\alpha\rangle=\sum_{\beta=1}^{\gamma} c_{\beta\alpha}|\hat{\Omega}^{\beta,\tau_\beta}\rangle+|\Omega^\alpha_{\perp\gamma}\rangle \qquad \text{for} \quad \alpha>\gamma \\ &\text{with} \quad c_{\beta\alpha}=\langle\hat{\Omega}^{\beta,\tau_\beta}|\Omega^\alpha\rangle \\ &\text{and} \quad \langle\hat{\Omega}^{\beta,\tau_\beta}|\Omega^\alpha_{\perp\gamma}\rangle=0\,. \end{aligned}\right\} \tag{3.92}$$

We have now to show how one can unambiguously proceed to the $(\gamma+1)$-st order. This makes it necessary to find the *next-key-factor* $|\hat{\Omega}^{(\gamma+1),\tau_{\gamma+1}}\rangle$ so that a)

$$\langle\hat{\Omega}^{\alpha,\tau_\alpha}|\hat{\Omega}^{\beta,\tau_\beta}\rangle=\delta_{\alpha\beta} \qquad \text{for} \quad \alpha,\beta=1,2,\ldots,(\gamma+1) \tag{3.93}$$

holds, and

b) that the $(\gamma+1)$-st order representation of $|u\rangle$ reads

$$\left.\begin{aligned} &|u\rangle=\sum_{\beta=1}^{\gamma+1} a_\beta|\hat{\Omega}^{\beta,\tau_\beta}\rangle+|u_{\perp(\gamma+1)}\rangle \\ &\text{with} \quad a_\beta=\langle\hat{\Omega}^{\beta,\tau_\beta}|u\rangle \\ &\text{and} \quad \langle\hat{\Omega}^{\beta,\tau_\beta}|u_{\perp(\gamma+1)}\rangle=0 \\ &\text{for} \quad \beta=1,2,\ldots,(\gamma+1) \end{aligned}\right\} \tag{3.94}$$

c) and that all socio-economic variables now can be decomposed as follows

$$\left.\begin{aligned}
&|\Omega^\alpha\rangle=\sum_{\beta=1}^{\alpha} c_{\beta\alpha}|\hat{\Omega}^{\beta,\tau_\beta}\rangle \quad \text{for} \quad \alpha\leq\gamma+1\\
&|\Omega^\alpha\rangle=\sum_{\beta=1}^{\gamma+1} c_{\beta\alpha}|\hat{\Omega}^{\beta,\tau_\beta}\rangle+|\Omega^\alpha_{\perp(\gamma+1)}\rangle \quad \text{for} \quad \alpha>\gamma+1\\
&\text{with} \qquad c_{\beta\alpha}=\langle\hat{\Omega}^{\beta,\tau_\beta}|\Omega^\alpha\rangle\\
&\text{and} \qquad \langle\hat{\Omega}^{\beta,\tau_\beta}|\Omega^\alpha_{\perp(\gamma+1)}\rangle=0\,.
\end{aligned}\right\} \tag{3.95}$$

Indeed we shall now show how $|\hat{\Omega}^{(\gamma+1),\tau_{\gamma+1}}\rangle$ can be constructed so that (3.93), (3.94) and (3.95) is fulfilled, given that (3.90), (3.91) and (3.92) are satisfied. The starting point of the construction is the residual utility $|u_{\perp\gamma}\rangle$ defined in (3.91) which must now be decomposed as

$$|u_{\perp\gamma}\rangle=a_{\gamma+1}|\hat{\Omega}^{(\gamma+1),\tau_{\gamma+1}}\rangle+|u_{\perp(\gamma+1)}\rangle \tag{3.96}$$

where $|\hat{\Omega}^{(\gamma+1),\tau_{\gamma+1}}\rangle$ must be orthonormal to all former key-factors in order to fulfill (3.93):

$$\langle\hat{\Omega}^{(\gamma+1),\tau_{\gamma+1}}|\hat{\Omega}^{\beta,\tau_\beta}\rangle=\delta_{\gamma+1,\beta}, \qquad \text{for} \quad \beta=1,2,\ldots,(\gamma+1) \tag{3.97}$$

and where $a_{\gamma+1}$ is chosen as

$$a_{\gamma+1}=\langle\hat{\Omega}^{(\gamma+1),\tau_{\gamma+1}}|u_{\perp\gamma}\rangle=\langle\hat{\Omega}^{(\gamma+1),\tau_{\gamma+1}}|u\rangle \tag{3.98}$$

so that

$$\langle\hat{\Omega}^{\beta,\tau_\beta}|u_{\perp(\gamma+1)}\rangle=0 \qquad \text{for} \quad \beta=1,2,\ldots,(\gamma+1)\,. \tag{3.99}$$

Inserting (3.96) into (3.91) one easily obtains the $(\gamma+1)$-st representation (3.94).

The new key-factor $|\hat{\Omega}^{(\gamma+1),\tau_{\gamma+1}}\rangle$ is now selected from the $|\Omega^\alpha_{\perp\gamma}\rangle$ with $\alpha>\gamma$ (see (3.92)). The $|\Omega^\alpha_{\perp\gamma}\rangle$ are the *residuals* of the *not yet used* socio-economic variables $|\Omega^\alpha\rangle$ with $\alpha>\gamma$.

By definition all $|\Omega^\alpha_{\perp\gamma}\rangle$ are orthogonal to all former key-factors $|\hat{\Omega}^{\beta,\tau_\beta}\rangle$, $\beta=1,2,\ldots,\gamma$ as required.

The same selection criteria as in step 1) can now be applied to find that $|\Omega^\alpha_{\perp\gamma}\rangle$ which fits best to the residual utility $|u_{\perp\gamma}\rangle$. The unambiguous result is $|\hat{\Omega}^{(\gamma+1),\tau_{\gamma+1}}\rangle$ after normalization and after renaming that $|\Omega^\alpha_{\perp\gamma}\rangle$ into $|\hat{\Omega}^{(\gamma+1),\tau_{\gamma+1}}\rangle$. To complete the transition from the γ-th to the $(\gamma+1)$-st order, it must be shown that also (3.95) holds. Indeed, for $\alpha=\gamma+1$ there follows

$$\left.\begin{aligned}
&|\Omega^{\gamma+1}\rangle=\sum_{\beta=1}^{\gamma+1} c_{\beta,\gamma+1}|\hat{\Omega}^{\beta,\tau_\beta}\rangle\\
&\text{with} \quad c_{\beta,\gamma+1}=\langle\hat{\Omega}^{\beta,\tau_\beta}|\Omega^{\gamma+1}\rangle
\end{aligned}\right\} \tag{3.100}$$

making use of (3.92) and taking into account that by choice of $|\hat{\Omega}^{(\gamma+1),\tau_{\gamma+1}}\rangle$

$$|\Omega_{\perp\gamma}^{\gamma+1}\rangle \equiv c_{\gamma+1,\gamma+1}|\hat{\Omega}^{(\gamma+1),\tau_{\gamma+1}}\rangle \; . \tag{3.101}$$

For $\alpha > \gamma + 1$ the decomposition

$$\left.\begin{aligned} &|\Omega_{\perp\gamma}^{\alpha}\rangle = c_{\gamma+1,\alpha}|\hat{\Omega}^{(\gamma+1),\tau_{\gamma+1}}\rangle + |\Omega_{\perp(\gamma+1)}^{\alpha}\rangle \\ \text{with}\quad &c_{\gamma+1,\alpha} = \langle\hat{\Omega}^{(\gamma+1),\tau_{\gamma+1}}|\Omega_{\perp\gamma}^{\alpha}\rangle = \langle\hat{\Omega}^{(\gamma+1),\tau_{\gamma+1}}|\Omega^{\alpha}\rangle \end{aligned}\right\} \tag{3.102}$$

can now be made, which leads to the rest of formula (3.95).

This completes our proof that by appropriate choice of the key-factor $|\hat{\Omega}^{(\gamma+1),\tau_{\gamma+1}}\rangle$ one can proceed from the γ-th order (defined by (3.90) to (3.92)) in an unambiguous way to the $(\gamma+1)$-st order (defined by (3.93) to (3.95)). Simultaneously we have found by progressing from order to order a sequence of key-factors $|\hat{\Omega}^{\beta,\tau_\beta}\rangle$ ranked according to their relevance, so that the utility vector $|u\rangle$ can be represented as their linear combination plus a small residual term.

In formal respect our approach consists of the Schmidt orthonormalization method combined with a selection principle which makes the successive choice of the basis vectors unambiguous.

The third step: Let us assume that the ranking procedure has been finished with the r-th order (because no further socio-economic variables are available or because the residual utility $|u_{\perp r}\rangle$ is very small rendering the $|\Omega^\alpha\rangle$ with $\alpha > r$ as irrelevant). Then we end up with the representation

$$\left.\begin{aligned} &|u\rangle = \sum_{\beta=1}^{r} a_\beta|\hat{\Omega}^{\beta,\tau_\beta}\rangle + |u_{\perp r}\rangle \\ \text{with}\quad &a_\beta = \langle\hat{\Omega}^{\beta,\tau_\beta}|u\rangle \; . \end{aligned}\right\} \tag{3.103}$$

The key-factors $|\hat{\Omega}^{\beta,\tau_\beta}\rangle$ are by construction an *orthonormal set of vectors*, but they are not identical with the original variables $|\Omega^\alpha\rangle$, $\alpha = 1, 2, \ldots, r$, which are neither orthogonal nor normalized.

It is however indicated to go back to these original variables since only they have a *direct socio-economic interpretation.*

The transformation of (3.103) into

$$|u\rangle = \sum_{\kappa=1}^{r} b_\kappa|\Omega^\kappa\rangle + |u_{\perp r}\rangle \tag{3.104}$$

is easily achieved because according to (3.92) or (3.95)

$$\left.\begin{aligned} &|\Omega^\kappa\rangle = \sum_{\beta=1}^{\kappa} c_{\beta\kappa}|\hat{\Omega}^{\beta,\tau_\beta}\rangle \quad \text{for} \quad \alpha = 1, 2, \ldots, r \\ \text{with}\quad &c_{\beta\kappa} = \langle\hat{\Omega}^{\beta,\tau_\beta}|\Omega^\kappa\rangle \end{aligned}\right\} \tag{3.105}$$

can be inserted in (3.104) yielding

$$|u\rangle = \sum_{\kappa=1}^{r}\sum_{\beta=1}^{\kappa} b_\kappa c_{\beta\kappa}|\hat{\Omega}^{\beta,\tau_\beta}\rangle + |u_{\perp r}\rangle \; . \tag{3.106}$$

The comparison of (3.106) and (3.103) gives

$$a_\beta = \sum_{\kappa=\beta}^{r} c_{\beta\kappa} b_\kappa \, . \tag{3.107}$$

From (3.107) the coefficients b_κ can be determined successively.

The form of the representation (3.104) agrees with (3.78b). But now the r variables $|\Omega^\kappa\rangle$, $\kappa = 1, 2, \ldots, r$ are linked directly via (3.105) to the orthonormal set of key-factors $|\hat{\Omega}^{\kappa,\tau_\kappa}\rangle$ $\kappa = 1, 2, \ldots, r$ whose successive ranking was found in progressing from order to order in finding representations of increasing quality of the utility vector in terms of key-factors.

II. Interregional Migration in Individual Countries

Synopsis of Part II

One main purpose of this book is the description and evaluation of interregional migration in several case studies on the basis of one and the same theoretical model, in order to secure comparability. In this sense Part II is the central part of the book.

The authors of the case studies of interregional migration in six countries, namely Canada, Federal Republic of Germany, France, Israel, Italy and Sweden, are well acquainted with the demographic, migratory and socio-economic situation of their countries. Furthermore they all have agreed to evaluate and interpret the migration process in their countries on the line of the model presented in Part I. Therefore the chapters of Part II, being independent analyses of the individual countries on the one side, coincide to some degree in their structure and are organized as follows.

In a first group of sections the regional system of the countries is introduced; furthermore, the population evolution, including natural growth as well as different types of migration, is discussed under historical and current points of views; finally, the sources of the data are noted.

In a second group of sections, the constitutive elements of the probability transition rates, the mobility and the regional utilities and preferences are evaluated making use of migratory data. The results are presented in various figures, which are interpreted in terms of the global and regional socio-economic situation creating the time-dependent migratory trends. The migratory stress is also introduced and interpreted within this frame.

The third group of sections comprises the choice of explanatory socio-economic variables and their classification and interpretation. Furthermore, the selection and ranking of key-factors out of the available socio-economic variables is implemented by the Ranking Regression Analysis of Chapter 3, ending with the representation of the global mobility and the regional utilities in terms of these key-factors.

Beyond this frame, complementary investigations enrich the general line of argumentation in several chapters. Examples are: The analysis of the regional inhomogeneity of the natural rate of growth per person in the case of Canada (see Section 5.1.3), the comparison between the present model and the gravity model in the case of France (see Section 6.7), the ranking of temporarily stable migration streams in the case of Israel (see Section 7.4), and the comparative representation of the mobility matrix in terms of geographical, or alternatively effective regional distances in the case of the FRG (see Section 4.2.2).

The sequence of the chapters of Part II has no relevance and is chosen alphabetically with the exception of the Federal Republic of Germany, whose place has been exchanged with Canada for technical reasons during the preparation of the book.

4 Federal Republic of Germany

Günter Haag, Martin Munz, Rolf Reiner and Wolfgang Weidlich

4.1 The Regional System and the Registration of Population and Migration

4.1.1 The Division of the Country into Federal States

The Federal Republic of Germany (FRG) is divided into 10 federal states and the region of Berlin (W) (Figure 4.1). The names of the federal states and the symbols used in graphical presentations of the results are listed in Table 4.1. Some of the states, such as Bayern, Bremen, Hamburg have a long history of their own. Others, such as Rheinland-Pfalz and Niedersachsen were established after the end of the Second World War. In 1952 Baden-Württemberg was formed after a plebiscite by fusion of the states Baden and Württemberg. Saarland joined the FRG since 1957, also after a plebiscite.

The federal states of the Federal Republic are ruled by their own government and administrations. The government (Landesregierung) of each federal state is elected in a four year turn by the population of that federal state and has well-defined responsibilities and fiscal resources. Intervention of the Federal Government (Bundesregierung) concerning states affairs takes place in exceptional cases only. In particular the local administration is responsible for territorial planning and for the regional development policies. The government of each federal state applies such development strategies according to the specific regional background and the political philosophy of the ruling parties. The differences of such political and economic strategies may lead to regional differentiations in the long run.

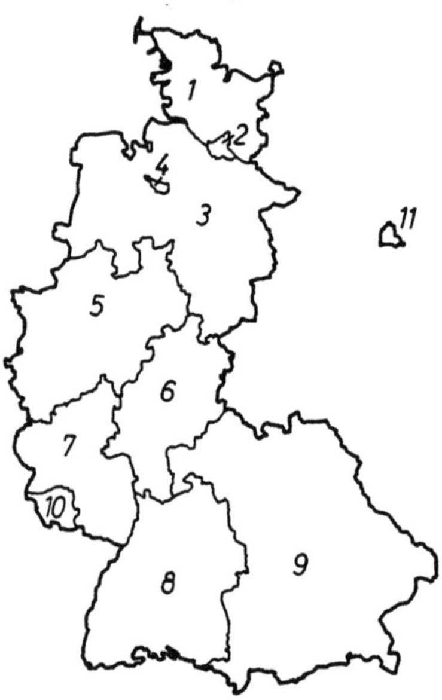

Fig. 4.1. The map of the Federal Republic of Germany and its subdivision into eleven federal states

Table 4.1. The federal states of the FRG. Numbers and symbols used in the graphical representation of regional quantities

symbol	region	name
+	1	Schleswig - Holstein
o	2	Hamburg
Δ	3	Niedersachsen
□	4	Bremen
◊	5	Nordrhein - Westfalen
·	6	Hessen
*	7	Rheinland - Pfalz
x	8	Baden - Württemberg
-	9	Bayern
>	10	Saarland
<	11	Berlin

With a population density of about 250 inhabitants per square kilometre, the FRG is one of the most densely populated countries in Europe. The urbanization process found all over the world is relatively advanced in the FRG. Economically strong centers such as Hamburg, Düsseldorf, Frankfurt, Stuttgart and München are more or less homogeneously distributed over the whole territory. The population of the FRG is relatively mobile. Between 1948 and 1978 more than eight million individuals changed their address each year. More than 50 per cent of these people crossed the border of a municipality during their move, and therefore they became migrants by definition.

4.1.2 The Volume and Registration of Migration

According to German law, everyone who permanently lives in the FRG should be registered in one of the parishes of the country. In the case of interregional migration (move over a state border), the move is recorded both in the state of origin and in the state of destination, and the time of move refers to the year of registration. An uncertainty effect results by a certain time lag in registration of migratory events. Because of this a certain, but small discrepancy exists with respect to population and vital events reported for any given year. Since multiple changes of residence can occur during the time period between two registrations the number of real relocations exceeds the number of registered migrants. However, this effect can be neglected on a state level and for the short time period of one year under consideration. The population numbers are corrected by census data about every 10 years [4.6].

The data of the parish registers were collected and published in parts in the Statistisches Jahrbuch [4.1]. Here, internal migratory moves between the federal states and the number of immigrants and emigrants are reported. The data are available on a yearly basis. Other important socio-economic data, belonging to the public service, labour market, housing market, and industry sector, are also published yearly for each region [4.1].

Nowadays the total volume of migration per year between parishes is about 3 million peoples. On the state level one observes about 1 million migrants per year (Figure 4.2). On the average, the inflow into a state is about 30 per cent higher than the number of births in that state and the outflow exceeds the number of deaths by about 10 per cent. This demonstrates the importance of a migration study for the forecasting of population on a state level. In Figure 4.2 the total volume of migration between the 11 federal states is shown. It should be stressed, that the total volume of migrants is the sum of all migratory events and should not be confused with the global mobility (compare Figure 4.2 and Figure 4.4). In the FRG both quantities show a similar time development. This is due to the fact that the utilities of the German states do not exhibit large variations with time. In general however it is not possible to obtain the global mobility simply by dividing the total volume of migrants by the total population size. For a detailed discussion of the total volume of migration see H. Birg [4.3]. Here, we restrict ourselves to the discussion of the global mobility.

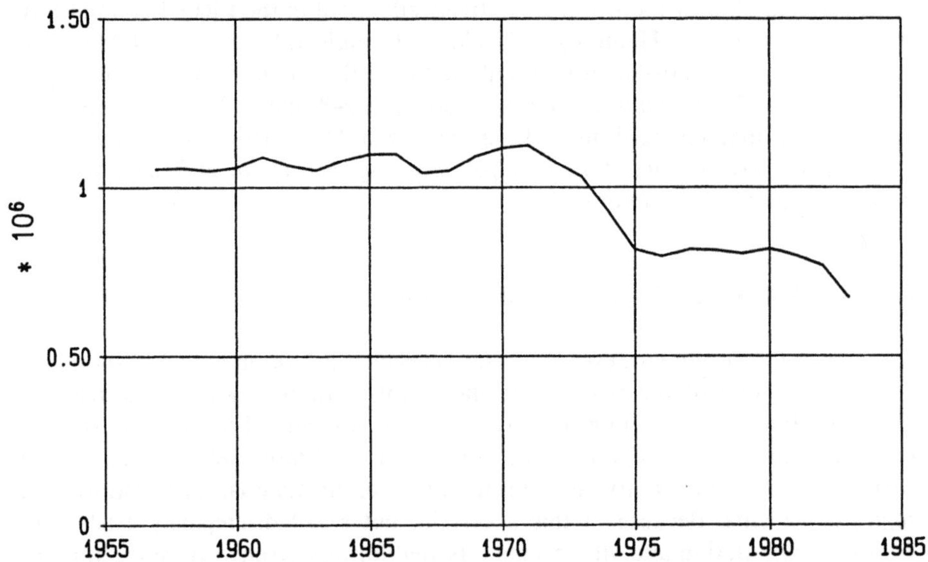

Fig. 4.2. The total volume $w_0(t)$ of migrants between the federal states of the Federal Republic of Germany

4.1.3 Choice of the Period of Evaluation

Before 1957 Berlin (W) was not considered at all in the migration matrices. Hence, only after 1957 a complete and continuous data set for migration flows is available. Therefore, we have restricted all empirical evaluations to the time period from 1957 to 1983.

The erection of the Berlin wall at August, 13th, 1961 of course fundamentally changed the migration statistics. Before this event, about thousand refugees per day from East Germany used Berlin (W) only as a transient stay for their flight to West-Germany. These migrants have been counted in the outflows of Berlin (W), whereas the number of refugees from East Germany were not registered as inflows in the migration table. This special politically induced postwar effect of course leads to a considerable distortion of the utilities with respect to Berlin before 1961. Another special case is the Saarland because this state belonged administratively to France until 1957.

Socio-economic data for all federal states of the FRG were collected systematically and of a comparable base since 1960. Therefore, our socio-economic analysis of trendparameters starts in the year 1960.

4.1.4 Assumption of One Homogeneous Population

Empirical inquiries have shown [4.2] that motivations for a change of residence are highly dependent on different personal characteristics, such as age, sex, family status and social status. It must also be expected, that the migratory behaviour of the native population and of guest workers differs considerably. Therefore it would be highly desirable to have data about sub-groups of the population which may migrate in a distinctly different way. Unfortunately such data are only partially available for the FRG and the other countries. The lack of data concerning a detailed information about personal characteristics causes further problems. Hence, we make the "coarse-graining" assumption of one homogeneous population, in particular, since it is the intention of this book to provide a comparative study of interregional migration for different countries. Simplifying the real situation we thus assume the same mean decision behaviour with respect to migration for all members of the population.

4.1.5 Total Population Growth

The total population growth of the FRG is governed by two sources: The balance of in-migration minus out-migration and the difference between the number of births and deaths. Indeed the forecasting of real population numbers requires detailed information of the rate of natural increase. Natural increase means, the difference between birth- and death rate (net growth rate) plus the difference between immigration minus emigration over the borders of the FRG. This rate is depicted for the total population in Figure 4.3 for the time period 1957 to 1983. The

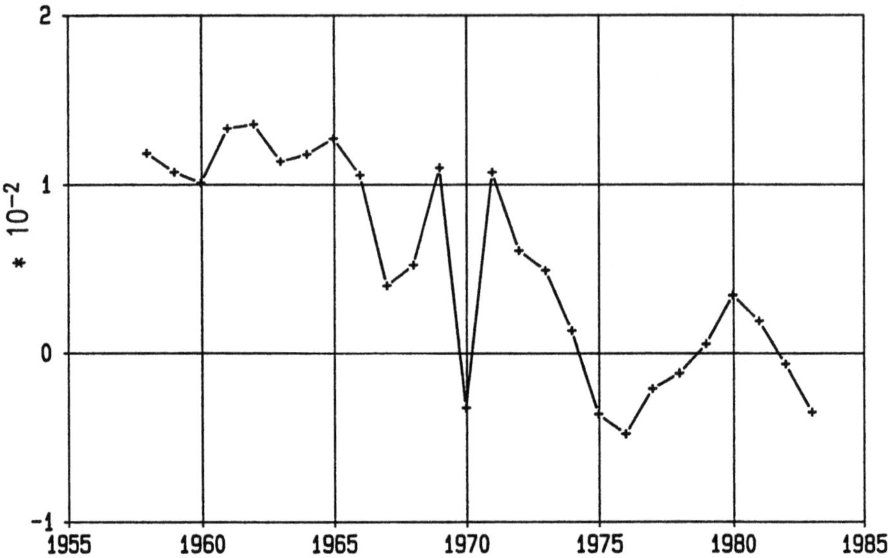

Fig. 4.3. The rate of natural increase of the total population of the Federal Republic of Germany

highly complicated structure of the rate of natural increase means that its forecasting is rather complicated. The death rate of the FRG on the other hand is spatially rather homogeneous and also slowly varying in time. Differences in the death rates of different years are mainly due to changes in the age structure of the population. In contrast, the birth rate shows a more complicated behaviour. This rate is higher in rural areas than in urban ones and is different for the native population and the guest workers. For a more detailed consideration of the demographical aspects of population growth in the FRG see [4.3, 4.7].

4.2 Transition Probabilities, Mobilities and Utilities

4.2.1 Form of Transition Probabilities

As introduced in Chapter 1, we now consider the form of the transition probability in terms of a symmetric mobility factor and a utility-dependent push-pull factor. We assume that the individual transition probability for a move of an individual from state j to state i has the form (1.39a).

$$p_{ij}(t)=v_{ij}(t)\exp\left[u_i(t)-u_j(t)\right] . \tag{4.1}$$

A multitude of socio-economic factors merges into the utilities $u_i(t)$ and the mobility matrix $v_{ij}(t)$. The numerical calculation of mobilities and utilities for the FRG in the period from 1957 to 1983 and their further analysis in terms of socio-economic key-factors is implemented in the following sections of this chapter.

4.2.2 Decomposition of the Mobility Matrix

The nonlinear estimation procedure for trendparameters developed in Section 3.1.2 leads to the best results, as the comparison of the quality of estimates in Section 3.1.3 has shown. Therefore we shall use the nonlinear estimation in the following evaluations for the FRG from 1957 to 1983.

Thus the evaluation of the utilities, the mobility matrix and its decomposition follows the scheme set up by eqs. (3.58), (3.60), (3.63) and (3.64). In particular, the non-decomposed mobility matrix $v_{ij}(t)$ can be computed making use of (3.60).

On the other hand in (1.28) a decomposition of the mobility matrix was introduced into a time-dependent global mobility $v_0(t)$, characterizing the frequency of moves, and a time-independent spatial deterrence function $f_{ij}(\boldsymbol{d})$, characterizing the distance and the spatial interdependence of regions.

$$v_{ij}(t) = v_0(t) f_{ij}(\boldsymbol{d}) \ . \tag{4.2}$$

The ansatz (4.2) for the mobility matrix could be substantiated to a high degree by the quality comparison of estimates in Section 3.1.3.

The global mobility is a measure for the frequency of residential changes across the borders of the states. The estimation of the global mobility makes use of eq. (3.63). The result is presented in Figure 4.4. We observe a long term decrease of the mobility since 1957. The mobility in 1983 is only 52 per cent of the value in 1957. The readiness of the population to move from one to another state, this means the acceptance of a far distance move, has dramatically reduced. An extraordinary decline of about 26 per cent is observed in the time interval inbetween 1971 and 1975. This decline of the mobility has enhanced the problem of structural unemployment.

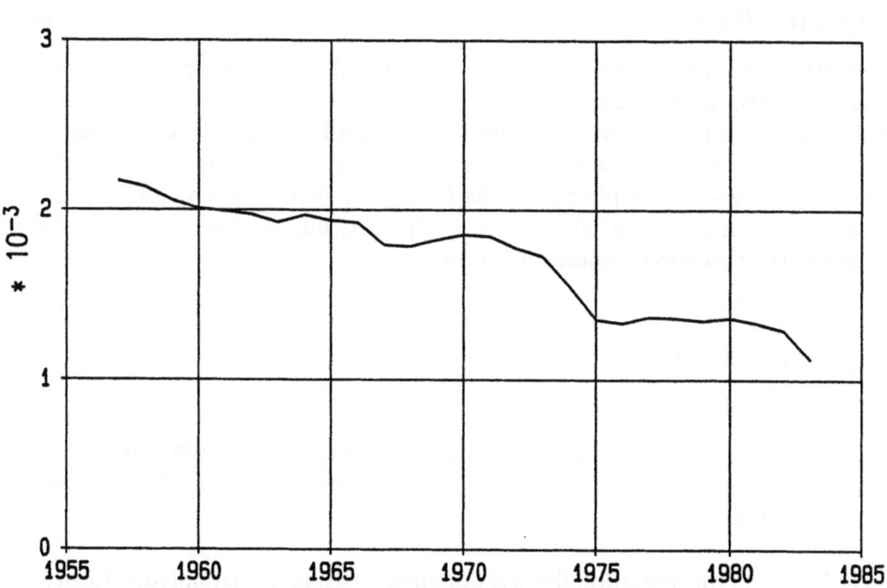

Fig. 4.4. The estimated global mobility $v_0(t)$ of the Federal Republic of Germany

Table 4.2. Estimated values of the deterrence function $f_{ij}(\boldsymbol{d})$ of the 11 federal states of the FRG

	1	2	3	4	5	6	7	8	9	10	11
1	0.00	5.95	1.80	0.86	1.08	0.52	0.34	0.59	0.48	0.11	0.69
2	5.95	0.00	2.21	0.63	0.71	0.48	0.21	0.47	0.40	0.08	0.68
3	1.80	2.21	0.00	3.69	2.46	1.12	0.50	0.84	0.66	0.15	1.12
4	0.86	0.63	3.69	0.00	0.57	0.33	0.17	0.32	0.24	0.06	0.31
5	1.08	0.71	2.46	0.57	0.00	1.61	1.93	1.33	1.12	0.39	1.15
6	0.52	0.48	1.12	0.33	1.61	0.00	2.12	1.67	1.37	0.43	0.84
7	0.34	0.21	0.50	0.17	1.93	2.12	0.00	1.70	0.75	1.80	0.46
8	0.59	0.47	0.84	0.32	1.33	1.67	1.70	0.00	2.79	0.61	0.92
9	0.48	0.40	0.66	0.24	1.12	1.37	0.75	2.79	0.00	0.27	0.78
10	0.11	0.08	0.15	0.06	0.39	0.43	1.80	0.61	0.27	0.00	0.16
11	0.69	0.68	1.12	0.31	1.15	0.84	0.46	0.92	0.78	0.16	0.00

The spatial deterrence function $f_{ij}(\boldsymbol{d})$ describes the influence of "distance" on the frequency to change location. The estimation of $f_{ij}(\boldsymbol{d})$ follows Section 3.1.2, eq. (3.64). The results for the FRG are listed in Table 4.2. In order to see how the deterrence function f_{ij} depends on the distance between state j and i, and whether the geographical distance is an appropriate measure for this distance effect, we first introduce the effective distance D_{ij} according to (1.31)

$$f_{ij}=L(L-1)\exp(-D_{ij})\,. \tag{4.3}$$

Obviously, one further advantage of our definition of the spatial deterrence function (4.3) is that an effective distance can easily be defined and computed. The effective distance D_{ij} is always positive (see Section 1.3) and is listed in Table 4.3.

In order to check on the other hand the role of the geographical distance we have made several assumptions for the functional dependence of the deterrence parameters f_{ij} on geographical distance d_{ij}. In particular we have investigated and compared the following distance functions

$$f_{ij}=a(d_{ij})^{-\alpha} \tag{4.4}$$

$$f_{ij}=a\exp(-\beta d_{ij}) \tag{4.5}$$

$$f_{ij}=a\exp(-\beta d_{ij}^2)\,. \tag{4.6}$$

The parameters a, α and β are estimated following the procedure described in Section 3.1.2, where in (3.51) $v_{ij}(t)$ has to be substituted by $v_0(t)f_{ij}$. Two important results are obtained:

- The best fit of the migration flows is obtained by using the exponential form (4.5). ($\beta=2.18\cdot10^{-3}$ per km, $a=1.73$)

- In general the deterrence function f_{ij} cannot be satisfactorily explained with the geographical distance only. (correlation $R^2=0.79$)

Before comparing the effective distance D_{ij} (Table 4.3) with the geographical distance d_{ij}, the latter must be scaled appropriately:

$$\tilde{D}_{ij}=\beta d_{ij}-\ln(a)+\ln(L(L-1)) . \tag{4.7}$$

The scaled geographical distances $\tilde{D}_{ij}$ are listed in Table 4.4. The comparison of the effective distance D_{ij} with the scaled geographical distance $\tilde{D}_{ij}$ now leads to the

Table 4.3. Effective distance D_{ij} between the 11 federal states of the FRG

	1	2	3	4	5	6	7	8	9	10	11
1	0.00	2.92	4.11	4.85	4.62	5.36	5.79	5.23	5.43	6.93	5.07
2	2.92	0.00	3.91	5.16	5.04	5.43	6.25	5.46	5.61	7.28	5.08
3	4.11	3.91	0.00	3.39	3.80	4.58	5.40	4.87	5.12	6.58	4.59
4	4.85	5.16	3.39	0.00	5.27	5.82	6.46	5.85	6.14	7.59	5.87
5	4.62	5.04	3.80	5.27	0.00	4.22	4.04	4.42	4.58	5.65	4.56
6	5.36	5.43	4.58	5.82	4.22	0.00	3.95	4.19	4.39	5.55	4.88
7	5.79	6.25	5.40	6.46	4.04	3.95	0.00	4.17	4.99	4.11	5.49
8	5.23	5.46	4.87	5.85	4.42	4.19	4.17	0.00	3.67	5.20	4.78
9	5.43	5.61	5.12	6.14	4.58	4.39	4.99	3.67	0.00	6.02	4.95
10	6.93	7.28	6.58	7.59	5.65	5.55	4.11	5.20	6.02	0.00	6.55
11	5.07	5.08	4.59	5.87	4.56	4.88	5.49	4.78	4.95	6.55	0.00

Table 4.4. Scaled geographical distance $\tilde{D}_{ij}$ between the 11 federal states of the FRG

	1	2	3	4	5	6	7	8	9	10	11
1	0.00	4.39	4.72	4.64	5.32	5.45	5.51	5.90	6.08	5.84	4.88
2	4.39	0.00	4.48	4.42	5.09	5.23	5.29	5.68	5.86	5.62	4.78
3	4.72	4.48	0.00	4.43	4.78	4.92	4.97	5.36	5.55	5.33	4.77
4	4.64	4.42	4.43	0.00	4.83	5.17	5.22	5.75	5.80	5.49	5.01
5	5.32	5.09	4.78	4.83	0.00	4.71	4.62	5.05	5.49	4.80	5.40
6	5.45	5.23	4.92	5.17	4.71	0.00	4.23	4.62	5.01	4.59	5.39
7	5.51	5.29	4.97	5.22	4.62	4.23	0.00	4.62	5.07	4.49	5.44
8	5.90	5.68	5.36	5.75	5.05	4.62	4.62	0.00	4.63	4.64	5.51
9	6.08	5.86	5.55	5.80	5.49	5.01	5.07	4.63	0.00	5.08	5.43
10	5.84	5.62	5.33	5.49	4.80	4.59	4.49	4.64	5.08	0.00	5.81
11	4.88	4.78	4.77	5.01	5.40	5.39	5.44	5.51	5.43	5.81	0.00

typical result, that for adjacent states with a long common border the effective distance is smaller than the geographical distance. As a consequence the migration flows between such states are higher than expected taking into account the geographical distance only. The effective distance thus turns out to be a better measure for the "migratory neighbourhood" than the pure geographical distance. Bayern (9) and Baden-Württemberg (8) may serve as an example. The strong socio-economic links between these adjacent states lead to strong migration between them and to a small effective distance.

4.2.3 Regional Utilities and Preferences

The regional utilities are estimated according to (3.58) and their time dependence is depicted in Figure 4.5 for 1957 to 1983. The symbols used for graphical representation of the utilities are listed in Table 4.1. By definition the utilities reflect the overall attractiveness of a federal state for a migrant.

At a first glance it can be seen that the utilities of all federal states remain relatively stable over time, with the exception of Berlin West (<). As already mentioned, the flow of refugees from Berlin to the FRG before the erection of the Berlin wall (1961)

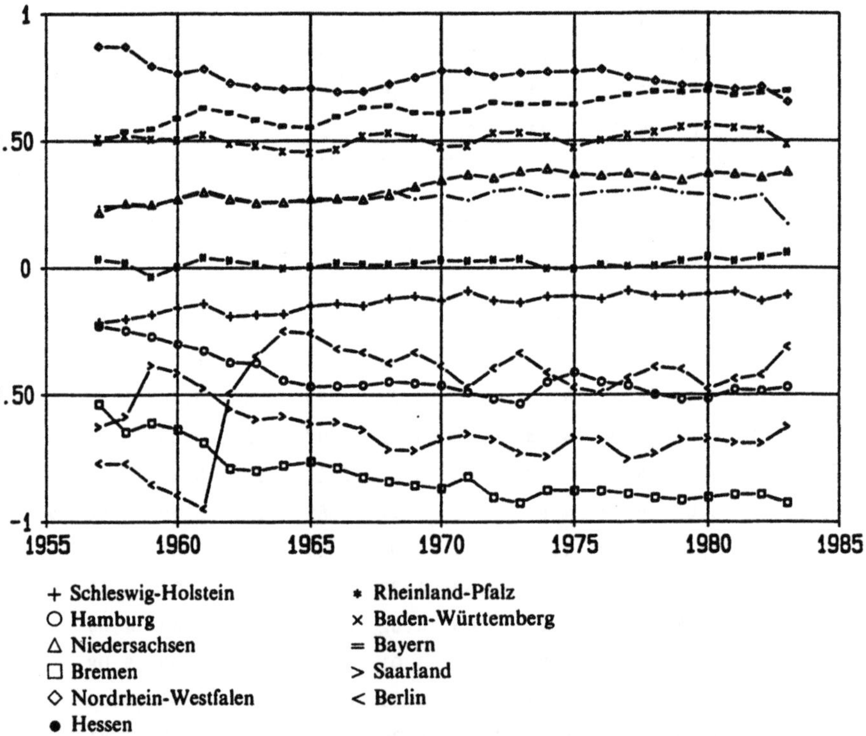

Fig. 4.5. The estimated regional utilities $u_i(t)$ of the Federal Republic of Germany

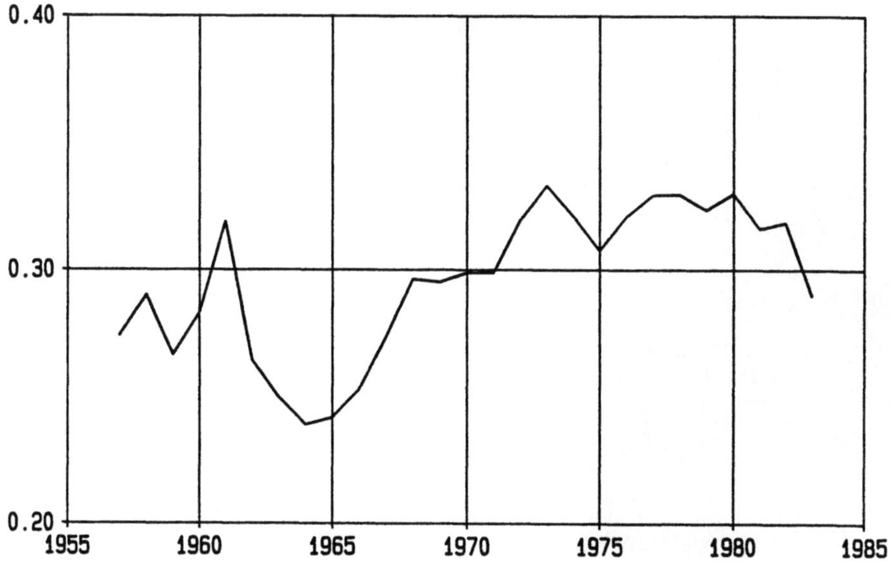

Fig. 4.6. The spatial variance of the regional utilities of the Federal Republic of Germany

leads to a formal underestimation of the "attractiveness" of Berlin in the period from 1957 to 1962. After 1961 the situation in Berlin stabilized, partially due to economic subsidies and an encouragement of cultural and industrial activities by the local administration and the government of the FRG.

The spatial variance of the utilities, (see (1.43)) is defined by

$$\sigma_u^2(t) = \frac{1}{L-1} \sum_{i=1}^{L} u_i^2(t) \tag{4.8}$$

and is depicted in Figure. 4.6.

This variance is a measure for the spread of utilities between the federal states. The dip from 1961 to 1965 in the temporal evolution of the variance essentially reflects the Berlin wall effect.

Another major effect can now be identified by inspection of Figure 4.5. The utility or attractiveness of a state to a large extent depends on its *size*. The states with a high population typically have higher utilities than those with lower population. This effect is illustrated by Figure 4.7 where the temporal mean values of the utilities of the states are depicted in the map of Germany.

Taking the population numbers as a proxy for size, it can be said, that states with large populations attract more people than states with small population numbers. This endogenous population size effect must now be distinguished from the size-independent effects, which also influence the utilities. It is therefore indicated to decompose the total regional utilities $u_i(t)$ as follows

$$u_i(t) = s_i(t) + \delta_i(t) \tag{4.9}$$

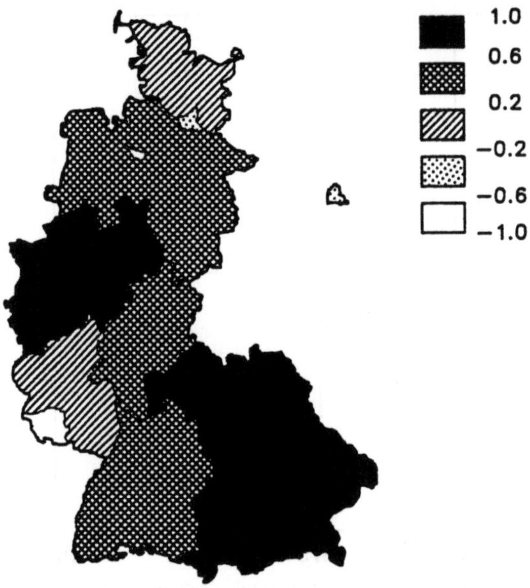

Fig. 4.7. Graphical repesentation of the temporal mean values of utilities of the states of the Federal Republic of Germany

where $s_i(t)$ is the size dependent part and $\delta_i(t)$ the residual size independent part of the utility of state i. The $\delta_i(t)$ are denoted as preferences.

It turns out, that the (empiric) population numbers $n_i(t)$ and the squared population numbers $n_i^2(t)$ are appropriate size-effect variables. Indeed, as explained in Section 4.3, the $n_i(t)$ and $n_i^2(t)$ are found to be the most important variables in a ranking of the influence of all socio-economic variables on $u_i(t)$. Therefore the size dependent part $s_i(t)$ of $u_i(t)$ can be represented as (see (1.44))

$$s_i(t) = \kappa \frac{(n_i(t) - \overline{n(t)})}{\overline{n(t)}} + \sigma \frac{(n_i^2(t) - \overline{n^2(t)})}{(\overline{n(t)})^2} \tag{4.10}$$

so that the form (4.10) is compatible with the constraints

$$\sum_{i=1}^{L} u_i(t) = 0; \qquad \sum_{i=1}^{L} \delta_i(t) = 0 \,. \tag{4.11}$$

The regression analysis of Section 3.2 now leads to the determination of the preferences, which are represented in Figure 4.8.

The preferences are measures of the size-independent attractiveness of the states. The mean values of these preferences for the period 1964–1968 are depicted in Figure 4.9, and those of the period 1979–1983 in Figure 4.10.

It turns out, that one group of states had a preference gain, another group remained practically stable, and a third group experienced a decrease in preferences. These changes are illustrated in Figure 4.11.

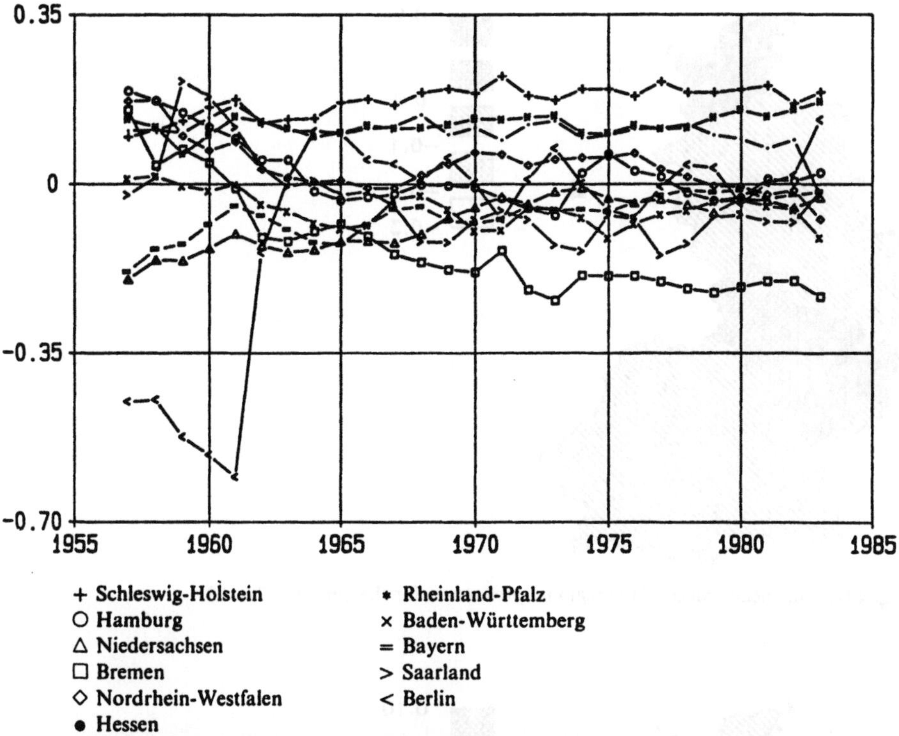

Fig. 4.8. The regional preferences of the federal states of the Federal Republic of Germany

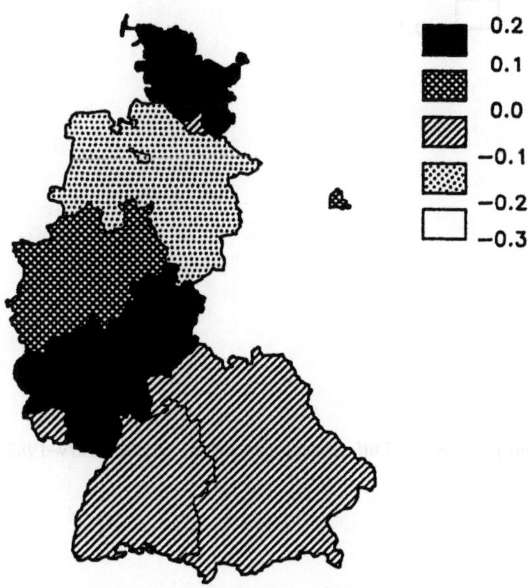

Fig. 4.9. The mean values of the regional preferences for the period 1964–1968

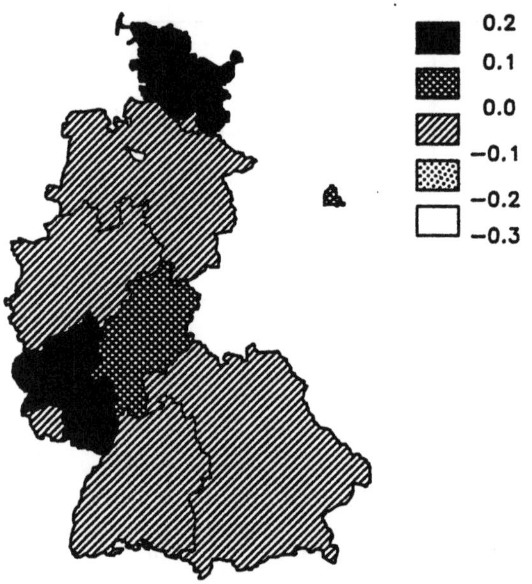

Fig. 4.10. The mean values of the regional preferences for the period 1979–1983

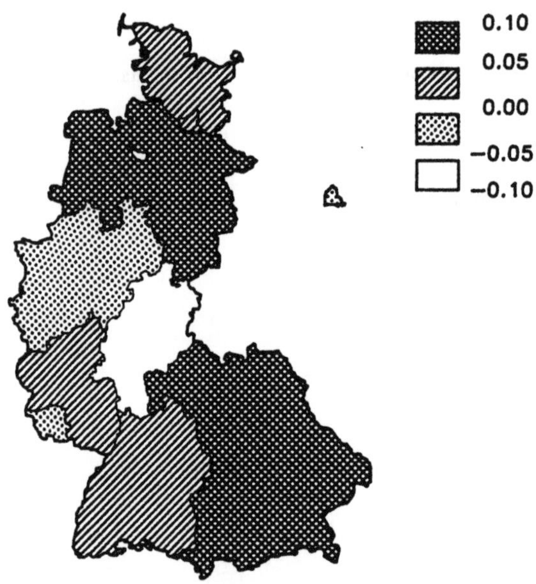

Fig. 4.11. Long term changes in the regional preferences: Differences between the periods 1979–1983 and 1964–1968.

In analogy to the spatial variance of the total regional utilities there can be introduced the variance of preferences (see (1.46))

$$\sigma_\delta^2(t) = \frac{1}{L-1} \sum_{i=1}^{L} \delta_i^2(t) \tag{4.12}$$

which is a measure for the size-independent inhomogeneity of the federal states. This variance is depicted in Figure 4.12.

The relevance of the size effect term $s_i(t)$ in explaining the differences between regional utilities is reflected in the small dispersion of the residual preferences: Comparing Figure 4.12 with Figure 4.6 it can be seen, that the variance of preferences is about 30 times smaller than the variance of utilities. This means: The differences between regional utilities are mainly due to size effects, whereas the states of the FRG are relatively homogeneous with respect to their size-independent preferences. This global situation has not changed with time in the period considered – apart from the Berlin effect, which shows up remarkably in the preferences, too, as expected.

Nevertheless a more detailed comparison of the size-independent preferences (Figure 4.8) with the total utilities (Figure 4.5) yields interesting insights: For example the population-rich Nordrhein-Westfalen (◇) loses its leading position and drops to position seven in the preferences. Rheinland-Pfalz (★) and Schleswig-Holstein (+) reached a leading position (in terms of preferences) in the mid-sixties and kept it until now. The city – state Bremen (□) is the tail – light in terms of utilities and preferences as well.

Some of the utilities are slowly increasing in the long run (Schleswig-Holstein (+), Niedersachsen (△), Bayern (=)), some remain practically constant (Hessen

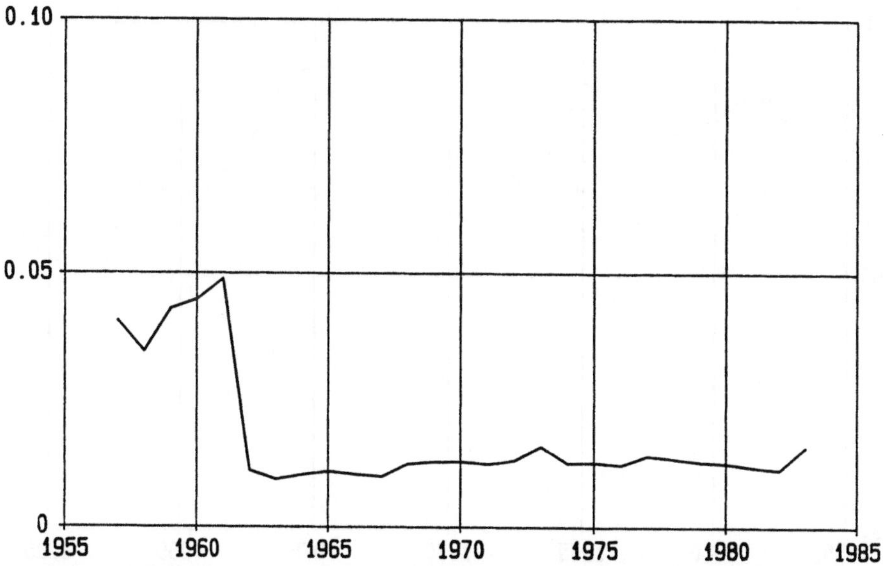

Fig. 4.12. The spatial variance of the regional preferences of the Federal Republic of Germany

(●), Rheinland-Pfalz (★), Baden-Württemberg (×)), and others show a long term decline (Bremen (□), Saarland (>), Hamburg (○), Nordrhein-Westfalen (◇)). Another effect is, that the small size regions, like Berlin (<), Bremen (□), Hamburg (○) and Saarland (>) exhibit considerable cyclic fluctuations in their preferences. In the next section we shall try to connect this spatio-temporal behaviour of utilities and preferences with the socio-economic situation in the federal states of the FRG.

4.2.4 Migratory Stress

The correlation coefficient (2.43) of the actual population size with the virtual equilibrium values of the migratory system of the FRG is shown in Figure 4.13. In Section 2.2 we have defined a migratory stress (2.46) using the deviation of the correlation coefficient from one. Since the migratory stress in the FRG is relatively small compared with other countries (Canada, Italy), gravity-type models may yield satisfactory results for the migration flows. This is due to the fact that the stationary solution (2.41) of our dynamic model is equivalent to a gravity-type model. However, gravity-type models fail if the system is not in equilibrium. In general a migratory system will be out of equilibrium.

Because of the low mobility of the system it would however need a relaxation time of a few hundred years to reach the stationary equilibrium state. Since political and economic processes will change the regional utilities on a much shorter time scale, the migratory system will always lag behind its virtual stationary state and practically never reach it. It should be mentioned that our dynamic migration

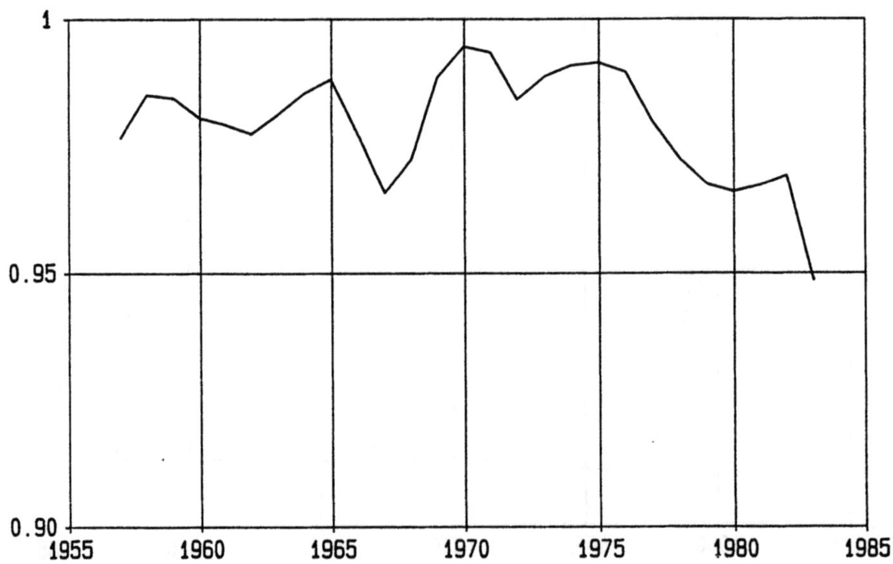

Fig. 4.13. The migratory stress of the Federal Republic of Germany

theory is equally well applicable to quasistationary situations near equilibrium as to situations far outside equilibrium with high migratory stress.

4.3 Choice of Socio-Economic Variables

The interaction of the migratory system with the socio-economic situation is now to be discussed following the line of Section 3.2. We begin with some qualitative considerations.

4.3.1 Identification of Influences on Global Mobility and Utilities

A first remark refers to the fact, that the interaction between migration and the socio-economic situation in a country is only *partial*. Not all political, social and economic changes within a country must necessarily influence the migratory behaviour of the population! On the other hand there exist influences on migration beyond socio-economics, such as climate and landscape of a state, which are "natural" and can be seen as constant for the time period under consideration. Some geographical aspects are taken into account by the deterrence factors f_{ij} and the concept of effective distance introduced in Section 4.2.

Secondly we must distinguish between *global* socio-economic developments affecting all regions homogeneously and socio-economic influences *differentiating* between the regions.

Global homogeneous effects can show up in the global mobility $v_0(t)$, but must not in the regional utilities $u_i(t)$. The latter statement follows from the definition of utilities, which are measures of the *relative* attractiveness of a region in comparison to all other regions, due to the constraint (4.11) and the fact, that only the differences of utilities enter the transition probabilities (4.1).

An example of such global effects are the two "oil shocks" 1972 and 1974 concerning all states of the FRG similarly. They may be one of the reasons for the observed dramatic decrease of the global mobility from 1972 to 1975. On the other hand this oil shock does not show up in the utilities.

Let us now go over to regionally differentiating effects: The structural change of the economy such as the decline of the steel-, coal-, heavy machine and ship-building-industry and on the other hand the dramatic growth of newly established electronic, computer and other high-tech industries does not concern all federal states in an equal manner!

Nordrhein-Westfalen, Saarland, Hamburg and Bremen have their specific structural problems with the decline of the mentioned "old" industries, whereas newly established high-tech industries mainly concentrate in federal states like Bayern and Baden-Württemberg.

It cannot be purely accidental, that this structural shift – which is somewhat coarsely characterized as the south-north gradient in the FRG – also shows up in the decline of the migratory regional utilities and preferences of the first group of states and in the increase or stability of utilities and preferences of the second group.

The following collection of socio-economic variables should comprise global economic indicators as well as regionally differentiating ones, so that their influence on the global mobility as well as on the utilities can be found out in the subsequent selection of key-factors.

4.3.2 Classification of Socio-Economic Variables

The time series of socio-economic variables used for our empirical analysis of the FRG are obtained from the publications of the Statistisches Bundesamt in Wiesbaden [4.1, 9, 10, 11, 12], the Statistischen Landesämter [4.13, 14], the Bundesanstalt für Arbeit in Nürnberg [4.15, 16], the Institut der deutschen Wirtschaft [4.17] and the Ifo-Institut [4.18, 19].

We have chosen the data set according to the following considerations:

- The data should be able to represent the influence of different socio-economic fields such as labour market, housing market, investment, public sector and living standard.
- All data should be available for a time period as long as possible on a yearly base.
- The data for the analysis of the regional utilities have to be spatially disaggregated at least on a state level and recorded in a comparable manner.

According to these criteria the following variables have been used in the regression analysis of the utilities and the global mobility. Of course, global data for the FRG can be easily obtained by aggregation of their state level. Abbreviations have been assigned to all variables turning out to be key-factors.

Size-Effect Variables:

- population numbers $n_i(t)$
- squared population numbers $n_i^2(t)$
- area of the federal states

Labour Market:

- total labour force (in thousands)
- labour force per capita
- employment rate (employed people per total population)
- unemployment rate $UR_i(t)$
- employment $E_i(t)$ (in thousands)
- rate of vacancies (number of open positions divided by the total labour force)
- vacancies $V_i(t)$ (in thousands)
- tertiary sector $TS_i(t)$ (percentage of total employment in the tertiary sector)
- labour market index (difference between open positions and unemployment divided by the total labour force)

Housing Market:

- housing stock (number of dwellings per capita)
- number of building permissions (per capita)
- number of completions of new dwellings (per capita)

Investment and Industrial Sector:

- investment structure index $Z(t)$ (expansionary minus rationalizing investment divided by total investment)
- plant investment (10^9 DM in prices of 1976)
- net production (10^9 DM in prices of 1976)
- gross domestic product (in 1000 DM per capita)
- total export (per capita and divided by the GDP)
- agricultural export
- industrial export
- export structure index $EI_i(t)$ (industrial minus agricultural export divided by the total export)

Public Sector:

- benefit payments (per capita)
- tertiary sector (see labour market)
- public expenditures (per capita in real prices)
- dwelling subsidies

Living Standard and Other Fields:

- real income $Y_i(t)$ (index 1976 = 100)
- weekly salary (in prices of 1976)
- total taxes (per capita)
- cars (per capita)
- overnight stays $OS_i(t)$ (per capita)

All economic variables are used in real prices (scaled in prices of 1976), even if it is not explicitly indicated above. It should be remarked that the confidence of the socio-economic data differs. Obviously, the number of cars is known very exactly, whereas the number of overnight stays may be more uncertain. On the other hand the regional data are recorded in most cases on the responsibility of the administration of the states. Therefore, there may occur slight differences in the confidence of the regional data. Nevertheless, the data base of the FRG is good compared with most other countries.

4.4 Selection of Key-Factors

In this section the selection of key-factors for utilities and the global mobility is considered.

Our procedure introduced in Chapters 3 and 15 differs from usually applied regression methods in some respects. The purpose of our algorithm is to select in an objective manner a few key-factors out of the above described set of socio-economic variables. Our analysis yields an "automatic" ranking of the relevance of key-factors. Since socio-economic variables have the tentative role of "causes" for migratory trends, the possibility of time-lags should be taken into account, because there can be some delay between changes of key-factors and their effects on migration flows. Of course, we can allow of positive time-lags only, since otherwise the imputed "causal" relation cannot exist. On the other hand our algorithm must not anticipate any interpretation of the results.

One important point is the sign of the coefficient of a key-factor which must be in agreement with the interpretation to be given. Another aspect is, that very often a high cross correlation between socio-economic variables can be observed. In the conventional regression analysis all these variables then simultaneously appear as relevant, whereas our procedure guarantees in this case the appearance of only one *representative* variable denoted as key-factor. This is achieved by the orthogonalization procedure described in detail in Chapter 3. For example the population size turns out to be highly correlated with other extensive socio-economic variables such as total labour force, number of households and total income. Therefore, it can happen that *only one of these extensive variables* appears as key-factor. In addition, other types of variables, for instance income per capita or labour force per capita may be relevant. This means that after a subtraction of the "size" effect the remaining variables should be intensive ones. In the following we consider the key-factor analysis for the global mobility and the regional utilities separately, also in view of the general remarks made in Section 4.3.

4.4.1 The Key-Factor Analysis of the Global Mobility

The global mobility is analyzed in terms of socio-economic variables following Chapter 15. Because of its global nature it is obvious that in this case only spatially aggregated variables have to be considered. For the FRG the temporal structure of the global mobility consists of a long term trend and superimposed short term variations of about 3 to 5 years. This relatively complicated structure of the global mobility is mainly comprehended by a set of four key-factors, namely real income per capita, number of open positions (vacancies), investment structure index and number of employed people. The detailed results of the regression analysis are presented in Table 4.5.

It is impressive to see that the labour market and the investment behaviour of the industry influences decisively the global mobility of the population in the FRG (see Figures 4.14–4.17). Long term as well as short term effects of the global mobility

Table 4.5. Results of the ranking regression analysis of the global mobility

variable*	timelag	coefficient	t - value	plotted in
Y	0	$-1.16\ 10^{-5}$	-49.3	Fig. 4.14
V	2	$3.52\ 10^{-7}$	12.7	Fig. 4.15
Z	0	$2.94\ 10^{-4}$	8.4	Fig. 4.16
E	2	$-1.09\ 10^{-7}$	-7.3	Fig. 4.17
constant	-	$5.41\ 10^{-3}$	-	-
correlation R^2 : 0.994				
corrected $\bar{R}^2$: 0.993				
F - test value : 858.7				
Durbin-Watson test : 1.71				

* abbreviations of the variables: Y: real income per capita, V: vacancies, Z: investment structure index, E: number of employed people

can be related to the evolution with time of the key-factors, which is represented in Figures 4.14 to 4.17. It turns out that a detrending of the mobility and thekey-factors is not necessary. This means we do not need an additional theory in order to "explain" the structure of the trend line. In other words: The key-factor of the first step (real income) is already more appropriate to describe the trend than a simple linear time function.

Because of its negative coefficient, somewhat surprisingly the *increase* of real income per capita in the FRG (see Figure 4.14) in the last three decades leads to a *decrease* of the global mobility. The interpretation may be, that the growing living standard and the resulting acquisition of homes and real estate by households reduces the mobility.

The vacancies depicted in Figure 4.15 appear as second important key-factor with a time-lag of two years. This means that an increase/decrease in the number of open positions is reflected in an increase/decrease of the global mobility two years later. This labour market indicator may influence the decisions of individuals to migrate in a sense that the social risks connected with a move are reduced by a high number of open positions.

The investment structure index (see Figure 4.16) is varying on a relatively short time scale. It has been shown [4.8] that this key-factor is most important to understand business cycles (Schumpeter cycles) in a quantitative manner. In so far it is not surprising that the investment structure index came as third important variable in the representation of the global mobility. Of course, an industrial bias to expansion gives rise to an increase of the global mobility, whereas a bias to rationalization reduces the mobility. During the interval 1955–65 the industry in the Federal Republic was nurtured by both post-war reconstruction of plant,

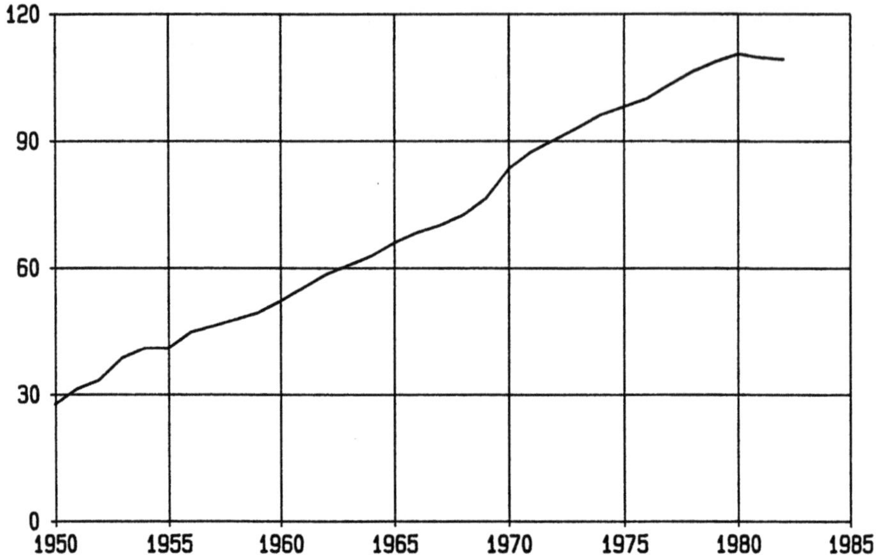

Fig. 4.14. The total income per capita in the Federal Republic of Germany

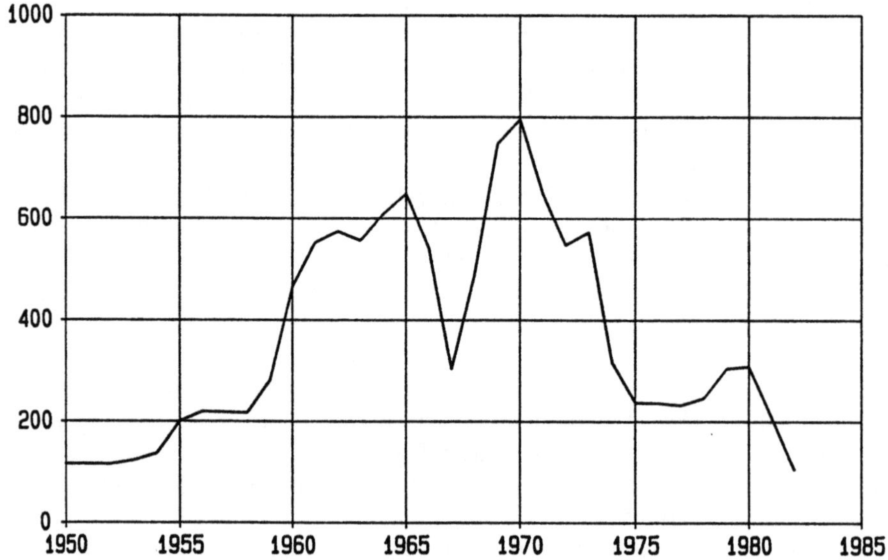

Fig. 4.15. The total number (in thousands) of open positions in the Federal Republic of Germany

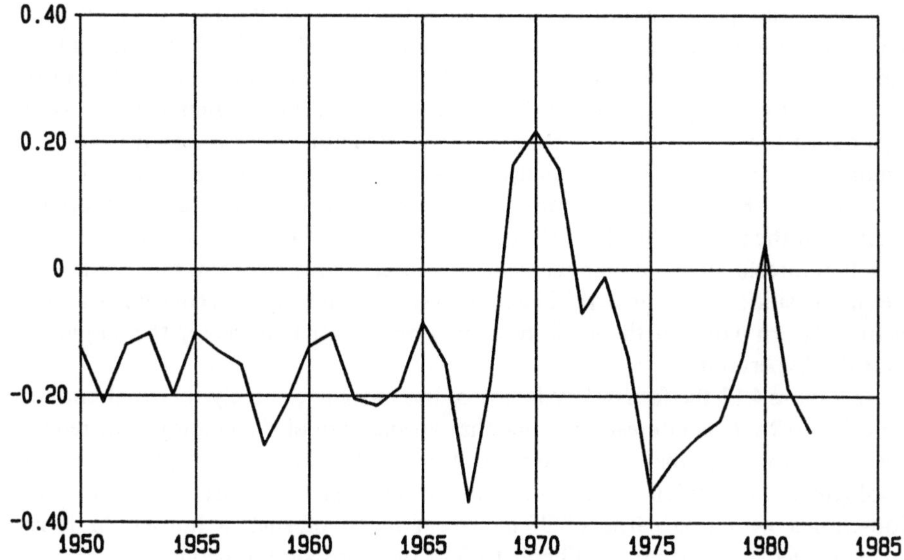

Fig. 4.16. The investment structure index of the Federal Republic of Germany

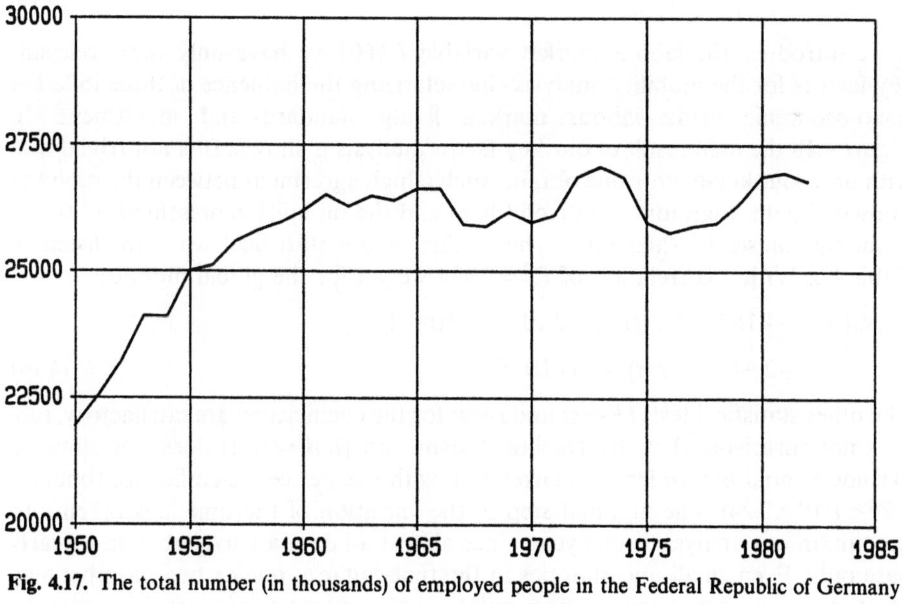

Fig. 4.17. The total number (in thousands) of employed people in the Federal Republic of Germany

equipment and infrastructure and by a high rate of product innovation in a number of growing international industries. The decline from 1965 to 1967 indicates a fairly general switch-over from product innovation to process innovation. During the late sixties, the newly formed "Große Koalition" resorted to a massive expansionary policy program. Its effect on the investment structure index is clearly visible as a remarkable peak in 1969/71 called the "Schiller Effect", after Karl Schiller, the accomplished Keynesian economist, who was minister of finance and economic affairs. For the period after 1971 the dramatic decline of Z may lend evidence to the view that forceful demand management only forces industrial firms more quickly to the supply side limits of growth. The two oil shocks altering the economic situation in the western world at the same time may be further reasons for the stagnation after the hyperboom.

In Figure 4.17 the fourth key-factor of the mobility, namely the total employment is shown. It is interesting to see that an increase of the employment reduces the mobility with a time-lag of two years. It is well known that a surplus increase of employment in FRG is mainly due to the fact that there are more and more households with two earners. Of course, a move of such a household is less probable because it is more difficult to find two jobs simultaneously.

Since both variables, vacancies and total employment appear with the same time-lag of two years the relevant variable for explaining the mobility is a combination of both key-factors

$$LM(t) = V(t) - 0.31\,E(t)\,. \tag{4.13}$$

If we introduce the labour market variable $LM(t)$ we have only three relevant key-factors for the mobility analysis characterizing the influence of three different socio-economic fields: labour market, living standard and investment. In Figure 4.18 the final result of our key-factor analysis is shown. It is not trivial that with only four key-factors one obtains such a high agreement between the mobility evaluated with migratory data (solid line) and the mobility represented by socio-economic causes (dashed line). The results of the statistical tests are listed in Table 4.5. With a correlation of $R^2 = 0.994$ we get for the global mobility

$$\begin{aligned} v_0(t) = & -1.16\,10^{-5}\,Y(t) + 3.52\,10^{-7}\,LM(t-2) \\ & + 2.94\,10^{-4} Z(t) + 5.41\,10^{-3}\,. \end{aligned} \tag{4.14}$$

The other statistical tests (F-test and t-test for the coefficients) are satisfactory, too. It is not surprising that the Durbin-Watson test ($DW = 1.71$) does not allow to exclude a serial first order autocorrelation with five per cent significance (bounds: $1.76 < DW < 2.24$). The minimal step in the variation of the time-lags taken into account in our analysis is one year, since almost all data are available on a yearly base only. Even small uncertainties in the time-lag (e.g. one or two months) may create a very high first order autocorrelation. Nevertheless, it is surprising that the obtained DW-test is inconclusive, this means that there is neither a confirmation nor an exclusion of a first order autocorrelation.

It should be mentioned that in the case of a nearly "perfect" fit the residuals become very small and therefore the evidence of the Durbin-Watson test, which considers the autocorrelation of the residuals, is insufficient.

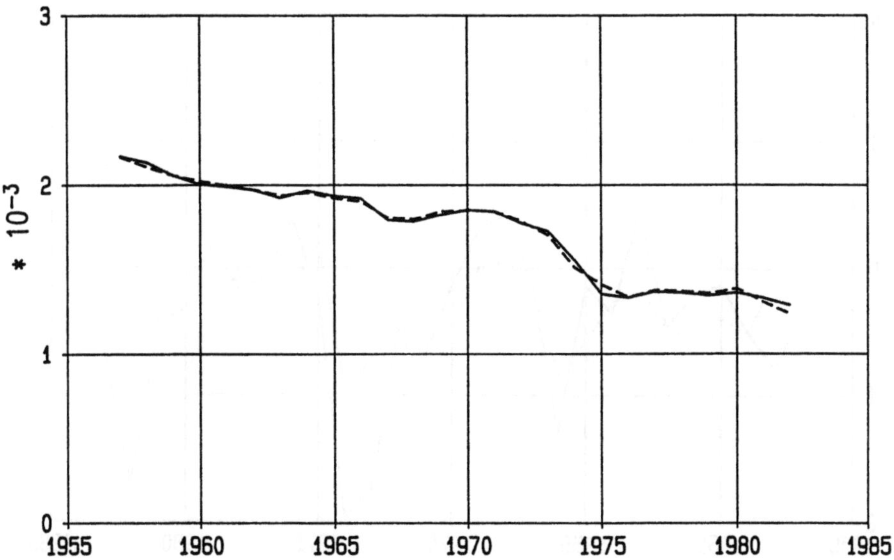

Fig. 4.18. Key-Factor analysis of the global mobility of the Federal Republic of Germany

Of course, the above described procedure yields a high correlation coefficient if there is a time trend (increase or decrease) in the time evolution of the global mobility. In order to avoid such a trivial correlation usually difference variables (for example $v_0'(t) = v_0(t+1) - v_0(t)$) are introduced [4.5]. It is obvious that the analysis of such detrended variables may yield another set of key-factors. In [4.5] it is supposed that the total volume of migration (Figure 4.2) shows a time evolution parallel to the gross national product. In Figure 4.19 the results of a linear regression using the difference variables obtained from the global mobility and the gross national product are given. In this case the correlation is $R^2 = 0.37$.

Using the key-factors depicted in Table 4.5 in the same detrended form one obtains a correlation of $R^2 = 0.76$. This result (Figure 4.20) is satisfactory, although we have used the original set of variables. It is remarkable that the coefficients of the key-factors differ only slightly from the coefficients listed in Table 4.5. In particular the signs of the coefficients are the same in both cases!

As already mentioned it is more appropriate for forecasting purposes of the migration flows to follow the method described in the beginning of this paragraph.

4.4.2 The Key-Factor Analysis of the Regional Utilities

We now continue in our key-factor analysis considering the regional utilities. The method is described in detail in Chapter 3. Of course, the socio-economic variables taken into account must be spatially disaggregated on a state level.

Since the spatial differences of the utilities are rather high compared with their temporal variations (see Figure 4.5), it is obvious that the explanation of their

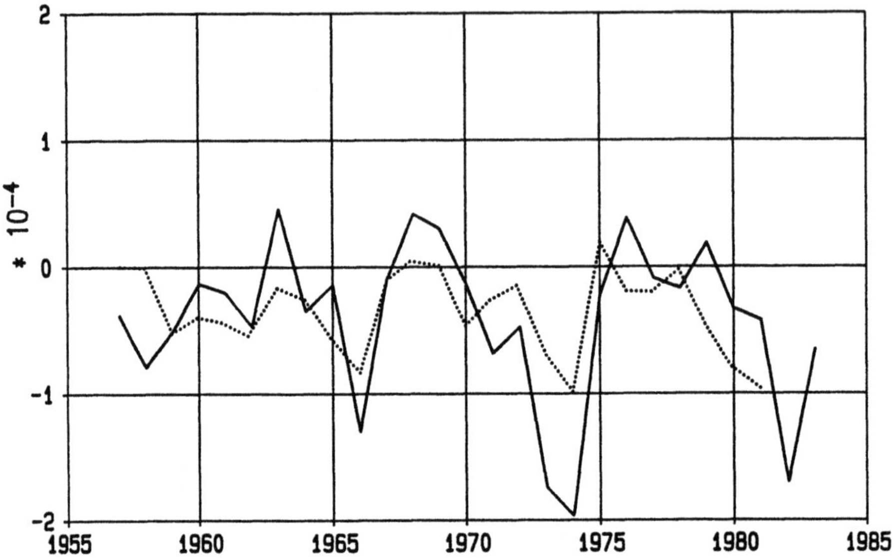

Fig. 4.19. Yearly changes of the global mobility (solid line) and of the gross national product (dotted line)

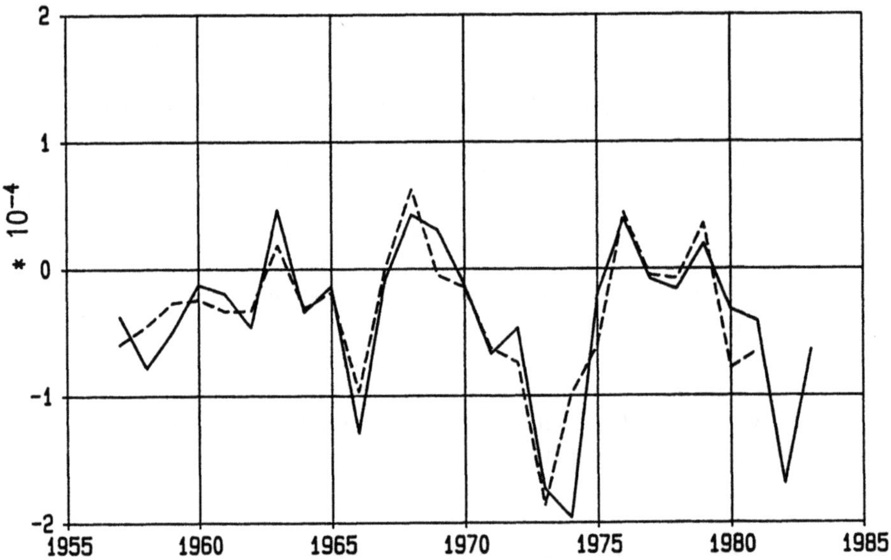

Fig. 4.20. Representation of yearly changes of the global mobility (solid line) by the four key-factors listed in Table 4.5 (dashed line)

spatial differences is more important. Nevertheless, at least the long term trends in the regional utilities should also be linked to key-factors.

It is not surprising that the "Berlin-effect" in 1961–63 does not have an explanation in terms of conventional economics. It cannot be expected, that such kind of political influence on the migratory system is reflected in the chosen set of socio-economic variables.

Our data set consists of 32 socio-economic variables and is described in Section 4.3. However, it is remarkable that only six key-factors are necessary to represent the complicated spatio-temporal structure of the regional utilities. The size effect is expressed by the dependence of the utilities on its regional population size (see (1.44) and (4.9), (4.10)) and saturation level. It is worthwhile to emphasize that population size $n_i(t)$ and squared size $n_i^2(t)$ are treated in the same manner as all other socio-economic variables. Nevertheless, the regression analysis always yields these size effect variables to be the first and second important key-factors. Their calculated coefficients have the expected sign: There is a positive correlation of the utilities with the population numbers, that means the agglomeration parameter κ is positive, whereas the coefficient σ of $n_i^2(t)$ is negative and describes the saturation behaviour. In other words the population prefers high density living, but only up to a certain level. This result is highly plausible and empirically confirmed. Subtracting this size effect from the regional utilities the regional preferences $\delta_i(t)$ are obtained (see (4.9) and Figure 4.8). As already mentioned only *size-independent socio-economic* variables should be found as key-factors for the preferences.

In the further steps of our analysis the *following four key-factors* have been obtained, namely the number of overnight stays per capita, the export structure index, the unemployment rate and the percentage of employment in the tertiary sector. The results of the regression analysis and the considered statistical tests are listed in Table 4.6.

Let us now start the discussion of the influence of the different key-factors. H. Birg [4.3] has used as a working hypothesis that the change of location of an individual is subject to:

- the possibility to earn a higher income
- a higher disposal of services, public and private goods
- a better environment and social structure
- a good infrastructure in order to use facilities to reach the other regions of the country.

The set of socio-economic variables chosen in Section 4.3 can be attributed to these four different fields of influence. It is remarkable that the ranking procedure automatically selects four intensive socio-economic key-factors which may be seen as representative variables of these fields. Of course, the classification of a variable with respect to these items is not in all cases possible in an unique way. Therefore different interpretations of the influence of the key-factors on the individual decision behaviour are possible. On the other hand a key-factor on the macrolevel of the migratory system may express the effect of different individual causes on the microlevel.

Table 4.6. Results of the ranking regression analysis of the regional utilities

variable*	timelag	coefficient	t - value	plotted in
n_i	0	1.037	205.8	Fig. 4.21
n_i^2	0	-0.180	-32.5	-
OS_i	0	0.156	17.7	Fig. 4.22
EI_i	2	0.474	9.2	Fig. 4.23
UR_i	0	-0.083	-5.7	Fig. 4.24
TS_i	0	0.193	5.0	Fig. 4.25

correlation R^2 : 0.984
corrected $\bar{R}^2$: 0.983
F - test value : 2332.0

* abbreviations of the variables: n_i: population number in region i, n_i^2: square of the population number in region i, OS_i: number of overnight stays in region i per capita, EI_i: export structure index in region i, UR_i: unemployment rate in region i, TS_i: quota of employment in the tertiary sector in region i.

In Figure 4.21 the scaled and dimensionless regional population numbers are depicted which turn out to be the most important first key-factor. According to (3.69) the scaling is done by going over to

$$\Delta n_i(t) = \frac{n_i(t) - \overline{n(t)}}{\overline{n(t)}} \tag{4.15}$$

with

$$\sum_{i=1}^{L} \Delta n_i(t) = 0 \tag{4.16}$$

where $n_i(t)$ is the number of individuals in region i, at time t, and $\bar{n}(t)$ denotes the spatial mean value of the population numbers. Therefore, $\Delta n_i(t)$ describes the relative deviation of the regional population from its spatial meanvalue. It can be seen (Figure 4.21) that there exists a stable ranking of the federal states with respect to their population size. Obviously, only the stable parts of their spatial differences and not the short term variations of the regional utilities $u_i(t)$ can be explained by this first key-factor. In the long run (since 1960) Bayern (=), Baden-Württemberg (×) and Hessen (•) have gained population at the cost of Nordrhein-Westfalen (◇), Berlin (<) and Hamburg (○). Such changes in the relative population share are mainly due to migratory effects.

As second important key-factor the squared population size $n_i^2(t)$ is obtained. The scaling of this variable is performed according to (1.44) and (4.10)

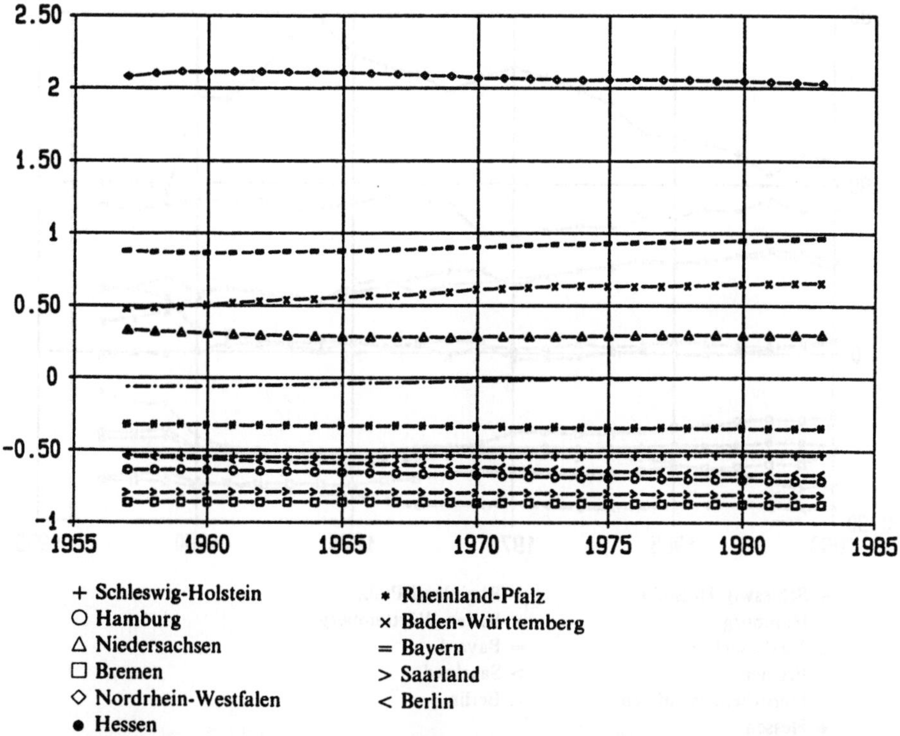

Fig. 4.21. The scaled regional population number of the Federal Republic of Germany

$$\Delta n_i^2(t) = \frac{n_i^2(t) - \overline{n_i^2(t)}}{(\overline{n(t)})^2} \tag{4.17}$$

with

$$\sum_{i=1}^{L} \Delta n_i^2(t) = 0 \, . \tag{4.18}$$

Since the coefficient of this variable, σ, is less than zero, this key-factor is responsible for saturation effects.

In Figure 4.22 the scaled numbers of overnight stays $OS_i(t)$ per capita are shown. This is the third key-factor. It may be seen as representative for describing differences in environment, e.g. recreation facilities as well as interregional and intraregional industrial activities. The more attractive regions from a touristic point of view are the coast-region of Schleswig-Holstein (+) and the southern part of Bayern (=) with its mountains. Of course, this influence is represented by a positive correlation of the preferences with the overnight stays per capita. Comparing Figure 4.22 with Figure 4.8 one sees that this key-factor $OS_i(t)$ is responsible at least in part for the ranking of the preferences.

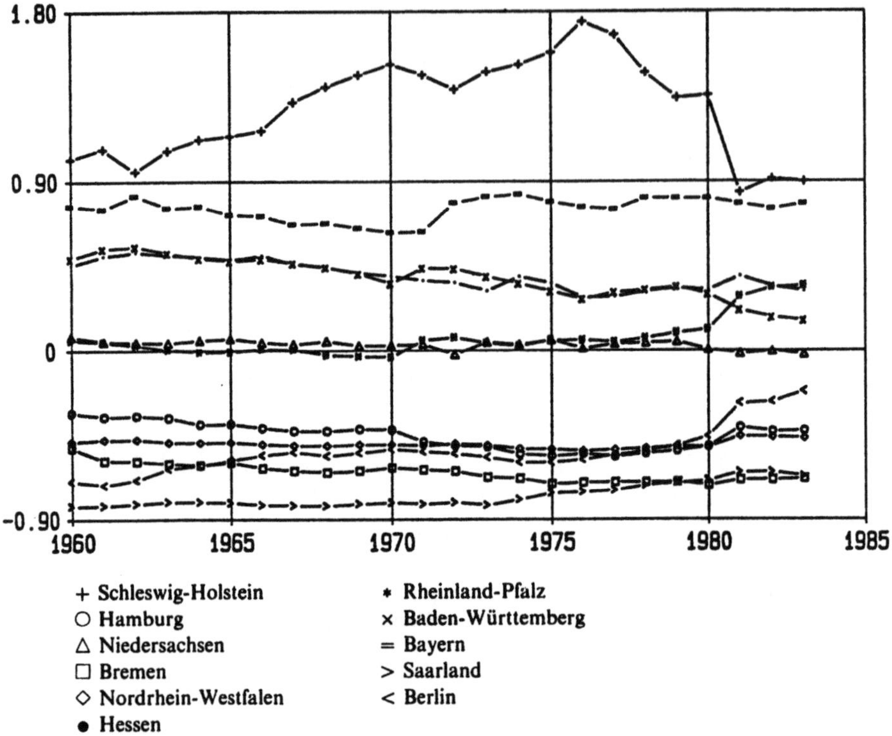

Fig. 4.22. The scaled regional numbers of overnight stays in the Federal Republic of Germany

The fourth key-factor is the export structure index $EI_i(t)$ as depicted in Figure 4.23. It represents the dependence of a region on the export share of agricultural and industrial products. It is obvious that since 1960 the regional differences between the federal states with respect to $EI_i(t)$ have enhanced. The considerable variations and fluctuations of this key-factor may partially be responsible for the short term variations of the preferences, because parts of the economic structure of a region are represented by it. The appearance of a time-lag of two years however indicates the inertia of the population in its reaction to changes of this key-factor. In the FRG, regions with a high export structure index are preferred against regions with a low $EI_i(t)$. The variation with time of the export structure index $EI_i(t)$ of Hessen (●), Saarland (>), Nordrhein-Westfalen (◇) and Baden-Württemberg (×) is very similar. This means that their time paths exhibit similar short term variations. An increase of $EI_i(t)$ of these states may be due to the development of higher specialized industrial products. The time evolution $EI_i(t)$ of the other states exhibits a more irregular structure.

The fifth-important key-factor is the regional unemployment rate $UR_i(t)$, depicted in Figure 4.24. The good economic conditions of the region of Baden-Württemberg (×) finds its expression in a low scaled unemployment rate during the whole period under investigation. From Figure 4.24 it is obvious that there

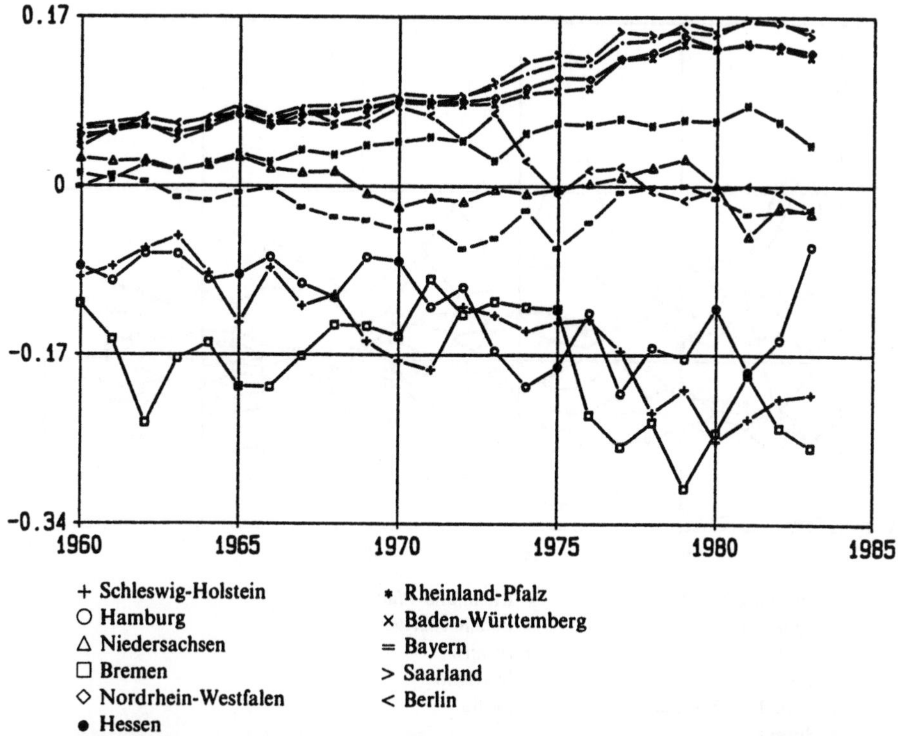

Fig. 4.23. The scaled regional export structure index of the Federal Republic of Germany

exists a north-south gradient in the unemployment rate. For example Bremen (□) and Niedersachsen (△) in the north of the FRG have a much higher unemployment than Bayern (=) and Baden-Württemberg (×). The regional differences of $UR_i(t)$ however are decreasing since 1960. In 1975 the spatial variance of $UR_i(t)$ is very small, because the total unemployment rate is increasing in all states of the FRG. This effect is caused by a global structural economic change affecting all federal states. After 1961 the high unemployment rate of Berlin (<) is rapidly decreasing to a medium level. The strong dependence of the regions of Bremen (□) and Saarland (>) on stagnating industries (steel and ship building industry) is reflected in their high unemployment rates.

As often observed small and medium size regions exhibit stronger variations of their socio-economic conditions. This fact can again be observed in Figure 4.25 where the regional quota of employment in the tertiary sector is depicted. Cyclical variations of $TS_i(t)$, especially of Berlin (<), Hamburg (○) and Bremen (□) can be recognized. This may be due to the fact that a higher share of the service sector is found in city – states of Germany. Apart from this fact, the other states are ordered mainly according to a north-south gradient. In this sense the states of Bayern (=) and Baden-Württemberg (×) have a very small share of employment in the tertiary sector. This key-factor appears in our classification of the socio-economic variables

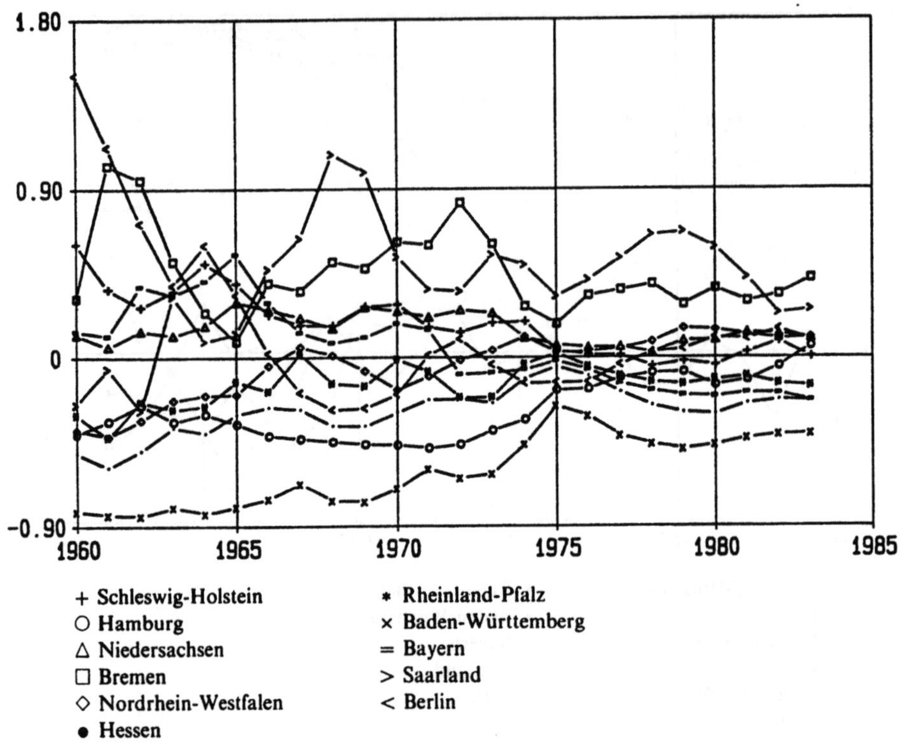

Fig. 4.24. The scaled regional unemployment rates of the Federal Republic of Germany

twice, since this variable not only represents the labour market but also the availability of services. The variable $TS_i(t)$ completes the series of six key-factors used in the representation of the utilities.

According to our ranking procedure the criterion for the ranking of these key-factors are the t-test values of these variables. The t-values of the key-factors are greater than the bounds for t with 0.1 percent error probability $t > t_b = 3.4$ (see Chapter 14). This is one reason for their significance. The effects of the different socio-economic key-factors are superimposed in the representation of $u_i(t)$ and yield a main part of the spatio-temporal variations of the regional utilities. The results of Table 4.6 inserted in (3.78a) with (4.15) and (4.17) finally yield

$$\begin{aligned} u_i(t) = {} & 1.037\,\Delta n_i(t) - 0.180\,\Delta n_i^2(t) \\ & + 0.156\,OS_i(t) + 0.474\,EI_i(t-2) \\ & - 0.083\,UR_i(t) + 0.193\,TS_i(t) \\ & + \delta_{0i}(t)\,. \end{aligned} \tag{4.19}$$

The remaining residuals of the regression analysis are denoted as $\delta_{0i}(t)$. The result of this regression of the regional utilities, listed in Table 4.6, lead to a correlation

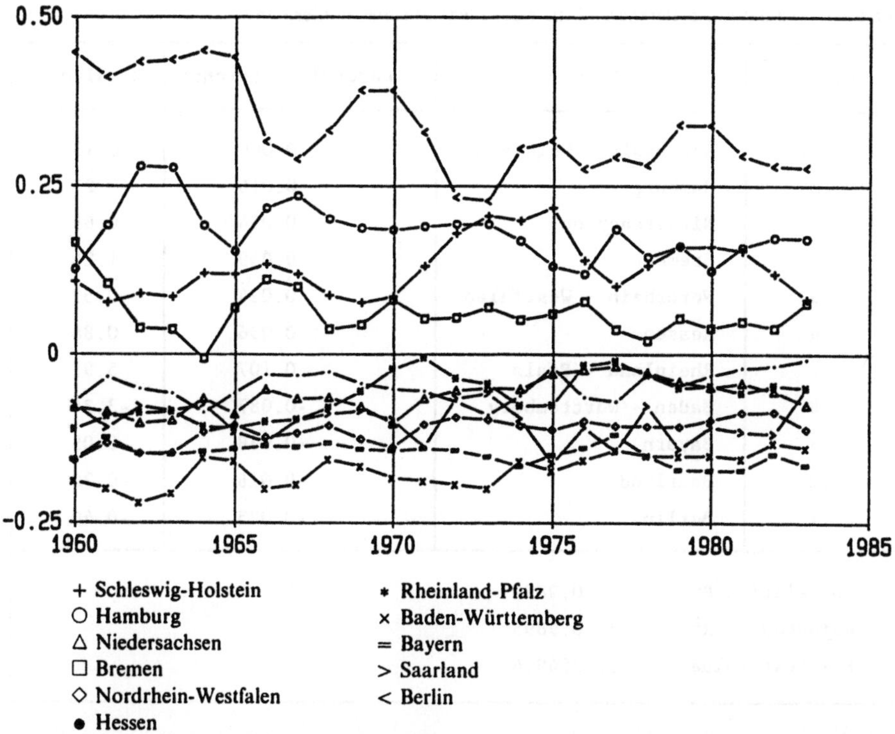

Fig. 4.25. The scaled quota of employment in the tertiary sector of the Federal Republic of Germany

coefficient $R^2 = 0.984$ and a test value $F = 2332$. These remarkable values of the statistical significance tests are promising for forecasting purposes.

The temporal meanvalues of the yearly residuals

$$\delta_{0i} = \frac{1}{T} \sum_{t=1}^{T} \delta_{0i}(t) \tag{4.20}$$

have been tested to be uncorrelated to the rest of socio-economic factors. Therefore, δ_{0i} can be seen as a measure for the "natural" preference of a federal state with respect to the migration of individuals. For the FRG these natural preferences are listed in Table 4.7.

The t-test for the natural preferences shows that δ_{0i} is significantly different from zero only for the regions of Rheinland-Pfalz ($t = 5.9$) and Baden-Württemberg ($t = -3.34$). Therefore the values of the natural preferences of the other states can be put equal to zero. Using these natural preferences δ_{0i} in (4.19) we obtain the test values listed in Table 4.7. Of course, the value of the correlation coefficient R^2 is now higher (0.990) since more parameters are involved in the analysis. But it is satisfactory that in our case also the *corrected* correlation coefficient is slightly improved (0.990) if the natural preferences are introduced. The F-test value of the utilities has increased remarkably, too ($F = 2849$).

Table 4.7. The natural preferences of the eleven federal states of Germany

region	state	natural preference	t-value
1	Schleswig - Holstein	0.006	0.13
2	Hamburg	0.016	0.27
3	Niedersachsen	0.034	0.69
4	Bremen	-0.059	-1.24
5	Nordrhein - Westfalen	0.024	0.59
6	Hessen	0.036	0.86
7	Rheinland - Pfalz	0.107	5.91
8	Baden - Württemberg	-0.087	-3.34
9	Bayern	-0.038	-2.05
10	Saarland	-0.016	-0.30
11	Berlin	-0.025	-0.40

correlation R^2	:	0.9899
adjusted $\bar{R}^2$	:	0.9895
F - test value	:	2849.4

In Figure 4.26 the estimated regional utilities (plotted with symbols according to Table 4.1) are compared with the regional utilities represented by a linear combination of the six socio-economic key-factors (solid lines) listed in Table 4.6. Obviously, the main feature, namely the spatial ranking of the regiotnal utilities is described excellently. The quality of the explanation of the details of the temporal evolution of the different utilities is somewhat lower. We see, that the development of the regional attractiveness of Rheinland-Pfalz (★), Bayern (=) and Baden-Württemberg (×) is rather well fitted, in contrast to Schleswig-Holstein (+), Berlin (<), Saarland (>) and Hamburg (○). It seems, that the key-factors are more able to describe the time path of the $u_i(t)$ of big states. Comparing Figure 4.22 with 4.26 it is indicated that the high disagreement between the migratory utility and the fitted utility of Schleswig-Holstein (+) in the period after 1974 is mainly caused by the evolution of the overnight stays. Our remarks about the uncertainty of the data in particular of small sized states should be kept in mind. Therefore it is not surprising that larger deviations in the time variations of the utilities of the smaller federal states exist.

4.4.3 Transition Rates in Terms of Key-Factors

In the above subsections the global mobility and the regional utilities have been analyzed separately in terms of socio-economic variables. These key-factors ex-

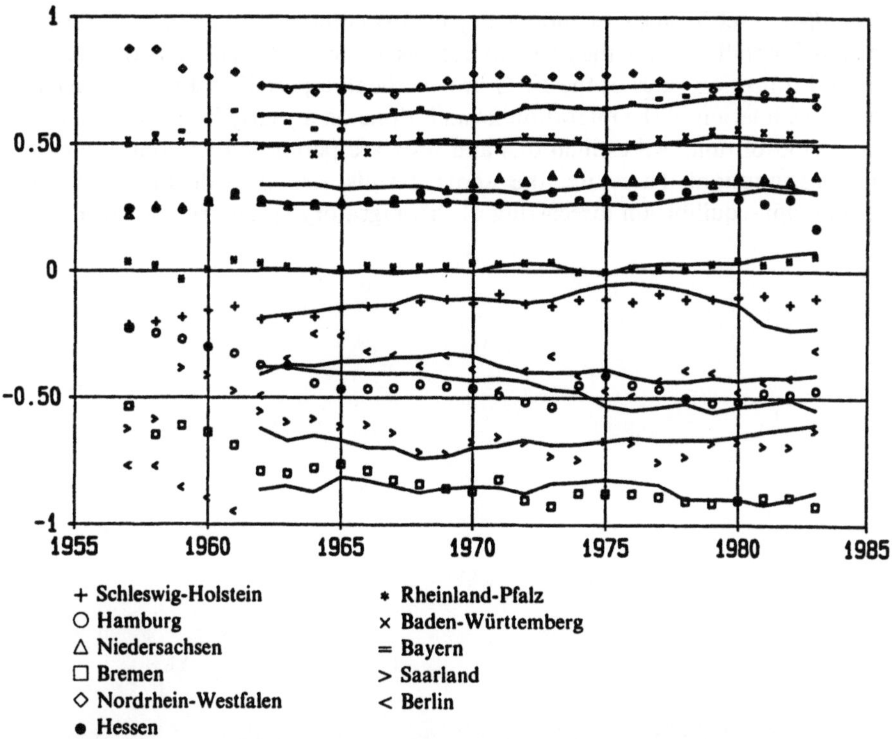

Fig. 4.26. Key-Factor analysis of the regional utilities of the Federal Republic of Germany

plain the spatio-temporal variations of these quantities. Therefore, the total transition rates $w_{ij}(t)$ can be composed by inserting (4.9), (4.3) and (4.12) in (1.40b). Hence, all 2310 interregional migration flows in the time period from 1962 to 1982 can be derived using eight exogenous socio-economic variables only. Taking into account the 55 spatial deterrence parameters f_{ij} and the two variables of the size effect, the test results for the total transition rates are listed in Table 4.8.

At the end of this investigation we come back to the fundamental assumption of our model, namely the decomposition (1.26) of the individual transition probability $p_{ij}(t)$ into a symmetric mobility factor $v_{ij}(t)$ and a push/pull factor G_{ij}. The two factors were assigned to different motivations in the individual decision behaviour, as explained in Section 1.3. The present analysis has confirmed this assumption

Table 4.8. Statistical tests for the total transition rates of the FRG

correlation R^2	0.983
adjusted $\bar{R}^2$	0.982
F - test value	1887.9

since *different* sets of socio-economic key-factors entered into the representation of the global mobility and of the regional utilities, respectively. The statistical tests for the total transition rates of the FRG are excellent, especially taking into account that the estimation of the migration flows is done on a yearly basis for a relatively long period of time. All estimations and tests are performed without any equilibrium assumption! Therefore, the above results also provide the basis for a dynamic non-equilibrium forecasting of the migratory system of the FRG.

5 Canada

Jacques Ledent

5.1 Introduction

In contrast to that of the other countries in this comparative study, the demographic development of Canada is fairly recent. Thus, prior to any analysis of current interprovincial migration patterns, it may be worthwhile to review some aspects of this development, particularly those pertaining to the evolution of the geographical distribution of the Canadian population.

5.1.1 Historical Patterns of Spatial Population Growth[1]

The growth of the Canadian population as well as its geographical redistribution were sparked by three successive waves of European immigrants. The first wave was composed of the increasing numbers of French settlers who followed the creation of New France in 1608. These settlers concentrated along the Saint Lawrence River in what has become the province of Quebec (see Figure 5.1), and they eventually gave rise to a population capable of sustaining itself through natural increase. After a change from a French to a British regime in 1763, British immigrants – the second wave – arrived in Canada. Many of these immigrants were British loyalists who, following the American revolution, fled the former British colonies. They settled not only in Lower Canada (Quebec), but also and predominantly in the Maritimes (Prince Edward Island, Nova Scotia, and New Brunswick) and Upper Canada (Ontario). Finally, the third wave was composed of an

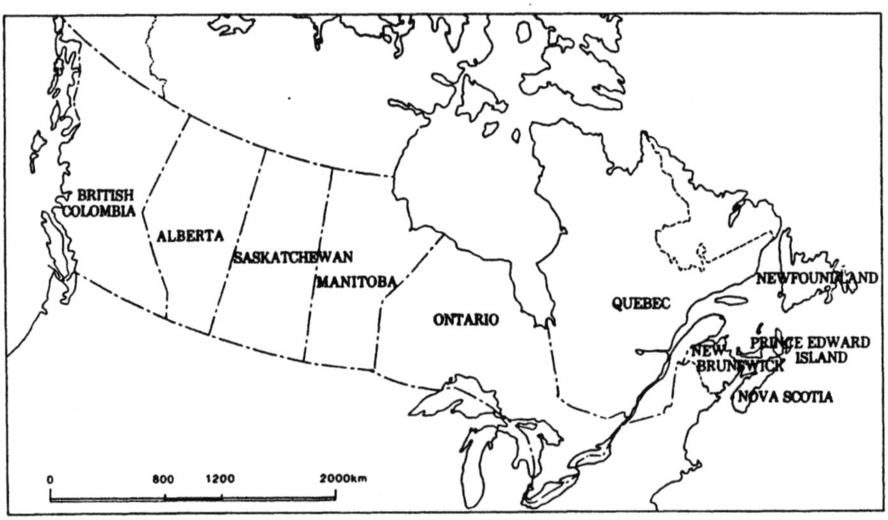

Fig. 5.1. The ten Canadian provinces

[1] This section is based on: W. Kalbach and W. McVey (1971) "Population growth in Canada". In: *The Demographic Bases of Canadian Society* Toronto, Ontario: McGraw-Hill.

additional immigration of Europeans during the post-Napoleonic era. By the mid-1800s, this had boosted the population of Upper Canada to a level that surpassed that of Lower Canada.

From that time, while growing under the sole influence of natural increase – net emigration was actually registered in the last four decades of the nineteenth century – the population of British North America pursued its westward expansion beyond Upper Canada. The penetration to the West, initiated by a gold rush to British Columbia and fueled by the gradual construction of the railway link from the Atlantic to the Pacific, led to the development of the Prairies (Manitoba, Saskatchewan, and Alberta) and the transformation by 1905 of the Dominion of Canada – created in 1867 by the union of Nova Scotia, New Brunswick, Quebec and Ontario – into the currently existing confederation of provinces and territories (except for Newfoundland which joined only in 1949).

In the first two decades of this century, European immigrants arrived again in great numbers, contributing to a substantial population increase in the Prairies region. Thus, by the 1921 census, the population of Canada had assumed a geographical distribution (according to provinces and territories) that has in the broad sense remained in force up to the present.

This is not to say that the breakdown of the national population by province (and territory) has not changed since the early twenties. In fact, the percentage share of the national population associated with most provinces has varied somewhat between 1921 and 1981 (see Table 5.1). Rather, the above observation should be interpreted as signifying that since 1921 immigration (or more exactly the differential capacity of the provinces to attract people from abroad) has been a much less important factor than before in shaping the spatial distribution of the Canadian population[2]. A corollary to this is that, in the last half century or so, natural increase rate differentials between provinces, and especially interprovincial migration, have been more influential than previously in breaking down the national population by province (and territory).

5.1.2 Current Patterns of Spatial Population Growth

Since assuming a distinctive geographical pattern broadly similar to that of today, the Canadian population, broken down by province, has continued to evolve slowly according to that pattern. Indeed, the variations exhibited in the recent past by the provincial percentage shares (see Table 5.1) reflect in most instances long-term tendencies that have been in effect since the early twenties.

In brief, the population shares of three provinces increased further between 1961 and 1981: Ontario, whose share reached 35.4 percent in 1981, up from 34.2

[2] In earlier days, immigrants often settled in the less populous provinces, contributing to their development, whereas in the twentieth century they elected in general to live in the largest metropolitan areas, that is, in the most populous provinces. Consequently, the population growth rate attributable to immigration, and thus the total population growth rate, have taken on values that after 1921 are more similar across provinces than before, resulting in a slower change in the breakdown of the national population according to province.

Table 5.1. Percentage distribution of the Canadian population, by provinces, 1921–1981

Province or territory	Year						
	1921	1941	1961	1966	1971	1976	1981
Newfoundland	-	-	2.5	2.5	2.4	2.4	2.3
Prince Edward Island	1.0	0.8	0.6	0.5	0.5	0.5	0.5
Nova Scotia	6.0	5.0	4.0	3.8	3.7	3.6	3.5
New Brunswick	4.4	4.0	3.3	3.1	2.9	2.9	2.9
Quebec	26.9	29.0	28.8	28.9	27.9	27.1	26.4
Ontario	33.4	32.8	34.2	34.8	35.7	36.0	35.4
Manitoba	6.9	5.5	5.1	4.8	4.6	4.5	4.2
Saskatchewan	8.6	5.9	5.1	4.8	4.3	4.0	4.0
Alberta	6.7	6.7	7.3	7.3	7.6	8.0	9.2
British Columbia	6.0	8.3	8.9	9.4	10.1	10.7	11.3
Yukon and North-west Territories	0.1	0.2	0.2	0.2	0.3	0.3	0.3
Canada[b]	100.0 (8.8)	100.0 (11.5)	100.0 (18.2)	100.0 (20.0)	100.0 (21.6)	100.0 (23.0)	100.0 (24.3)

a. Data from the 1981 Census of Canada.
b. The figures in parentheses give the corresponding population level in million of inhabitants.

percent in 1961, and the two westernmost provinces (Alberta and British Columbia), whose combined shares amounted to 20.5 percent, up from 16.2 percent in 1961. The population shares of six other provinces pursued long-term declines: the four Atlantic provinces (Newfoundland, Prince Edward Island, Nova Scotia, and New Brunswick), whose combined shares decreased from 10.4 percent in 1961 to 9.2 percent in 1981, and two of the Prairies provinces (Manitoba and Saskatchewan), whose combined shares went down over the same time span from 10.2 to 8.2 percent. In fact, the population share of only one province, Quebec, took on a divergent course in the recent past. Instead of increasing as it did in the previous decades, Quebec's population share experienced in the mid-sixties a reversal that pushed its value down to 26.4 percent in 1981, as against 28.8 percent in 1961.

It should be stressed here that the recent evolution of the geographical distribution of the Canadian population just observed has largely resulted from significant population movements within Canada. Among these, two movements are particularly prominent:

1) Increased migration flows into the two westernmost provinces caused by the rise of economic opportunities, especially in Alberta.

In the late seventies, following the rapid development of mineral extraction, Alberta experienced an economic boom that attracted many people from the rest of the country, including from the distant Atlantic provinces[3].

2) Increased (decreased) migration flows out of (into) Quebec.

In the early sixties, a "quiet revolution" began in Quebec, the only province with a predominantly French-speaking population, a product of the first wave of European settlers. This political unrest eventually led to rule by an "independentist" party and was instrumental in driving many anglophones out of Quebec (as well as keeping many anglophones from moving to Quebec)[4].

5.1.3 Spatial Differentials in Residual Population Growth

Given the substantial changes observed since the early 1960s in the distribution of the Canadian population by province, the model used throughout this book appears to be an interesting basis for analyzing such shifts and, more precisely, the interprovincial migration patterns responsible for such shifts.

The implementation of this model, however, relies on an assumption – namely, that population redistribution is the result of only interregional migration – that does not generally hold in the multiregional population systems that are commonly observed. It may be valid in the first approximation for some systems, but it is probably unrealistic for those systems in which regional differences in other sources of population growth (natural increase and international migration) are substantial.

Is the hypothesis of equal residual growth rates made in Chapter 2 (that is, of equal rates of population growth from sources other than interprovincial migration) applicable to the Canadian system of provinces and territories? To answer this question, let us consider, for each region i, a residual growth index $\tilde{\rho}_i$ defined as the ratio of the residual growth rate $\rho_i = g_i - m_i$; where $g_i = \frac{1}{n_i}\frac{dn_i}{dt}$ is the total growth rate of province i and $m_i = \frac{1}{n_i}\sum_j{}' (w_{ij} - w_{ji})$ is the provincial growth rate due to interprovincial migration, to the national growth rate $\hat{g} = \sum_i x_i g_i = \frac{1}{N}\sum_i \frac{dn_i}{dt}$, with $x_i = \frac{n_i}{N}$, $\sum_i x_i = 1$ – that is,

$$\tilde{\rho}_i = \frac{g_i - m_i}{\hat{g}} . \tag{5.1}$$

[3] In the wake of the 1981 economic crisis that hit Canada especially hard, the huge population gains experienced by British Columbia and Alberta through internal migration turned rapidly into population losses.

[4] Most of the anglophones moving out of Quebec have been bound for Ontario, which explains to a large extent the diverging evolutions of the population shares associated with these two provinces.

A simple way to assess the spread of $\tilde{\rho}_i$ across provinces is to calculate the spatial variance of this index, i.e.,[5]

$$\sigma_\rho^2 = \frac{1}{L-1}\sum_i x_i(\tilde{\rho}_i - 1)^2 \tag{5.2}$$

where L is the number of provinces[6], or rather its standard deviation (square root of the preceding variance). Between 1961 and 1983, $\tilde{\rho}_i$ registered a mean average of 0.2643 with a corresponding standard deviation of 0.1124 (see first column of Table 5.A1 in the Appendix), thus suggesting a spread covering a wide interval with extreme values of 0.2643 ± 0.1124 (that is, 0.1519 and 0.3767).

Although it may appear surprising, this result summarizes nicely a reality in which each year $\tilde{\rho}_i$ takes values that differ substantially from one[7]. Thus, for the year 1979–1980, which has the most uneven distribution of the $\tilde{\rho}_i$-values (see Figure 5.A1 in the Appendix), $\tilde{\rho}_i$ ranges from 0.25 in the case of Prince Edward Island to 2.23 in the case of Alberta (corresponding to a residual growth rate ranging from 0.31 to 2.75 percent since $\tilde{\rho} = 1.24$ percent). Even for the year 1963–1964 which had the most even distribution of the $\tilde{\rho}_i$-values, $\tilde{\rho}_i$ ranges from 0.66 in the case of Nova Scotia to 1.15 in the case of Prince Edward Island (corresponding to a residual growth rate ranging from 1.24 to 2.87 percent since the mean average value of $\tilde{\rho}_i = 1.88$ percent).

The Canadian population thus exhibits growth rates from sources other than interprovincial migration that differ substantially across provinces. This result contradicts the explicitly retained assumption of homogeneous residual growth rates[8]. Fortunately, this finding bears no impact on estimation, in Section 5.2, of the model's main indices (global mobility and regional utilities), as their values depend solely on the interregional migration process. It may, however, influence assessment in Section 5.3 of these indices in terms of the key socio-economic factors. Substantial biases in the residuals between the estimated and observed values may appear that do not reflect levels of or changes in socio-economic conditions but rather result from the assumption of homogeneous growth rates.

---

[5] The average value of $\tilde{\rho}_i$ is $\tilde{\rho} = \frac{1}{\hat{g}}\left(\sum_i x_i g_i - \sum_i x_i m_i\right) = \frac{\hat{g}}{\hat{g}} = 1$, since $\sum_i x_i m_i = \frac{1}{N}\sum_{j,i}{}' (w_{ij} - w_{ji}) = 0$.

[6] Because of the small size of the population in the territories, the system of regions considered below is reduced to the 10 provinces.

[7] Such wide differences between the various provincial values naturally reflect:

- provincial differences in the rates of natural increase;
- a different capacity to attract immigrants which favors the provinces with larger percentage shares of their population in metropolitan areas; and
- provincial differences in an artificial population change due to the inconsistency of Statistics Canada's estimates of population levels and components of change.)

[8] Figure 5.A1 in the Appendix suggests that the values of $\tilde{\rho}_i$ have diverged rather than converged over the years – a trend confirmed by a single linear regression of the standard deviation of $\tilde{\rho}_i$ against time (see first column of Table 5.A1 in the Appendix) – so that the hypothesis of uniform values of $\tilde{\rho}_i$ is, in the case of Canada, a possibility that becomes more distant.

5.2 Global Mobility, Utilities and Preferences

Central to the stochastic theory applied here is the instantaneous rate p_{ij} of transition from region j to region i. This rate is assumed to have the following specification:

$$p_{ij} = v_{ij} \exp(u_i - u_j) \tag{5.3}$$

where v_{ij} is a mobility factor reflecting the ease of moving from region j to region i (as well as from region i to region j, for it is also assumed that $v_{ij} = v_{ji}$) and $u_j(u_i)$ is the "utility" of region $j\,(i)$.

In practice, from any given matrix of interregional migration streams, a nonlinear procedure (see Chapter 3 for details) allows one to estimate

1) the symmetric matrix of the mobility factors v_{ij} as well as a summary index of these factors, or global mobility,

$$v_0 = \frac{1}{L(L-1)} \sum_{i,j}{}' v_{ij} \tag{5.4}$$

where L is the number of regions in the system

2) the set of the regional utilities u_i constrained to be such that

$$\sum_i u_i = 0 \tag{5.5}$$

since (5.3) defines u_i up to an additive factor.

This procedure was applied repeatedly to the matrix of the migration streams between the 10 Canadian provinces estimated annually by Statistics Canada from the file it maintains on family allowance recipients (Statistics Canada, Cat. 91–208, various issues)[9]: 22 observations of this matrix were used from 1961–1962 to 1982–1983.

5.2.1 Global Mobility

At first glance, the time variations of the global mobility index v_0 (depicted in Figure 5.2) suggest a tendency for this index to decrease with time. Such a visual impression is confirmed by a simple linear regression of v_0 against time: see the second column of Table 5.A1 in the Appendix where it is shown that the coefficient of the time variable has a negative sign while being significant.

[9] For each annual period starting June 1 of a given year and ending May 31 of the following year, this matrix refers to interprovincial migrations (i.e., all moves observed between each pair of provinces during this period) rather than to migrants (i.e., all changes of province observed between the beginning and the end of the observation period).

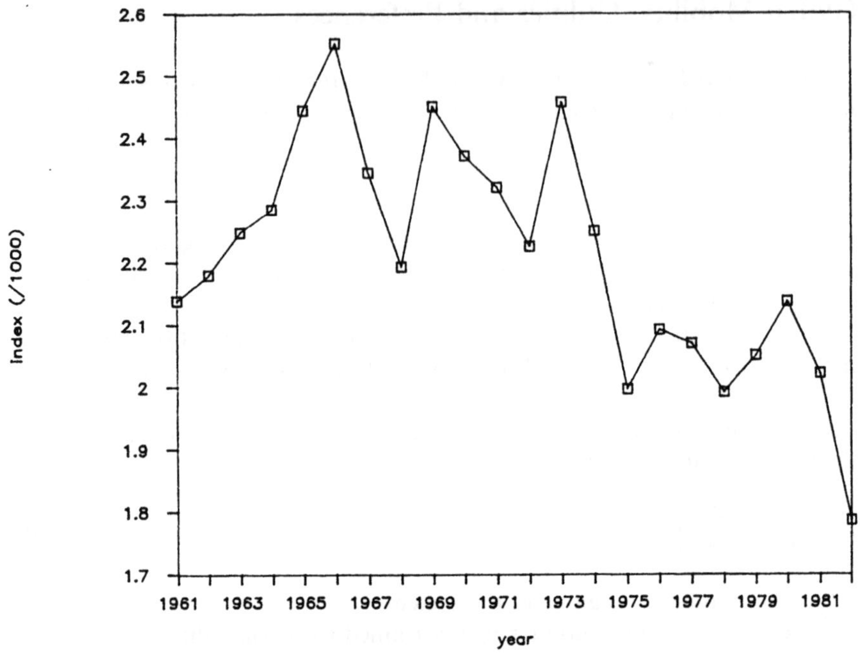

Fig. 5.2. Global mobility, Canada, 1961–62 to 1982–83

A closer look at Figure 5.2, however, allows one to identify three successive periods over the 22-year span:

1) an initial period of increasing mobility (up to 1966),
2) an intermediate period of stationary mobility (between 1966 and 1973), and
3) a final period of decreasing mobility (since 1973).

This more refined visual impression is somehow supported by a quadratic regression of v_0 against time: see again the second column of Table 5.A1 in the Appendix, where it appears that the coefficients of the linear and quadratic time terms are positive and negative, respectively, and that both are significant.

In other words, the trend of increased mobility that has long prevailed in Canada halted in the late sixties and subsequently reversed following the 1973 oil crisis.

5.2.2 Regional Utilities

Comparison of the mean average values taken by their respective utility indices (see first column of Table 5.2) allows one to classify the 10 provinces into four groups:

1) a group of provinces with high utility: Ontario;
2) a group of provinces with moderately high utility: Quebec, British Columbia, and Alberta;

Table 5.2. Estimates of linear and quadratic regressions of the regional utilities against time

Province	Regional utilities		Regression type[b]				
	Mean average[a]	Standard deviation	Linear		Quadratic		
			t	R^2	t	t^2	R^2
Newfoundland	-0.488 (9)	0.096	0.0076 (2.66)[c]	0.26	0.0119 (1.06)	-0.0002 (-0.40)	0.27
Prince Edward Island	-1.150 (10)	0.092	0.0017 (0.52)	0.01	0.0221 (1.93)	-0.0010 (-1.85)	0.16
Nova Scotia	-0.210 (7)	0.053	0.0016 (0.87)	0.04	0.0173 (2.90)	-0.0007 (-2.73)	0.31
New Brunswick	-0.324 (8)	0.065	0.0021 (0.98)	0.05	0.0198 (2.63)	-0.0008 (-2.43)	0.27
Quebec	0.607 (2)	0.197	-0.0276 (-9.77)	0.83	-0.0395 (-3.67)	0.0006 (1.15)	0.84
Ontario	0.960 (1)	0.108	-0.0102 (-3.50)	0.38	-0.0062 (-0.54)	-0.0002 (-0.36)	0.38
Manitoba	-0.117 (5)	0.063	-0.0053 (-2.90)	0.30	-0.0247 (-4.51)	0.0009 (3.67)	0.59
Saskatchewan	-0.136 (6)	0.154	0.0112 (2.39)	0.22	-0.0222 (-1.34)	0.0016 (2.08)	0.37
Alberta	-0.325 (4)	0.107	0.0124 (5.18)	0.57	0.0048 (-0.52)	0.0004 (0.85)	0.59
British Columbia	0.534 (3)	0.129	0.0065 (1.53)	0.11	0.0168 (1.02)	-0.0005 (-0.65)	0.12

a. The rank of each province according to decreasing values of the regional utility index is shown in parentheses.
b. The coefficient of determination of each regression equation appears under the column heading «R^2».
c. The figure in parentheses associated with each coefficient estimate is the corresponding t-statistic.

3) a group of provinces with moderately low utility: the Prairies provinces minus Alberta and the Atlantic provinces minus Prince Edward Island; and
4) a group of provinces with low utility: Prince Edward Island.

This picture of utility differentials between the provinces is not an average picture that conceals diverging evolutions of the various provincial utilities. As suggested by Figure 5.3, which displays the time variations of the utility index associated with each province, it is a somewhat robust picture of differentials that prevailed more or less throughout the entire observation period. Column 2 of Table 5.2, which shows the standard deviation of the utility index associated with each province, points to annual variations in utility that are far from uniform across provinces. Some provinces (Nova Scotia, Manitoba, and New Brunswick) experienced comparatively smaller variations, while others (British Columbia, Saskatchewan, and Quebec) experienced comparatively wider variations. Given their commonly small magnitude, however, these nonuniform variations in utility did not affect substantially the relative positions of the provincial utility indices between the beginning and end of the observation period.

The only modifications identifiable from an examination of Figure 5.3 concern the three provinces with comparatively higher annual variations in utility. But, they appear to have a permanent nature in only one instance. On the one hand, the utility indices for Saskatchewan and British Columbia exhibit U-shaped variations (oriented downward for the former, upward for the latter) such that they take on, by the end of the period, values that are close to their corresponding values at the beginning of the period (as is typically the case with the other provincial indices with average or comparatively smaller annual variations). On the other hand, Quebec's utility index declined almost continually over the observation period, from a value close to that of Ontario in the early sixties to one that is now comparable to those of Alberta and British Columbia.

The set of linear regressions that plot each provincial utility against time (see columns 3 and 4 of Table 5.2) confirms the decline of Quebec's utility index from a statistical view point – as the *t*-statistic associated with the time variable amounts to -9.77. It also reveals significant linear trends that could not be identified visually, namely the upward orientation of the utility index for Alberta ($t=5.18$), Newfoundland ($t=2.66$), and Saskatchewan ($t=2.39$), and the downward orientation of the utility index for Ontario ($t=-3.50$) and Manitoba ($t=-2.90$).

Use of a quadratic rather than of a linear function of time brings out significant results for three of the four remaining provinces (see columns 5, 6, and 7 of Table 5.2) – the utility indices for Prince Edward Island, Nova Scotia, and New Brunswick exhibit variations that can be depicted by U-shaped curves oriented downward – as well as improved results for two of the provinces with significant linear trends (Manitoba and at a lesser degree Saskatchewan). Thus, only British Columbia appears to have an utility index that exhibits no systematic variations with time[10].

[10] The time variations visually identified for British Columbia and to a lesser extent for Saskatchewan thus are not easily picked up by simple (linear or quadratic) functions of time.

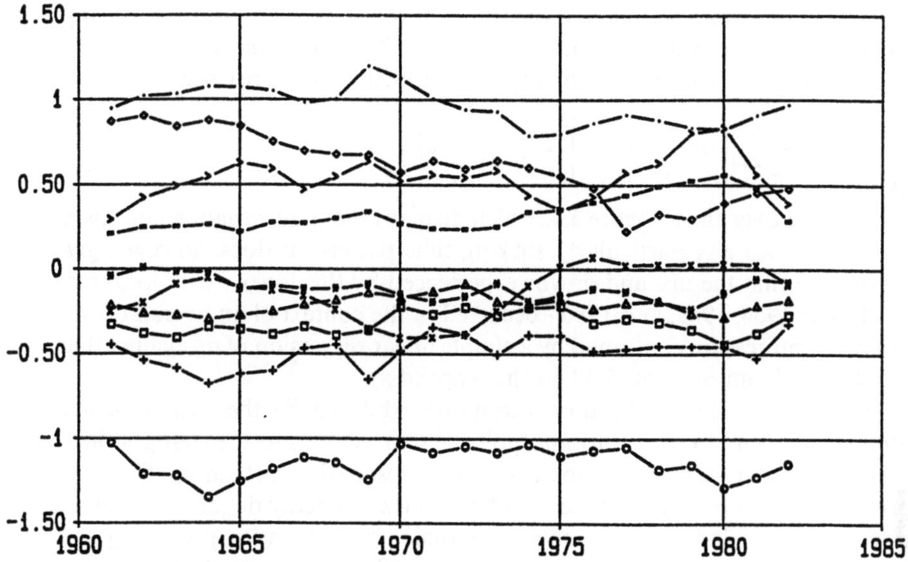

+ Newfoundland
○ Prince Edward Island
△ Nova Scotia
□ New Brunswick
◇ Quebec
● Ontario
* Manitoba
× Saskatchewan
= Alberta
> British Columbia

Fig. 5.3. The regional utilities of Canada, 1961–62 to 1982–83

Table 5.3. Interpretation of the symbols used in Figures 5.3, 5.5 and 5.9

symbol	province
+	Newfoundland
o	Prince Edward Island
Δ	Nova Scotia
□	New Brunswick
◊	Quebec
·	Ontario
*	Manitoba
x	Saskatchewan
-	Alberta
>	British Columbia

Have such time variations contributed to narrowing or to widening the differentials between the various utility indices? To answer this question, one must examine how the variance of the regional utilities across provinces

$$\sigma_u^2 = \frac{1}{L-1} \sum_i \left[u_i - \frac{1}{L} \sum_j u_j \right]^2 = \frac{1}{L-1} \sum_i u_i^2 \tag{5.6}$$

has evolved over time. Figure 5.4, which displays the time variations of this index, does not reveal any particularly striking time pattern. It does, however, give the impression that the ups and downs experienced by this index – with local maxima reached in 1964, 1969, and 1980 – occurred in the context of an overall downward trend (an impression confirmed by a simple linear regression of σ_u^2 against time: see the third column of Table 5.A1 in the Appendix).

Having ascertained the time variations exhibited by the various provincial utilities, we should now try to interpret them in relation to the changes that have affected the socio-economic climate of the corresponding provinces. This does not prove feasible, however, because the utility index implicitly defined by (5.3) is not a true composite index of the socio-economic characteristics prevailing in each province. To a large extent, the utility index u_i reflects the population level of region i as can be seen from the similar rankings obtained by ordering provinces according to increasing values of their utility index u_i and of their population level n_i. The only substantial difference in the two rankings concerns Quebec, a province with more than twice as many inhabitants as Alberta and British Columbia but

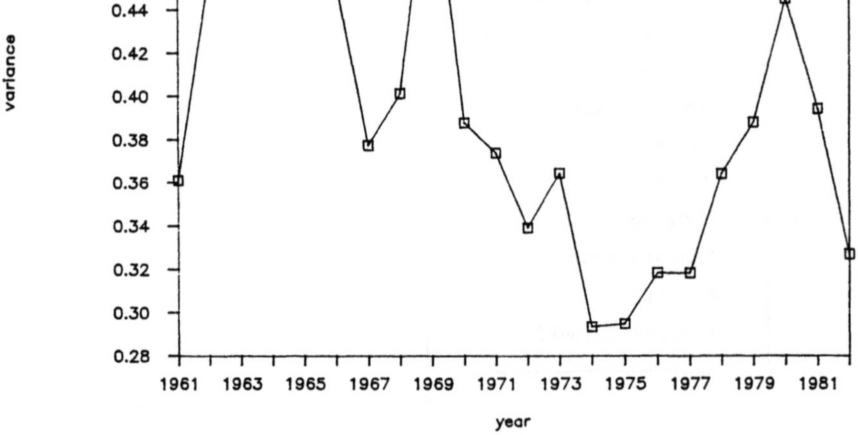

Fig. 5.4. Spatial variance of utilities, Canada, 1961–62 to 1982–83

with about the same utility index as these two provinces. A control for population size is thus in order if one wishes to make useful inference about the utility indices determined above.

5.2.3 Regional Preferences

According to the statistical procedure described in Chapter 3, a desirable control for population size can be carried out at the same time that are determined the socio-economic factors that influence the mobility behavior of individuals. More specifically, the utility index u_i is regressed against a set of various socio-economic indicators of region i to which are added two variables linked to the population level of the same region. Then another index δ_i, rid of any size effect and labelled regional preference, is obtained as the difference between the regional utility index u_i and the sum s_i of the two population terms in the statistical explanation of u_i. By construction, the set of the regional preferences δ_i is constrained in the same manner as that of the regional utilities, i.e.

$$\sum_i \delta_i = 0 . \tag{5.7}$$

The first column of Table 5.4, which displays the mean average values taken by the various preference indices, suggests a simple but sharp contrast between two provinces – Quebec and Prince Edward Island – and the remaining eight. On the one hand, Quebec and Prince Edward Island have a preference index with a highly negative mean average value. This is synonymous with (i) a weak attraction exerted on the residents of other provinces, and (ii) a strong repelling effect exerted on their own residents. Such a finding reflects, in the case of Quebec ($\delta_i = -0.43$), the strong language and cultural barrier that stands between this predominantly French-speaking province and the predominantly English-speaking rest of Canada. In the case of Prince Edward Island ($\delta_i = -0.39$), this finding reflects the relative geographic isolation of that province and its ensuing lack of economic opportunities.

On the other hand, the other eight provinces have a preference index with a non-negative mean average value that falls in a rather narrow interval ranging from 0.07 (Manitoba) and 0.14 (British Columbia) – except for Newfoundland (−0.01)[11] and Ontario (0.24). This seems to suggest that on average six of the eight provinces considered here exert a similar attraction, whereas Ontario (Newfoundland) appear to exert, on average, a comparatively higher (smaller) attraction.

Even though the time variations exhibited by the various preference indices are somewhat more acute and pervasive than those exhibited by the various utility indices (see Figure 5.5), the contrast just uncovered between two subsets of the Canadian provinces is one that was sustained quite clearly over the whole observation period.

---

[11] It is negative but almost equal to zero in the case of Newfoundland.

Table 5.4. Estimates of linear and quadratic regressions of the regional preferences against time

Province	Regional preference		Regression type[b]				
	Mean average[a]	Standard deviation	Linear		Quadratic		
			t	R^2	t	t^2	R^2
Newfoundland	-0.009 (8)	0.099	0.0086 (3.03)[c]	0.31	0.0084 (0.75)	0.0000 (0.01)	0.31
Prince Edward Island	-0.394 (9)	0.090	0.0020 (0.64)	0.02	0.0191 (1.68)	-0.0008 (-1.56)	0.13
Nova Scotia	0.092 (4)	0.061	0.0050 (2.77)	0.28	0.0174 (2.69)	-0.0006 (-1.99)	0.40
New Brunswick	0.076 (6)	0.069	0.0044 (2.03)	0.17	0.0213 (2.82)	-0.0008 (-2.32)	0.35
Quebec	-0.433 (10)	0.216	-0.0305 (-10.49)	0.85	-0.0463 (-4.27)	0.0008 (1.50)	0.86
Ontario	0.242 (1)	0.086	-0.0050 (- 1.80)	0.14	0.0075 (0.72)	-0.0006 (-1,23)	0.20
Manitoba	0.073 (7)	0.057	-0.0003 (- 0.17)	0.00	-0.0224 (-3.98)	0.0010 (4.06)	0.47
Saskatchewan	0.082 (5)	0.181	0.0185 (3.96)	0.44	-0.0127 (-0.75)	0.0015 (1.91)	0.53
Alberta	0.139 (2)	0.072	0.0029 (1.22)	0.07	0.0082 (-0.87)	-0.0002 (-0.58)	0.09
British Columbia	0.127 (3)	0.132	-0.0051 (- 1.15)	0.06	-0.0021 (-0.12)	-0.0001 (-0.18)	0.06

a. The rank of each province according to decreasing values of the regional preference index is shown in parentheses.
b. The coefficient of determination of each regression equation appears under the column heading «R^2».
c. The figure in parentheses associated with each coefficient estimate is the corresponding t-statistic.

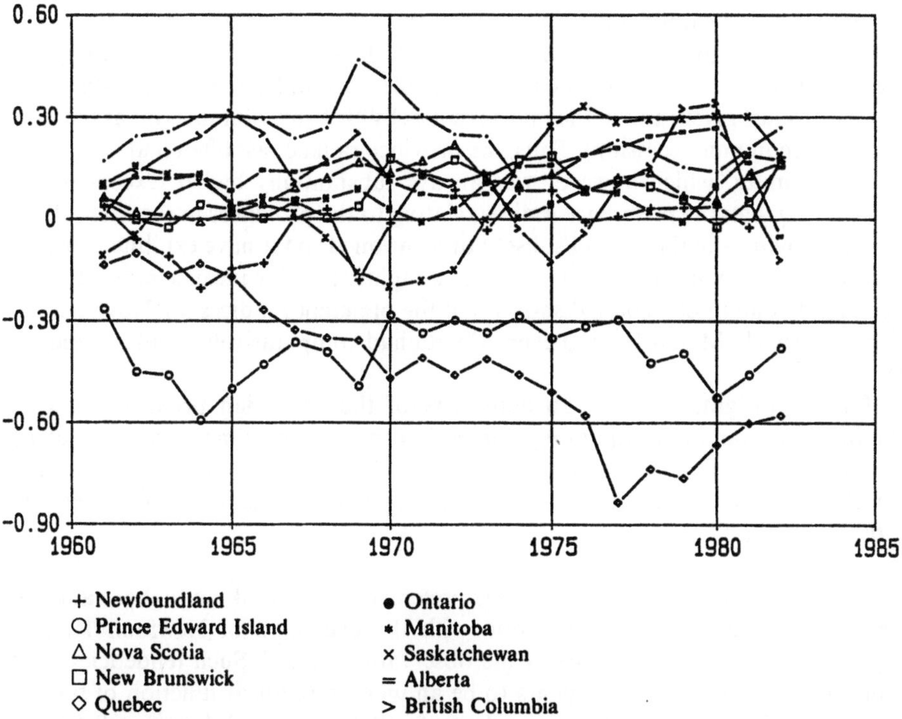

Fig. 5.5. The regional preferences of Canada, 1961–62 to 1982–83

A closer look at the time variations exhibited by the preference indices relating to the provinces in the first subset indicates that, while the preference index for Prince Edward Island registered ups and downs around its mean average value, the corresponding index for Quebec declined substantially before reversing in the late seventies. The latter finding comes as a surprise given that the 1976 takeover of the provincial government by the Parti Québécois, an "independentist" party, contributed to increasing the propensity of Quebec anglophones to leave this province and to erode the attraction exerted by Quebec on anglophones living in other provinces. Despite this reversal, the long-term tendency suggests a substantial decline in the relative attractiveness of Quebec. Thus, Quebec has replaced Prince Edward Island as the least attractive province.

Figure 5.5 reveals that the relative positions of the provinces in the other subset (provinces with non-negative values of their mean average preference index) have varied somewhat over time. Thus, Ontario – on average the most attractive province – did not always rank first. Ontario occupied first position through the sixties, lost that position after the 1973 crisis, and retrieved it when the 1981 economic recession hit hard the two westernmost provinces. In between, first position was occupied, rather surprisingly, by Saskatchewan. That agricultural province had by the mid-seventies recovered from a local recession that took place about 1970 as a result of depressed agricultural prices. First position was also held briefly by British Columbia in the last two annual periods of the seventies.

Similarly, Newfoundland – on average the least attractive province of the subset – occupied the last position only through the sixties and at times during the late seventies. Otherwise, this position was filled around 1970 by Saskatchewan (for reasons just given) and in the mid-seventies by British Columbia (hit at the time by a local economic recession). Thus, the province ranked last often had a negative preference index, which however was always substantially higher (except in the early sixties) than the preference indices relating to the provinces in the first subset.

The provinces in the second subset not mentioned so far have exhibited smaller variations of their preference index around a slightly positive mean average. Indeed the second column of Table 5.4 shows that the preference indices for Nova Scotia, New Brunswick, Manitoba, and even Alberta had comparatively smaller standard deviations.

Further insights into the attractiveness of the Canadian provinces can be obtained by examining separately for each province the time variations of its preference index. This can be done visually (see Figures 5.A2 – All in the Appendix) as well as by means of a simple regression analysis (see righthand side of Table 5.4). It turns out that:

1) In all four Atlantic provinces, the preference index took on higher values in the seventies than in the sixties, but, with the exception of Newfoundland, it apparently fell in the latter part of the observation period. Such tendencies were confirmed statistically as δ_i appears to be an increasing linear function of time in the case of Newfoundland but a quadratic function (oriented downward) for the other three provinces.

2) The long-term decline of Quebec's preference index mentioned earlier is a reality substantiated by the high performance of a linear regression of the corresponding δ_i against time.

3) In two of the Prairies provinces (Manitoba and Saskatchewan), the preference index took on comparatively higher values at the extremities than in the middle of the observation period, so that δ_i for these provinces comes out as a quadratic function of time (oriented upward).

4) Finally, for Ontario, Alberta, and British Columbia, no clear pattern of variations was obtained using a linear or a quadratic function of time. Their respective preference indices registered brief variations (Ontario in the late sixties, Alberta in the late seventies, British Columbia in the mid-seventies) that cannot be picked up by such simple functions of time.

Another variable of interest here is the variance of the preference indices across provinces:

$$\sigma_\delta^2 = \frac{1}{L-1}\sum_i\left(\delta_i - \frac{1}{L}\sum_j \delta_j\right)^2 = \frac{1}{L-1}\sum_i \delta_i^2 \tag{5.8}$$

Its time variations, displayed in Figure 5.6, reflect a pattern of change which, unsurprisingly, is very similar to that observed in the case of the variance of the utility indices (note the coincidence of the years in which the extremes of the two curves concerned occurred). Nevertheless, there is a substantial difference between the time variations in Figures 5.4 and 5.6. Whereas the time variations of the spatial variance associated with the utility indices were oriented downward, those of this spatial variance associated with the preference indices are oriented upward: σ_δ^2 is an increasing linear function of time (see Table 5.A1 in the Appendix). In other words, once controlled for population growth, the indices reflecting the attractiveness of the various provinces appear to diverge rather than to converge!

5.2.4 Migration Stress

An alternative and perhaps more sensible way, from a demographic viewpoint, to examine how the relative attractiveness of the various provinces varies over time is (with reference to each annual period over the 22-year span)

1) to derive the stationary population distribution that would result if the transition intensities in the multiregional population system were governed indefinitely by preference indices with an unchanged structure, and

2) to calculate the correlation coefficient $r(n, n_{st})$ given by formula (2.43) which expresses the similarity of this stationary population distribution with the current one.

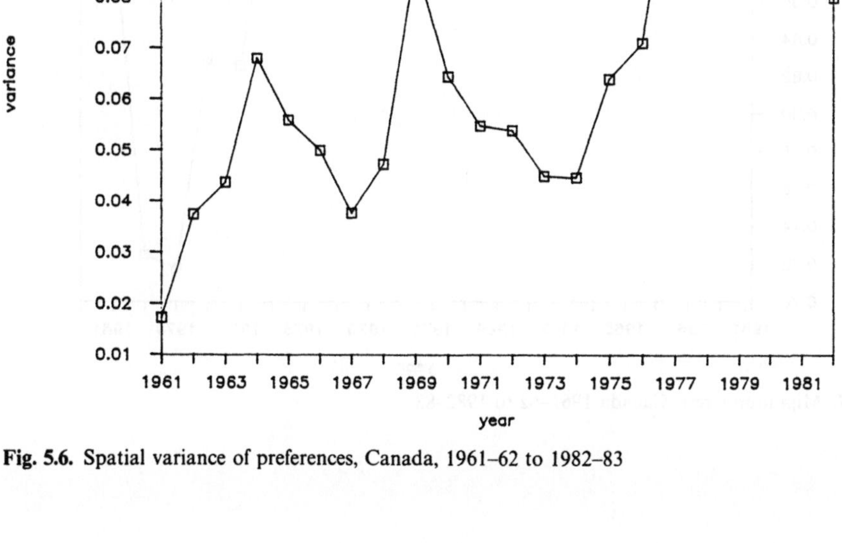

Fig. 5.6. Spatial variance of preferences, Canada, 1961–62 to 1982–83

As indicated by Figure 5.7, this coefficient, determining migration stress $s=\frac{1}{2}[1-r(\boldsymbol{n}, \boldsymbol{n}_{st})]$ registered some interesting variations over the observation period. After an initial decrease from a value close to 1 in the early sixties to about 0.9 in the late sixties, this correlation coefficient experienced about 1970 a reversal that brought its value back to roughly 0.95 in the mid-seventies. It then went on in the late seventies to fall dramatically to a value just above 0.7 before returning to a more traditional value (about 0.9) in the last two annual periods. Despite these fluctuations, the overall tendency clearly points to a decreasing value of the correlation coefficient, which means an increasing value of migration stress over time (see the last column of Table 5.A1 in the Appendix).

We thus conclude that the system of the Canadian provinces that was close to equilibrium in the early sixties has, in the last quarter of the century, moved gradually away from its equilibrium state, although two events contributed temporarily to a larger discrepancy between current and ultimate distributions:

1) the excellent economic performance of Ontario in the late sixties (which pushed up its preference index), and

2) the economic boom in the late seventies in British Columbia and especially Alberta, which motivated a larger proportion of migrants to move across the country.

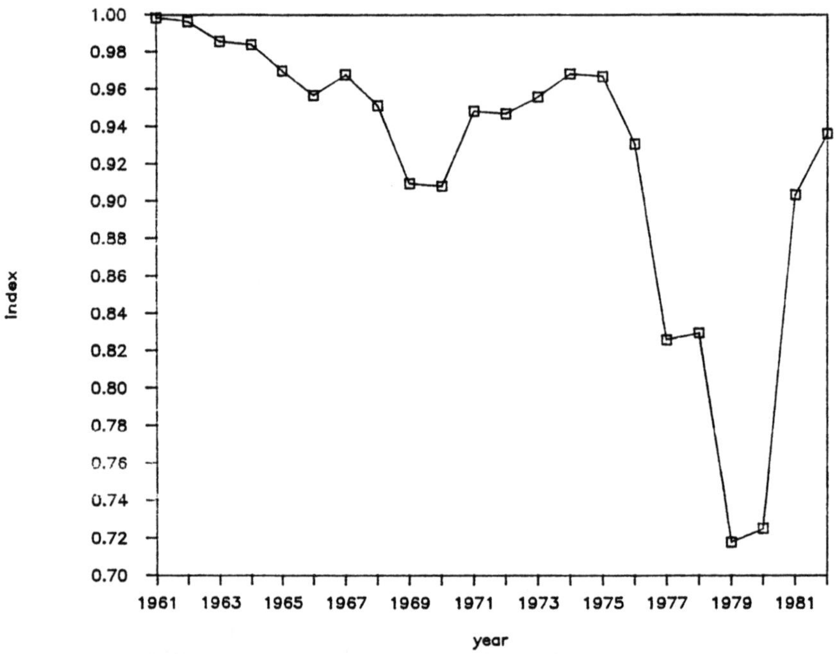

Fig. 5.7. Migration stress, Canada 1961–62 to 1982–83

5.3 Socio-Economic Analysis

The analysis now shifts to determining the socio-economic factors that influence the set of transition rates p_{ij}. Initially, a certain number of variables – contained in a data base used by this author in the context of an earlier econometric analysis of interprovincial migration[12] – were selected for having a potential influence,

1) in the case of their national values, on the global mobility index, and
2) in the case of their provincial values, on the regional utility indices.

These variables, which relate to population, employment, income, etc., are listed in Table 5.5 along with respective sources. Population was not selected for its own sake. Rather, it was used as deflator of the three income variables considered, thereby giving rise to as many income variables on a per capita basis.

5.3.1 Analysis of the Global Mobility

In case of the global mobility index, determination of the explanatory variables was carried out using the nonlinear regression method outlined in Chapter 15. It turns out that three factors have a significant influence on v_0: see Table 5.6. They are in decreasing order of importance (based on the associated t-statistics):

1) the level of per capita income (obtained as the ratio of total income to the population level) which tends to depress v_0 quite significantly – the associated t-statistic is extremely high – but with a two-year lag;

2) the level of economic activity (measured by an index of employment level in the largest companies) in the current period which affects v_0 positively and still quite significantly (as the associated t-statistic is higher than six); and

3) the unemployment rate lagged one year which also has a positive although barely significant influence on v_0.

In all, the three factors help explain about 87% of the variance regarding the global mobility index, but, as is obvious from Figure 5.8, they are incapable of duplicating satisfactorily its actual variations. The estimated values of v_0 (that is, the values calculated from the regression equation just discussed) lead to a somewhat smoother curve than the curve linking the annual observed values. Some turning points, especially between 1965 and 1975, are missed.

[12] See J. Ledent (1984) *Demoeconomic modeling of interprovincial migration in Canada: The case of time-series.* Paper prepared for presentation at the 8th Annual meeting of the Canadian Regional Science Association, Guelph, Ontario, May 31–June 1, 1984.

Table 5.5. Socio-economic factors selected as independent variables in the explanation of (1) the global mobility index and (2) the regional utility indices

Type	Factor	Source
Population	Population level	Statistics Canada, Cat. 91-201, Estimates of population for Canada and the provinces
Employment	Employment level	Statistics Canada, Cat. 71-201, Historical labour force statistics actual data, Seasonal factors, Seasonally adjusted data
	Employment index	Statistics Canada, Cat. 72-206, Employment, earnings and hours survey
	Unemployment rate	Statistics Canada, Cat. 71-201, op. cit.
	Average weekly salary	Statistics Canada, Cat. 72-206, op. cit.
Income	Total income	Statistics Canada, Cat. 13-201, National income and expenditure accounts
	Disposable income	Statistics Canada, Cat. 13-201, ibid.
	Wages, salaries, supplementary labor income	Statistics Canada, Cat. 13-201, ibid.
Others	Consumer price index	Statistics Canada, Cat. 62-010, Consumer prices and price indexes
	Benefit payments	Statistics Canada, Cat. 73-001, Statistical report on the operation of the unemployment insurance act
	Housing starts	Statistics Canada, Cat. 64-002, Housing starts and completions
	Retail sales	Statistics Canada, Cat. 63-005, Retail sales

Table 5.6. Regression equation for the global mobility index

variable	time lag	coefficient	t-value
per capita income	2	-7.13×10^{-7}	-34.8
employment index	0	2.90×10^{-5}	6.2
unemployment rate	1	6.58×10^{-5}	2.3
correlation R^2 = 0.871; Durbin-Watson statistics = 2.39			

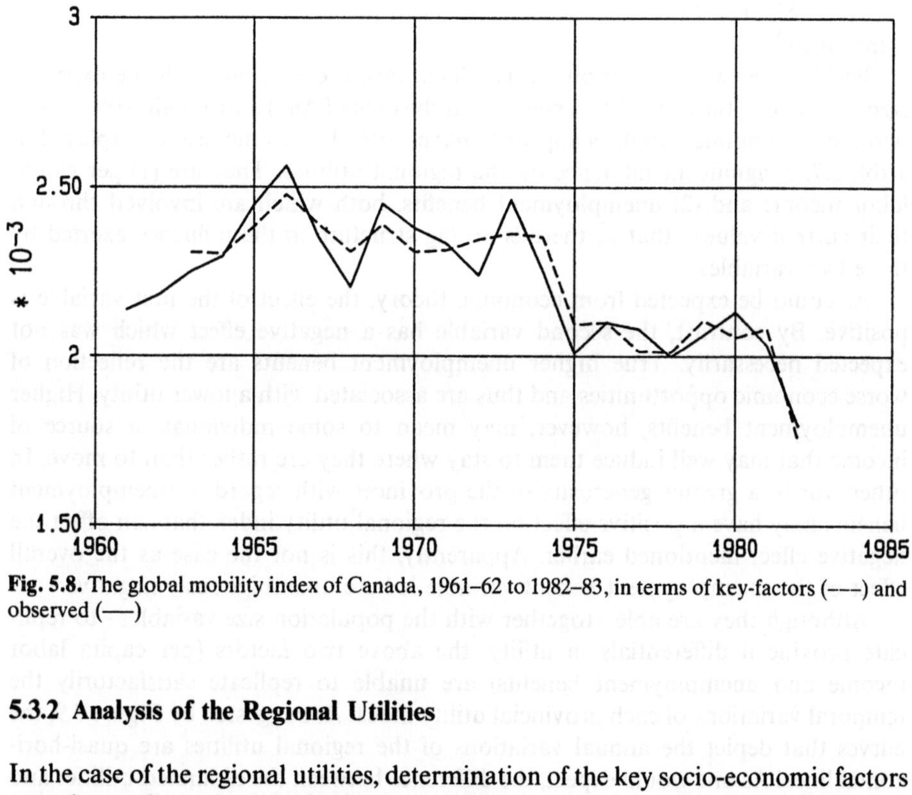

Fig. 5.8. The global mobility index of Canada, 1961–62 to 1982–83, in terms of key-factors (---) and observed (—)

5.3.2 Analysis of the Regional Utilities

In the case of the regional utilities, determination of the key socio-economic factors was also performed on the basis of a nonlinear regression method (see Chapter 3). As was hinted before when dealing with the regional preference indices, this method involves adding to the explanatory variables $\Omega(i, t)$ (see (3.69))

$$\Omega(i, t) = \frac{\tilde{\Omega}(i, t) - \frac{1}{L}\sum_i \tilde{\Omega}(i, t)}{\frac{1}{L}\sum_i \tilde{\Omega}(i, t)} \tag{5.9}$$

associated with each socio-economic factor $\tilde{\Omega}(i, t)$ listed in Table 5.5 two variables Δn_i and Δn_i^2 linked to the population level n_i by

$$\Delta n_i = \frac{n_i - \frac{1}{L}\sum_j n_j}{\frac{1}{L}\sum_j n_j} \tag{5.10}$$

and

$$\Delta n_i^2 = \frac{n_i^2 - \frac{1}{L}\sum_j n_j^2}{\left(\frac{1}{L}\sum_j n_j\right)^2} \tag{5.11}$$

respectively[13].

Besides the two population terms which have a coefficient with the expected sign (positive in the case of Δn_i, negative in the case of Δn_i^2) and a high t-ratio, only two socio-economic variables appear to have, according to the results displayed in Table 5.7, a significant influence on the regional utilities. They are (1) per capita labor income and (2) unemployment benefits, both which are involved through their current values – that is, there is no lag structure in the influence exerted by these two variables.

As could be expected from economic theory, the effect of the first variable is positive. By contrast, the second variable has a negative effect which was not expected necessarily. True higher unemployment benefits are the reflection of worse economic opportunities and thus are associated with a lower utility. Higher unemployment benefits, however, may mean to some individuals a source of income that may well induce them to stay where they are rather than to move. In other words, a greater generosity of the provinces with regard to unemployment benefits may have a positive effect on the regional utility index that can offset the negative effect mentioned earlier. Apparently, this is not the case as the overall effect of the unemployment benefit variable is here unambiguously negative.

Although they are able – together with the population size variables – to replicate provincial differentials in utility, the above two factors (per capita labor income and unemployment benefits) are unable to replicate satisfactorily the temporal variations of each provincial utility index. As suggested by Figure 5.9, the curves that depict the annual variations of the regional utilities are quasi-horizontal lines! As one would expect, the difference between the estimated utilities and the "observed" ones is greater for the provinces whose utility experienced a large

[13] The use of $\Omega(i, t)$, Δn_i and Δn_i^2 rather than of $\tilde{\Omega}(i, t)$, n_i and n_i^2 ensures that the explanatory variables satisfy constraints similar to the one that linked the dependent variables, namely, $\sum_i \Omega(i, t) = 0$, $\sum_i \Delta n_i = 0$ and $\sum_i \Delta n_i^2 = 0$.

Table 5.7. Regression equation for the regional utilities

variable	time lag	coefficient	t-value
population Δn_i	0	1.304	138.1
squared population Δn_i^2	0	-0.227	-17.0
per capita labor income	0	0.877	12.3
benefit payments	0	-0.231	-5.6
correlation R^2 = 0.932; Fishers F = 731			

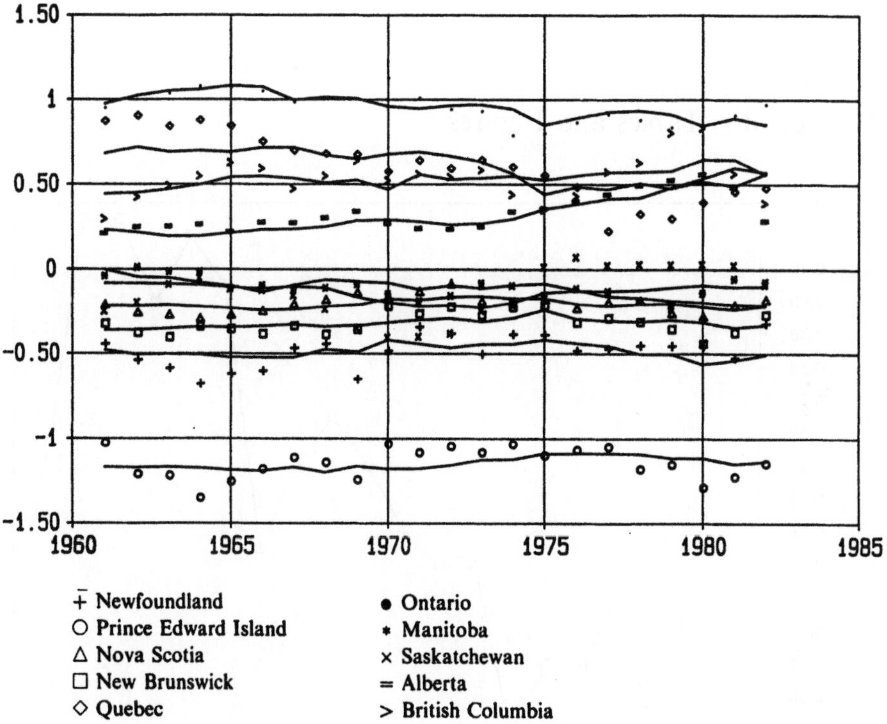

Fig. 5.9. The regional utilities of Canada, 1961–62 to 1982–83, in terms of key-factors (——) and observed

standard deviation (Quebec, Saskatchewan, and British Columbia), especially in the late seventies. Thus, the rather surprising reversal of the regional preference index uncovered earlier in the case of Quebec may not reflect reality. Rather, it is perhaps an artificial result of the above socio-economic analysis.

5.4 Conclusion

Application of the model to the case of Canada has enabled us to establish some interesting features of interprovincial migration through a proper interpretaion of the levels of and changes in the "observed" values of the model's main indices (global mobility, regional preferences). By contrast, it has been somewhat less successful in identifying the socio-economic variables that are responsible for these features. It can be conjectured that the latter finding follows less from the unrealistic assumption of homogeneous growth rates (which underlies implementation of the model) than from the lack of consideration of relevant socio-economic factors, which have been found in other studies available in the literature to have a significant effect on Canadian interprovincial migration. It is thus suggested that the global mobility and regional utility indices – estimated in Section 5.2 – be fitted, in future runs of the socio-economic analysis, to a more complete set of explanatory factors.

5.5 Appendix: Tables and Figures

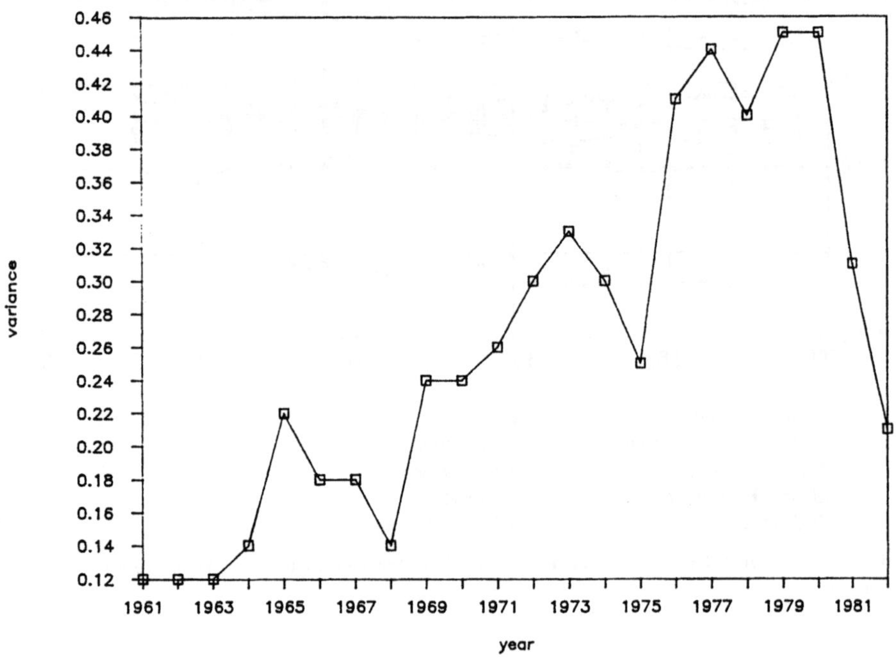

Fig. 5.A1. Spatial variance of pop. change res, Canada 1961–62 to 1982–83

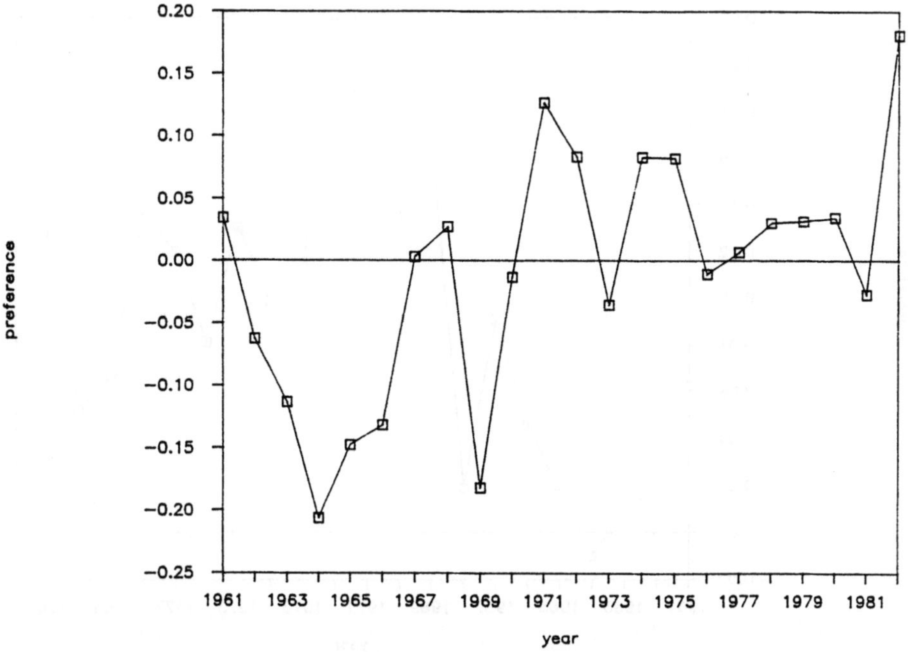

Fig. 5.A2. Regional preference, Newfoundland, 1961–62 to 1982–83

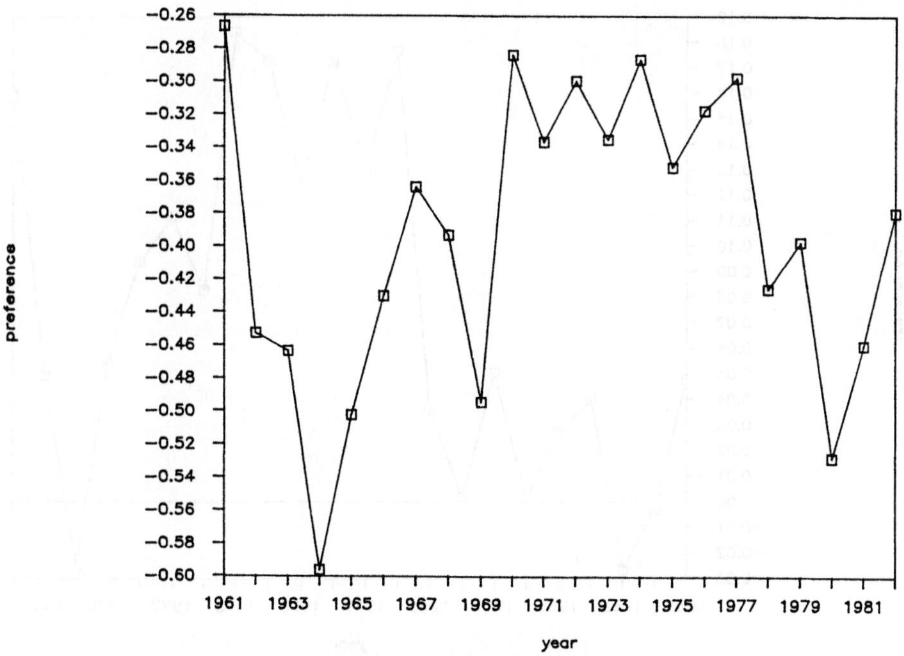

Fig. 5.A3. Regional preference, P.E.I., 1961–62 to 1982–83

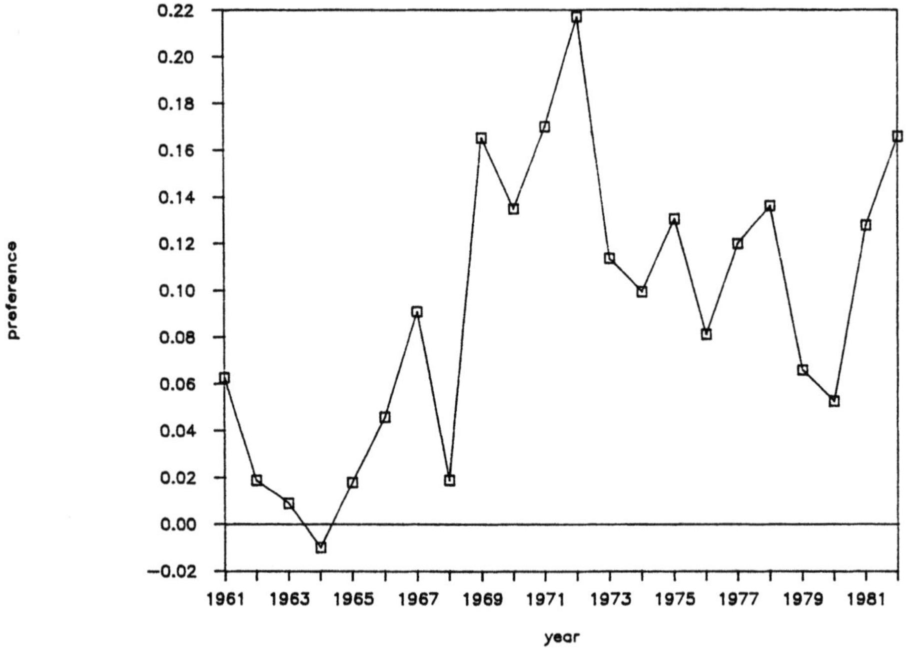

Fig. 5.A4. Regional preference, Nova Scotia, 1961–62 to 1982–83

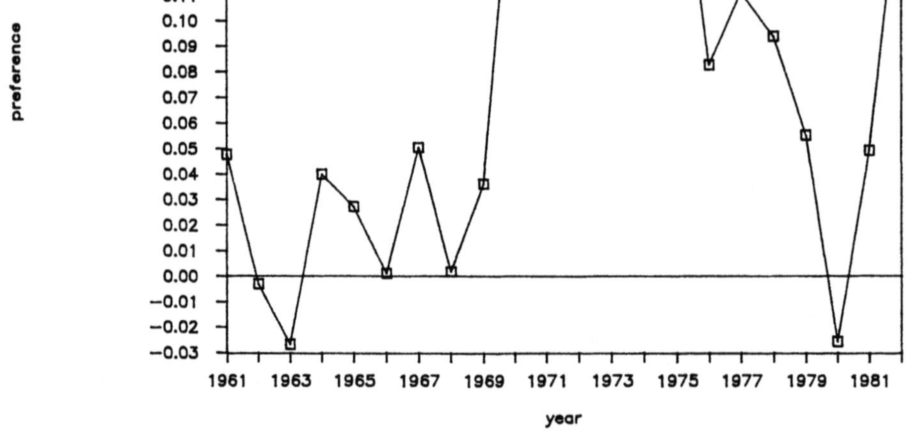

Fig. 5.A5. Regional preference, New Brunswick, 1961–62 to 1982–83

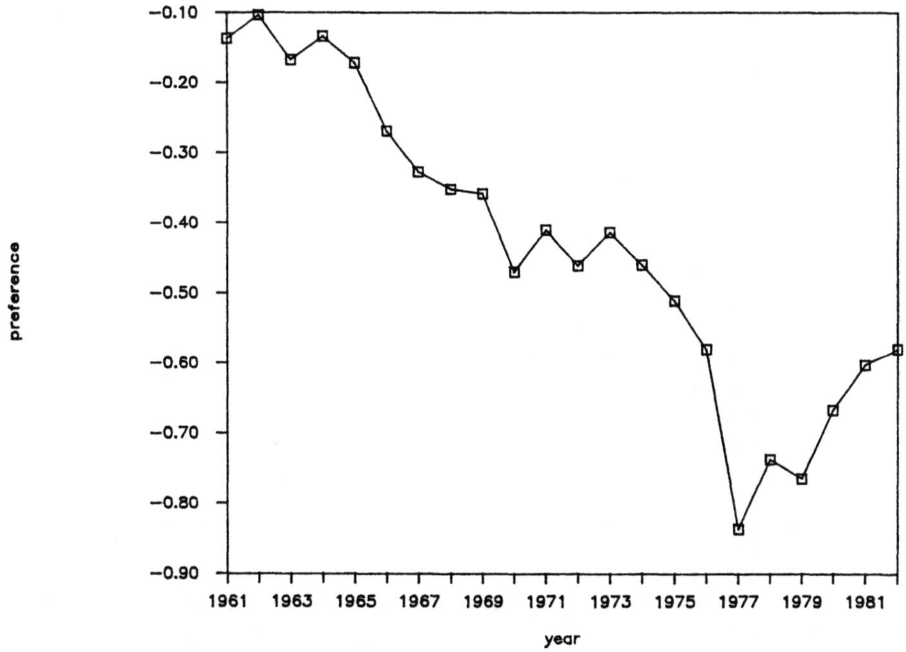

Fig. 5.A6. Regional preference, Quebec, 1961–62 to 1982–83

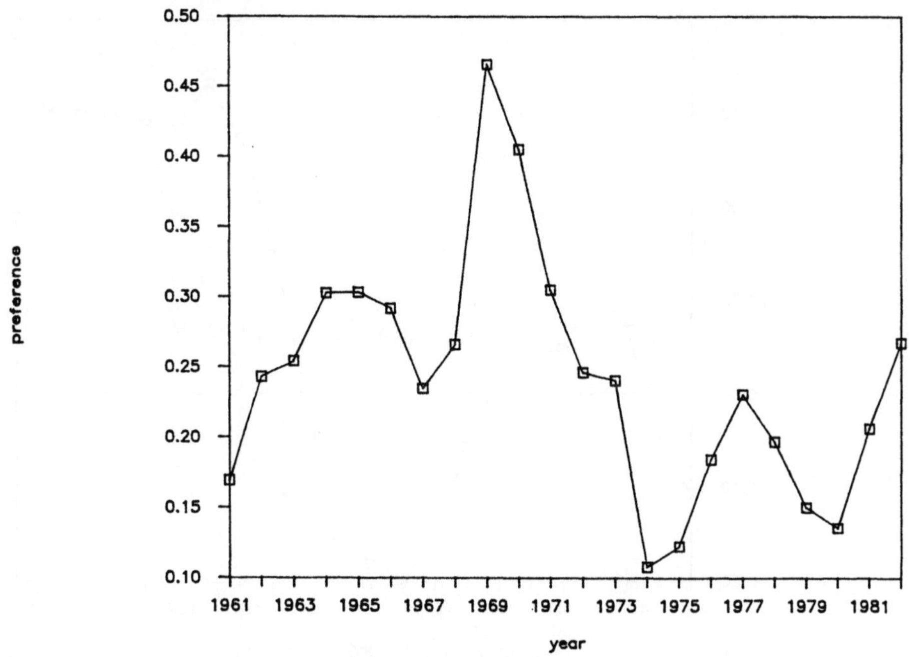

Fig. 5.A7. Regional preference, Ontario, 1961–62 to 1982–83

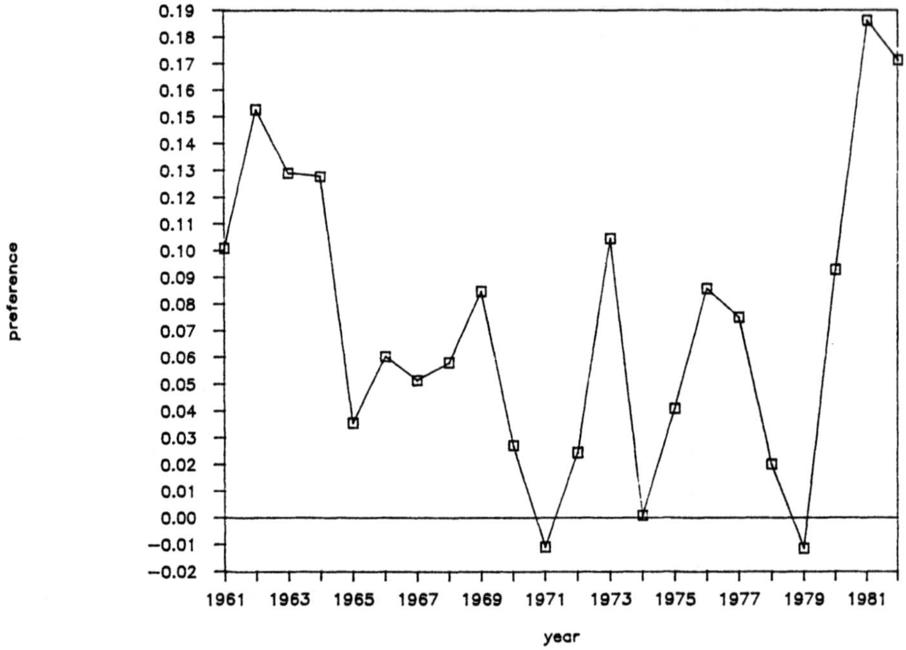

Fig. 5.A8. Regional preference, Manitoba, 1961–62 to 1982–83

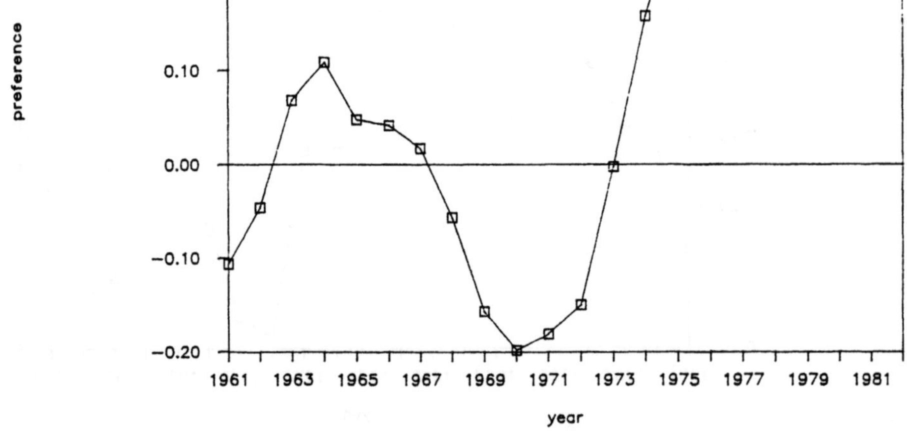

Fig. 5.A9. Regional preference, Saskatchewan, 1961–62 to 1982–83

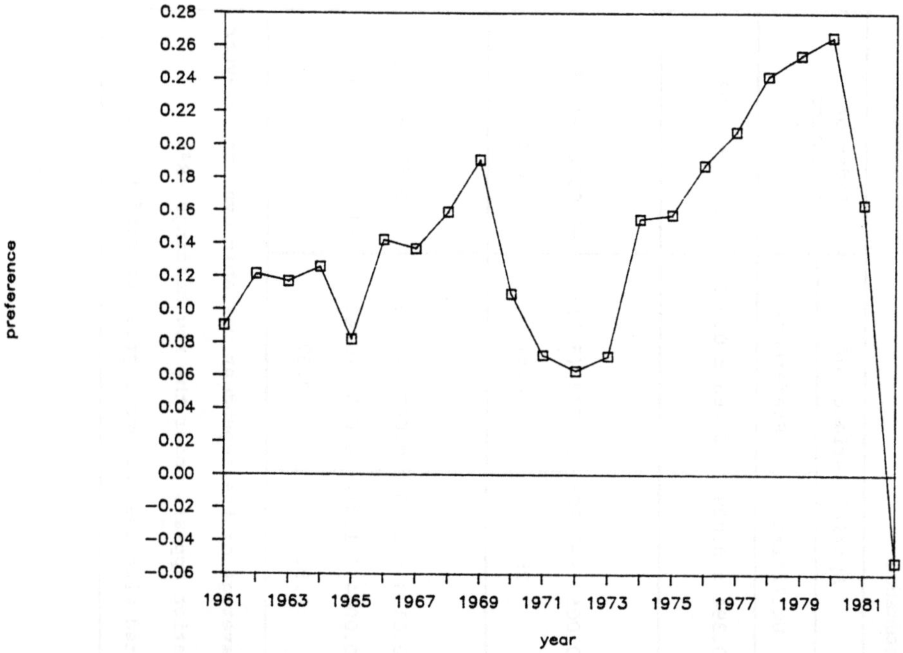

Fig. 5.A10. Regional preference, Alberta, 1961–62 to 1982–83

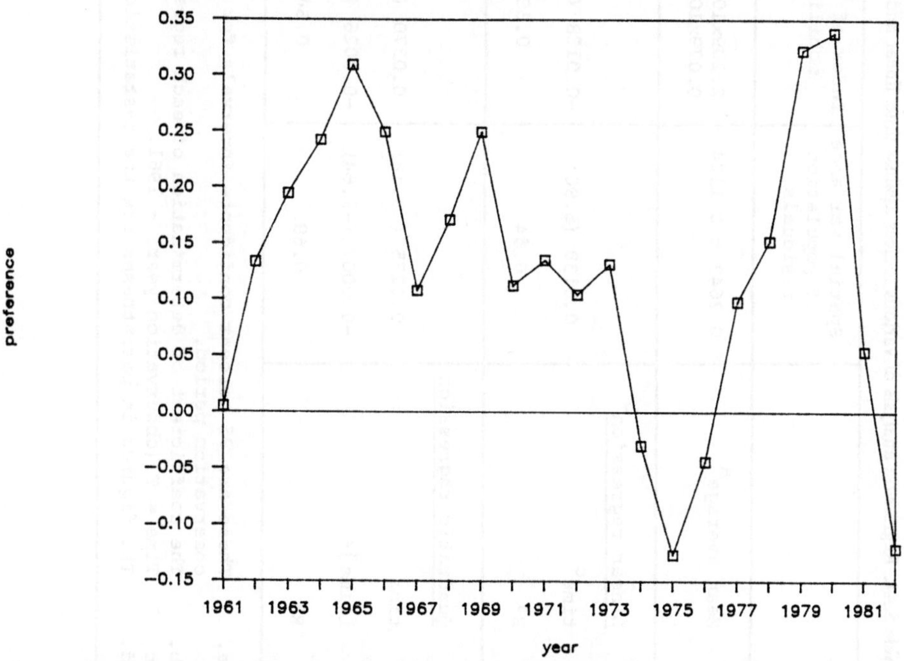

Fig. 5.A11. Regional preference, British Columbia, 1961–62 to 1982–83

Table 5.A1. Regression analysis of various indices against time: linear and quadratic specifications

	Spatial variance of population residuals	Index of global mobility	Spatial variance of		Index of migration stress
			Utilities	Preferences	
Mean average[a]	0.2643 ± 0.1124	$2.209\times10^{-3} \pm 0.079\times10^{-3}$	0.394 ± 0.029	0.066 ± 0.012	0.922 ± 0.033
<u>Linear regression</u>[b]					
time[c]	0.0138 (5.90)	-0.0178 (-3.5)[d]	-0.0057 (-2.78)	0.0032 (5.11)	-0.0082 (-4.08)
R^2	0.64	0.38	0.28	0.57	0.45
<u>Quadratic regression</u>					
time	0.0275 (3.18)	0.0370 (2.4)	-0.0133 (-1.70)	0.0021 (0.85)	-0.0053 (-0.67)
$[time]^2$	-0.0007 (-1.64)	-0.0026 (-3.7)	0.0004 (1.01)	0.0001 (0.47)	-0.0001 (-0.37)
R^2	0.68	0.64	0.32	0.57	0.46

a. These are 95 percent confidence intervals for the mean average of the dependent variable over the 22-year observation period.
b. The coefficient of determination of each regression equation appears under the row headings «R^2».
c. Time = T (observation year) - 1961.
d. The figures in parentheses are the t-statistics associated with the various regression coefficients.

6 France

Denise Pumain

6.1 Population and the Regional System

A few specific features of French population and geography, which may be of consequence for migrations, have to be recalled first. France is a medium-sized country of 54 millions inhabitants for 550 000 km^2. Its density of population is particularly low, when compared to most of the neighbouring European countries: 100 inhabitants per squared kilometer. But this population is rather unevenly distributed over the French territory: about 18% is concentrated in the single urban agglomeration of Paris, which has 8.7 millions inhabitants, whereas the second largest urban unit, Lyon, has only 1.2 millions. Many rural or mountainous areas are sparely populated: almost 60% of the French Communes (22 000 among 36 000) have less than 500 inhabitants each, which means, over a surface equivalent to half of the territory, a global density of less than 20 inhabitants per squared kilometer.

Those contrasts in settlement density are widely smoothed by the subdivision in 21 administrative "régions de programme" (the 22^d region, Corse's island, has been excluded from the study because of its spatial discontinuity with the rest of the country) (see Figure 6.1 and Table 6.1). Since those regions were designed recently (1955), and mainly for planning purposes, they are more homogeneous in size than the regions of other countries in this book: the smallest, Limousin, has 0.7 million inhabitants and the second largest, Rhône-Alpes, has about 5 millions, while Ile-de-France which includes Paris, has 10 millions. They do not exactly correspond to functional regions, but most of them are roughly centred on their largest city. Their borders almost never separate a large town (the worst case is for Avignon belonging to Provence-Côte d'Azur, which has 175 000 inhabitants, and among them 10% belonging to the neighbouring region of Languedoc-Roussillon).

The regions are still rather highly differentiated by their economic structure and socio-cultural characteristics: the average gross regional product per inhabitant of the poorest region (Limousin) is half of the richest (Ile-de-France). The main geographical features are a north/south opposition for demographic indicators (higher fertility and younger population in northern regions than in southern ones), and a north-east/south-west opposition (along a line Le Havre-Grenoble) for economy: the regions in northern and eastern part, where the industrial revolution of the last century took place, are much more industrialized than the other. The urbanization has followed the industrialization but is also well developed in the region Provence-Côte d'Azur, where the tertiary sector is essential.

During the period under study (between the two censuses of 1954 and 1982), very significant changes have occurred in the geographical distribution of population. In 1954, France was still a rather rural country with only 57% of its population living in cities. In thirty years, the labour force engaged in agriculture has been divided three times (5.1 millions of working people in 1954, 1.75 in 1982), and the average urbanization rate raised to 73%. During the same time, population and economy were growing rapidly, at least until the early seventies. This growth and the substitutions in activities (decrease in employment of concentrated industries and growing importance of the tertiary sector, from 38% to 56% of the labour force), induced large redistributions of activities and population between

Fig. 6.1. The map of France with the 21 "régions de programme"

regions. After a first stage of increasing concentration in Paris and the most industrialized regions, reverse movements appeared and were encouraged by national public agencies, namely the decentralization from the Paris region since the beginning of the sixties. Population growth has then shifted from the "central regions" around Paris, the north, the east and also Rhône-Alpes, to the "periphery", in the western part, south-eastern and even south-western recently. During the last ten years, the population of the most populated regions started to decrease, and the "shift to the south" has become more sensible.

Interregional migrations are then of particular importance to understand the recent evolution of French regions and to elaborate about their future. After a presentation of the available data and of the results given by the model, an attempt will be made to underline its specific contribution to the analysis of French migrations, as compared with a classical gravity-type model.

6.2 The Data

Since France does not have any register of population, information about spatial mobility is available from population censuses. For the first time in 1962 a question was asked concerning the place of residence at the date of the previous censuses in 1954. The same question was asked by the following censuses in 1968, 1975 and 1982. Answers were exploited to produce interregional migration tables, in an exhaustive way for the 1954–1962 period and from samples for the others (with various rates: 1/4 in 1962–68, 1/5 in 1968–75 and 1/20 in 1975–82; source: [6.1]).

Table 6.1. Symbols used in the figures for the utilities and the preferences of France

symbol	region	name
1. Regions 1 - 11		
+	1	Ile de France
o	2	Champagne Ardenne
△	3	Picardie
□	4	Haute Normandie
◊	5	Centre
·	6	Basse Normandie
*	7	Bourgogne
x	8	Nord Pas de Calais
-	9	Lorraine
>	10	Alsace
<	11	Franche Comte
2. Regions 12 - 21		
+	12	Pays de Loire
o	13	Bretagne
△	14	Poitou Charente
□	15	Aquitaine
◊	16	Midi Pyrenees
·	17	Limousin
*	18	Rhone Alpes
x	19	Auvergne
-	20	Languedoc Roussilon
>	21	Provence Cote d'Azur

This kind of statistical source does not allow to study year by year the variations in spatial mobility. Moreover, it provides information only about *migrants* (people having changed their residence between two dates and surviving at the last date – with their children counted as migrants even if they were born after the move) but not about *migrations* – that is, the total number of residential changes that really occurred during the period: multiple migrations, migrations of people who died between the two censuses and migrations ending by a return to the place of departure are neglected. Thus, there is a systematic underestimation of spatial mobility when studied from this source of data, and the intensity of underestimation increases with the length of the period between two censuses (eight, six or seven years). A model has been proposed by Courgeau [6.2] to estimate annual migration rates and to allow comparisons between the data from different periods. Its results are consistent with estimations of annual mobility derived from other sources (like surveys on the mobility of labour force). We will then use this model to provide some figures for comparison with the other countries. But there is no model allowing to correct this effect for each flow of migrants appearing in the interregional tables, which have been used as they were for the analysis.

6.3 Evolution of Global Mobility

The level of residential mobility in France, with about 10% of people changing their housing per year, is very near of the figures observed in other European countries or in Japan, but much less than the mobility rates of Canada or United States, which are twice as high.

The general level of mobility in France, which was increasing since the last century (Tugault [6.3]) started to decrease between 1975 and 1982 (Courgeau, Pumain [6.4]). This decrease was more pronounced for medium- and long-distance migrations (changes in department or region of residence) than for short-distance moves (changes in commune or housing). As it appeared in every age group, even for retired persons, this decrease in mobility cannot entirely be explained by the reduction in job opportunities due to the economic crisis. Despite the crisis may have accentuated the phenomena, it has more likely to be related to the general decrease in mobility levels which were observed sooner in other countries (in the sixties for United States, since 1962 for Belgium and 1970 for Scandinavia).

Figure 6.2 shows the evolution of a global mobility parameter calculated by the model between 1954 and 1982 at the regional level and Table 6.2 allows its comparison with estimations derived from the same source by the model of Courgeau [6.2]. Both illustrations underline the rapid increase in interregional mobility between 1968 and 1975 and its decrease after this date.

One hypothesis upon which our migration model is based is that this general mobility level should be considered as homogeneous for the whole country. However, this can only be hold as a first approximation: Tables 6.3 and 6.4, show

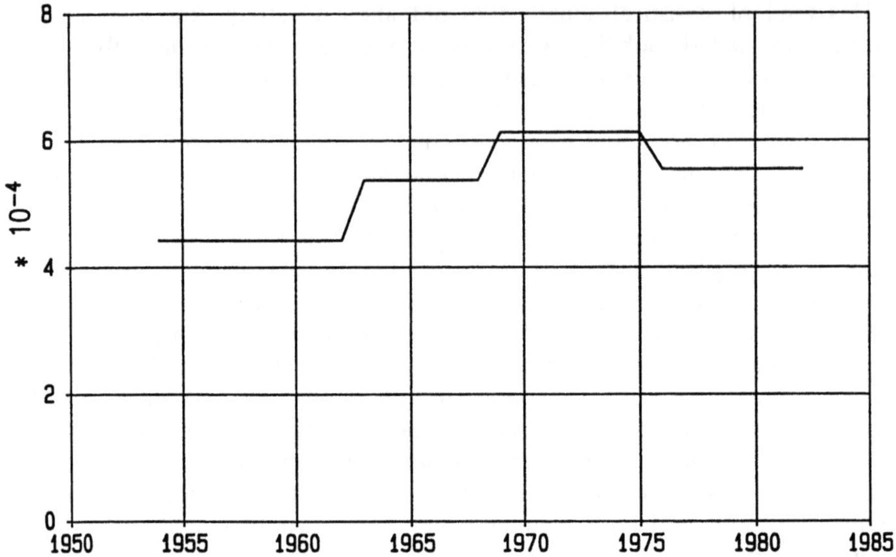

Fig. 6.2. The estimated global mobility $v_0(t)$ of France

Table 6.2. Evolution of global mobility from 1954 to 1982 at the regional level. (1) Source: Courgeau, Pumain [6.4]

Period	Annual rate of interregional migration (%) (1)	Global mobility parameter estimated by the model
1954-1962	1.42	4.41×10^{-4}
1962-1968	1.59	5.37×10^{-4}
1968-1975	1.90	6.12×10^{-4}
1975-1982	1.72	5.53×10^{-4}

significant variations in interregional mobility for groups of people differing by age, with a strong decrease of the mobility level after 40 years, and by socio-professional status [6.5]. Long-distance migrations are about twice as frequent for professionals or managers and executives (and of course for categories liable to mobility as army or police) than for entrepreneurs, and three times when compared to the mobility of workers – the farmers being an even less mobile category. Interregional variations of a mobility index, as shown on Figure 6.3, are nevertheless of smaller amplitude than inter-group ones, but cannot be totally explained either by regional differences in age structure or in social composition of population.

Table 6.3. Variations in mobility rates according to the age of population. Source: Population et Sociétiés, 1984, no. 179

Age (at census)	Proportion of people having changed their region of residence between two censuses (%)	
	1968-75	1975-82
less than 20 years	10.1	9.1
20-29 years	14.9	13.1
30-39 years	12.0	11.1
40-59 years	5.6	4.8
60 and over	4.5	4.3

Table 6.4. Variations in mobility rates according to the socio-professional status (after correction of the age effect). Source: INSEE, Survey on residential mobility [6.5]

Socio-professional status of head of household	Proportion of households having changed between 1973 and 1978:	
	their housing	their region of residence
Farmers	10.3	1.9
Agricultural workers	31.2	6.7
Entrepreneurs	26.3	14.4
Professional or managers	25.3	28.3
Executives	36.4	21.4
Employees	32.5	17.5
Workers	29.5	9.2
Domestic services	33.5	11.0
Army police	43.0	26.7
Non actives	32.0	15.6

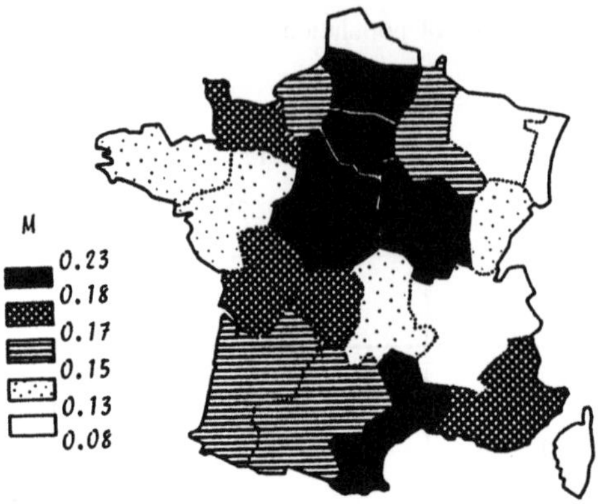

M = (I+O)/P
I = interregional immigrants (1975–1982)
O = interregional outmigrants "
P = population of the region

Fig. 6.3. Inequalities in the interregional mobility level

6.4 Regional Utilities

Figures 6.4a and 6.4b show the temporal variation in regional utilities. Figure 6.5 maps the spatial distribution of these utilities for 1954–1962 (a) and 1975–1982 (b) and of their variations between the two periods (c).

During the fifties, the differences in utilities reflected mainly the inequalities in population size between regions. The most populated Ile-de-France, Rhône-Alpes, Nord-Pas-de-Calais and Provence-Côte d'Azur had the highest values, while at the other end Limousin, Franche-Comté, Champagne-Ardenne, Basse-Normandie and Auvergne, which were the five less populated areas, received four of the five smallest values. However, there are exceptions to this general relationship between utilities and population size: Alsace, Bourgogne, Lorraine seem to be more attractive than one would expect from their size and on the reverse Bretagne, Poitou-Charentes and Pays de la Loire look less attractive. For this 1954–1962 period, utilities express then the two main characteristics of spatial mobility in France after the war: the attraction towards large metropolis on the one hand, leading to an accentuation of the spatial concentration of population, and on the other hand the shift from the less industrialized zones of the western part towards the industrialized regions of the eastern part.

The graphs of Figure 6.4 and map c of Figure 6.5 illustrate clearly the two main tendencies in the evolution of these regional utilities between 1954 and 1982. First, the relationship between utility and population size of the regions has become looser. The three most populated regions undergo a continuous decrease in their utility level: rather strong for Ile-de-France and Nord-Pas-de-Calais, for which it has even become negative, a little less rapid but still sensible for Rhône-Alpes. It is a

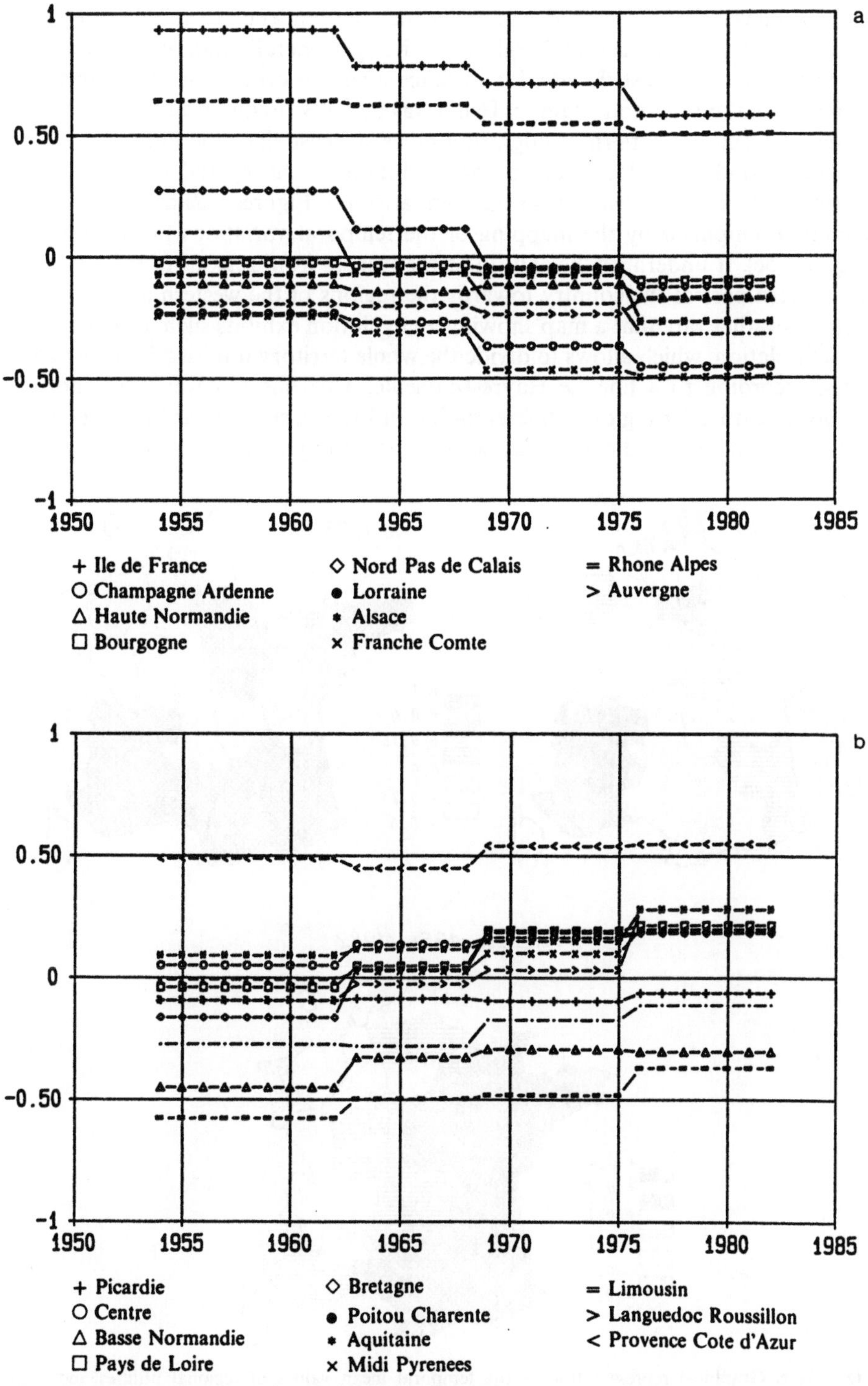

Fig. 6.4a,b. The temporal variation of regional utilities $u_i(t)$

major achievement of the model to show that the tendency towards population deconcentration from "central" regions to the "peripheral" ones, which was perceptible only in the last decade, for instance after the evolutions of the regional shares of total population (Pumain [6.6]) was actually predictible after the design of the interregional pattern of migration flows, at least since the fifties.

The second main tendency in the evolution of the spatial distribution of regional utilities is deducible from the comparison of Figures 6.5a and 6.5b, and is strongly exemplified by the mapping of the temporal variations in utility levels (Figure 6.5c). It underlines the shift in attractiveness from regions of the northern and eastern part of the country towards the regions of the western and southern part. It is rather rare that a map showing an evolution exhibits such a clear spatial autocorrelation, which allows to divide the whole territory into two homogeneous parts, according to a line Le Havre-Marseille. This line is nevertheless not an unknown feature for regional differentiation in France, since it has for a long time be used to summarize the nowaday secular dichotomy between the north-eastern

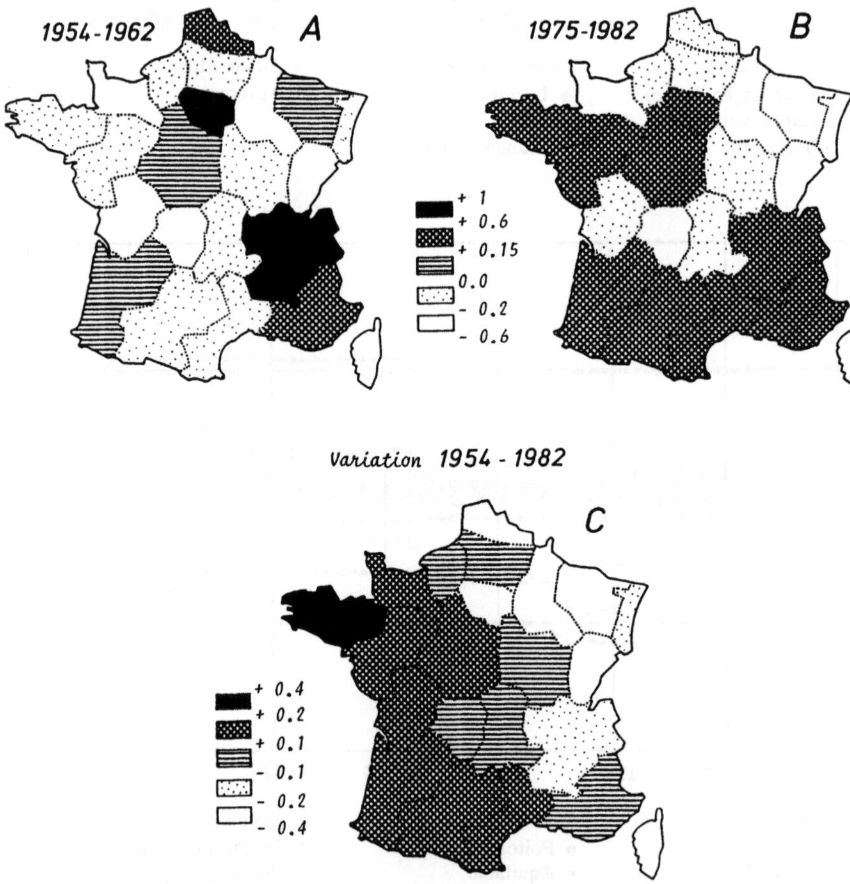

Fig. 6.5a–c. Graphical representation of the temporal mean values of regional utilities and their variation between 1954 and 1982

industrialized part of the territory, which was attracting population since the middle of nineteenth century, and the south-western part, still more engaged in agriculture and loosing population in an absolute and relative way until recently. The variations in regional utilities indicate then a tendency to a reversal in interregional gradients in attractiveness. One may wonder when this reversal started: just after or just before the second-world-war?

Those two main tendencies did not affect all regions in the same way. We noticed already that Rhône-Alpes resisted better than Ile-de-France and Nord-Pas-de-Calais to the decrease in utility of largest regions. Provence-Côte d'Azur, despite highly populated and urbanized, maintained also and even increased slightly its utility level. While most of the regions undergo a continuous increase or decrease of their utility level, a few of them exhibit fluctuations. They are mainly regions with negative values, as Picardie, Haute- and Basse-Normandie, Alsace, Auvergne. Among the regions of the western and southern part, if Bretagne, Pays de la Loire, Midi-Pyrénées and Languedoc-Roussillon succeeded in passing from a negative to a positive utility level, other remained with more or less negative values, as Basse-Normandie, Poitou-Charentes, Auvergne and Limousin.

Figure 6.6 shows that the inequalities in regional utilities, which are actually not very strong, did not vary much during those last thirty years. After a decrease between 1962 and 1968, they are increasing again but very slowly. As a matter of fact, the range of utility values has considerably decreased, with values ranking from −.57 to .93 in 1962, and from −.49 to .57 in 1982. This can be explained by the general spatial redistribution of utilities, with a regression towards the mean of the most extreme values corresponding to long-established attractive or repulsive situations, whereas new and less stronger disparities appeared among the other regions.

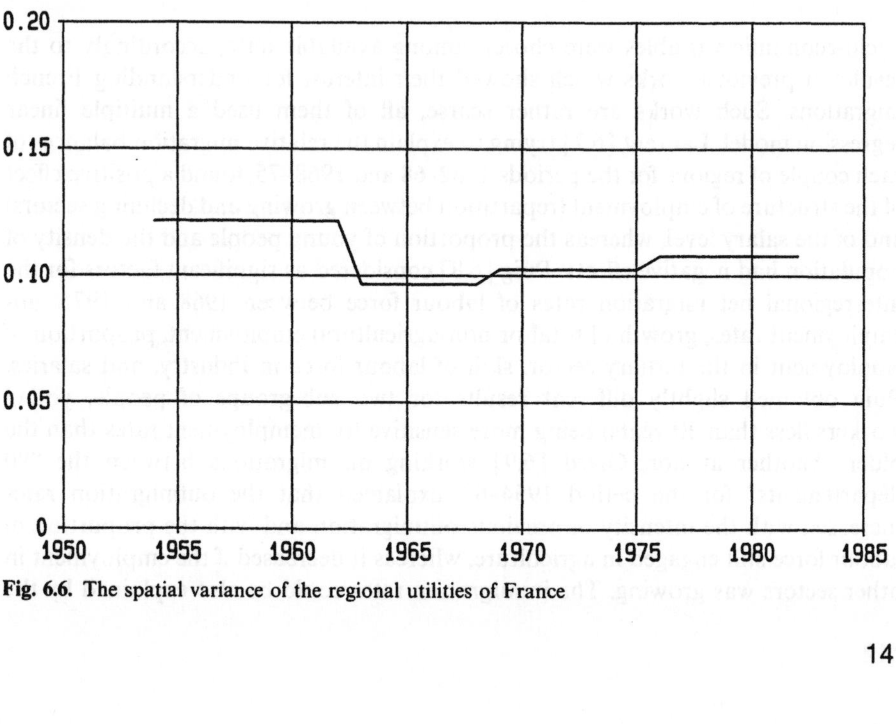

Fig. 6.6. The spatial variance of the regional utilities of France

6.5 Regional Preferences

Maps of the Figure 6.7 show the spatial distribution of preferences, which are derived from the utilities after excluding the size effect of the regions (they are residuals of a regression of utilities against population size and its square put as a saturation effect). Those maps lead to precise the general interpretation of migratory trends, which globally remain the same as previously described.

From Figure 6.7a it can be seen that, in the early sixties, migratory flows tended to avoid all the western regions from Basse-Normandie to Midi-Pyrénées as well as the Nord and Champagne-Ardenne. Provence-Côte d'Azur was the most preferred region, followed by Ile-de-France, Rhône-Alpes, Bourgogne and Alsace. In 1982, western and southern regions (the two Normandies excepted) are systematically preferred, while with the exceptions of Picardie and Bourgogne all the northern and eastern regions are less preferred, including Ile-de-France and Rhône-Alpes. As for utilities, the evolution of preferences between 1954 and 1982 has been most of the time a continuous one, attesting the deep roots of those tendencies in population redistribution over the French territory.

This last remark may be exemplified by Figure 6.8 showing a continuous increase in the "migratory stress", that is the correlation between the actual distribution of population among regions and the one which would be observed if the preferences detected for each period were maintained over time. The tendency out of equilibrium has been stronger since 1975, when the previous tendencies towards a reversal in migration flows became more obvious to all observers.

6.6 Socio-Economic Key-Factors

Socio-economic variables were chosen among available data, accordingly to the results of previous works which showed their interest for understanding French migrations. Such works are rather scarse, all of them used a multiple linear regression model. Laurent [6.7] trying to explain the relative migration balances of each couple of regions for the periods 1962–68 and 1968–75, found a positive effect of the structure of employment (repartition between growing and declining sectors) and of the salary level, whereas the proportion of young people and the density of population had negative effects. Puig [6.8] considered as significant factors for the interregional net migration rates of labour force between 1968 and 1975: unemployment rates, growth of total or non-agricultural employment, proportion of employment in the tertiary sector, skill of labour force in industry, and salaries. Puig obtained slightly different results for two sub-groups of people, young workers (less than 30 years) being more sensitive to unemployment rates than the older. Another author, Giard [6.9] working on migrations between the "90 départements" for the period 1954–62, explained that the outmigration rates increased with the intensity of previous outmigration and with the proportion of labour force still engaged in agriculture, whereas it decreased if the employment in other sectors was growing. The immigration rates could be also explained by the

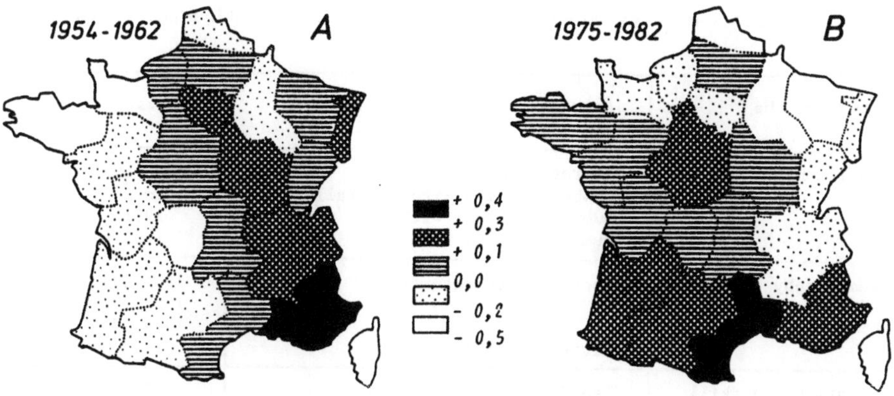

Fig. 6.7a,b. Graphical representation of the temporal mean values of regional preferences between 1954 and 1982

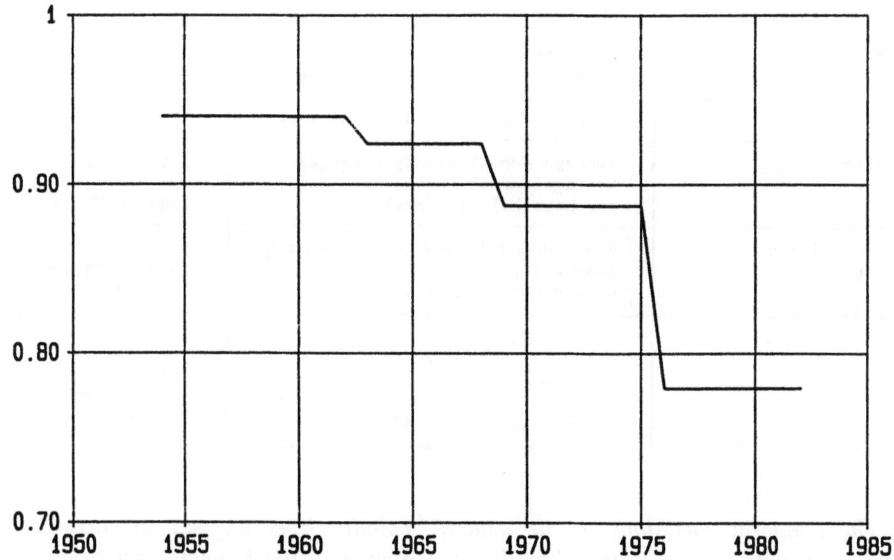

Fig. 6.8. The correlation coefficient $r(n, \hat{n})$ of France (migratory stress $=(1-r(n, \hat{n}))/2$)

intensity of previous in and outmigrations. But this model failed to predict with a sufficient precision further migrations of the period 1962–68.

We collected then indicators related to the factors quoted by the two first studies. We added also a few variables which were put forward to explain the most recent migratory moves (for instance in [6.10], [6.11] and [6.12]) as environmental amenities (indicators of climate) and education level of the population (see Table 6.5 for the labels of variables).

Table 6.5. Selected socio-economic variables for regions. Source: Statistiques et Indicateurs des Régions Françaises

Nature of the indicators	Actually measured variables	Years
Environment	Average temperature Number of days with rainfall Number of hours with sun Number of days with frost	average 1931-1960 average 1931-1960 1981 1981
Demography	Reproduction rate (%) " " " " Birth rate (%)	1961-1963 1967-1969 1974-1976 1982
Education and skill of labour force	People aged over 15 years with diploma higher than baccalaureat (%) % of managerial staff in production % of managerial staff in services (among salaried persons employed in establishments whose size are over 10 salaried persons)	1968 1975 1982 1973,1979 1973,1979
Employment	Activity rate (%) % employed in tertiary sector	1954,62,68 1975,82 1962,68,75,82
Housing construction	Number of housings per capita completed at the end of the year	each year 1959 to 1983
Income	Average annual salary (francs) Average annual income per household (francs)	1963,67,75 1963,70,75,79
Productivity	Gross domestic product per unit of labour force weekly duration of work	1962,70,75,80 1975,1980
Unemployment	Number of unsatisfied job demands/labour force - monthly average - at the end of the month of march Unemployment rate (%)	 1964 to 1970 1971 to 1983 1954,62,68,75

The temporal variations in the general mobility level happen to show no correlation with any of those variables, excepted with the housing construction rates (number of housings completed each year). This relationship mainly refers to the interaction between the long distance moves of people and the building activity. It would be hard to decide about a causal effect: does mobility increase when more housings are constructed, or does one construct more housings when more people move? More probably, the rate of housing construction and the mobility level are both tied to deeper changes in the location of population due to the conversion of economic activity (from agriculture to industry and then to services) and to changes in the settlement system (rural depopulation and urbanization). The intensity of those changes was rather high in France after world-war II and was then strongly reduced from the middle of the seventies: the depletion of the rural "reservoir" of labour force, the urbanization rate coming near to its maximum level, the decelaration in the demographic growth (due to inferior natural increase rates and to decrease of the call to external immigration) and the economic crisis which reduces the number of available jobs and increases the uncertainty of moving, have all been put forward as factors of the recent decrease in mobility level, and may all have also affected in the same way the rate of housing construction. Two other factors acting since the beginning of sixties may also have had effects with longer delays (about ten years), and could contribute to this relative stabilization of populations: one is the tendency for industrial activity to locate nearer of the available labour force, because its costs were smaller in remote cities or in rural areas than in already industrialized regions; the other refers to the public regional policies which strived towards the decentralization of industry (since the early sixties) and then of some services (around 1970) from the Ile-de-France region towards the other regions, and mainly the most disadvantaged ones, in order to correct the disequilibrium between "Paris and the French desert" as quoted by Gravier [6.13]. Those measures tended then to reduce large interregional migrations in stopping the concentration process and in trying to bring nearer the jobs and the resident population.

However, our data, which are known only for a few large periods of time, do not allow to conclude more firmly about the lack of temporal association between the general mobility level and the socio-economic indicators that we have selected.

The regional differences in utility level are more directly explainable from those indicators. Table 6.6 shows the results of a multiple regression of regional utilities against those variables and the size of the regions (absolute number of inhabitants and square of this number). As we noted earlier, the size of the regions is the main factor explaining the utilities: the larger is a region and the higher is its utility for migrants. As expected, this attraction is not infinite but exhibits a saturaton effect, as showned by the negative coefficient associated to the square of the size. It is as everywhere the second explanatory factor in the equation of regression, but for France its value is rather low as compared to other countries: the repulsiveness of the most populated regions is not (not yet?) very sensible.

Among the socio-economic variables, only five proved to be of some significance to describe the differences between the regional utilities. They all appear with the sign that one could intuitively expect: the housing construction rate, the environmental amenity, the proportion of workers engaged in the tertiary sector and the level of education tend to increase the regional utilities, whereas unemployment has negative effects.

Unemployment plays the major role. Since 1968 it has ceased to strike mainly the rural regions and it grows more rapidly industrial and urban ones (Ile-de-France, Nord-Pas-de-Calais, Provence-Alpes-Côtes d'Azur and Lorraine) and more recently in western regions. But the significance of the relationship between the unemployment level and the utility for migrants is not simple: for instance attractive regions of the southern part of France as Languedoc-Roussillon or Provence-Côte d'Azur have as high unemployment rates as the repulsive Nord-Pas-de-Calais or Lorraine: in the first group of regions the population increases more rapidly than the jobs are created, whereas in the second the population

Table 6.6. Results of the regression of utilities against socio-economic variables

variable	time lag	coefficient	t-value
Size (1)	0	0.913	181.6
Square of size	0	-0.181	-40.0
Unemployment rate	0	-0.228	-25.4
Housings completed per capita	0	0.269	16.6
Environmental amenity (2)	0	0.679	15.7
Tertiary sector (3)	0	0.788	14.6
Education (4)	0	0.634	7.7

correlation R^2 = 0.92
Fishers F = 139.7

(1) Number of inhabitants
(2) Average annual temperature
(3) Proportion in labour force
(4) Proportion of people age over 15 years with diploma higher than baccalaureat

decreases but less rapidly than the creations of jobs. Unemployment rate is then together a cause of the low utility of some regions and a consequence of the high utility of the others.

As for the general mobility level, the construction of housings can be seen as an indicator which accompanies the growth of population in regions with higher attractiveness (see Figure 6.9), but cannot be interpreted as a direct determinant of global mobilities: this causal interpretation would be more plausible at the geographical scale of an urban area, between center and suburbs, but it can hardly hold for interregional migrations.

A variable which is paradoxically more "explanatory" in character in the environmental amenity, as measured here by the average annual temperature: this component contributes to the higher utility level in southern regions, despite of their sometimes high level of unemployment. This "shift towards the south" has received many interpretations. It is true that considerations about the quality of the environment, and especially climatic ones, can express themselves when economic activities are more mobile and when managers can influence their choice in location. But to be really effective, the climatic advantages of the south had to be reinforced in a more complex process leading to improve the general "image" of

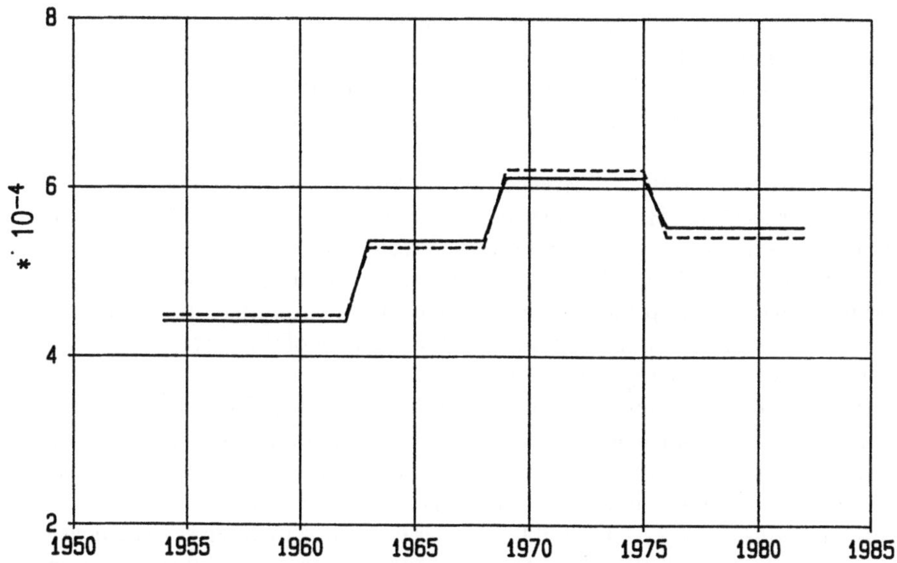

Fig. 6.9. Fitting of the global mobility of France with housing constructions as key-factor

those regions in the mental representations of entrepreneurs and of potential migrants. Advertising compaigns like the one for "Montpellier la surdouée" illustrate well the need to enlarge the basis of attraction and to give a more up-to-date image to this southern part, which was until recently a little backward, economically.

It is also not surprising to find as a fourth explanatory variable the relative importance of the tertiary sector. The share of the labour force employed in this type of activity has grown from 38% in 1954 to 58% in 1982. Ile-de-France and Provence-Côte d'Azur keep the highest proportions, larger than 67%, whereas it varies in the other regions between 47 and 60%, according to their still agricultural or industrial prevalent character. Despite a slight reduction in those interregional functional disparities, the relative positions of the regions have remained about the same over time and this can explain why the more tertiarized ones, where many jobs were still created in this sector of activity, had a greater utility for migrants.

The educational level of the population is more and more invocated as a location factor for the economic activities using the more advanced technologies. As it appears here in the equation, it plays only a minor role in the regional utilities. Perhaps will it become still more effective in the future. Perhaps also, as the French gradient for educational level is roughly spatially distributed from northern regions, with low values (Ile-de-France excepted), towards the southern ones, and is then rather highly correlated with the average annual temperature.

Shall one be surprised, not to find any effect of differences in salaries or income level in the explanation of the regional utilities? At a first sight, this could appear as a refutation of one of the theoretical proposal which is very frequently put forward to explain migrations: the search for a better salary. But such an interpretation would fall in a trap usually called the "ecological fallacy": a relationship or an explanation may exist at a given geographical scale (here, for the individuals) and many not appear at another geographical scale (here, the regions). As a matter of fact the interregional differences for the average salary level are not very high in France: the value for Ile-de-France is 50% higher than the one for the poorest regions, but there is only a difference of 10% among the 20 regions where Ile de France is excepted. Higher salaries are found in more urbanized regions (Rhône-Alpes, Provence-Côte d'Azur) and in regions which are together urbanized and industrialized (Haute-Normandie, Lorraine, Alsace, Nord-Pas-de-Calais). During the whole period those differences have been strongly reduced: the raising up of the lowest salaries and the industrial development of the western regions have contributed to this homogeneization.

So the lack of relationship between the salary levels and the regional utilities reflects again the main tendency of the evolution of economic development and migration flows during the last twenty years (at least since 1962), which has been a trend of territorial reequilibration towards the western and southern regions. Economic private interests and public policies combined themselves to achieve this ambition of territorial planners. However, one may wonder if this evolution as illustrated by the reversal in regional preferences between 1954–62 and 1975–82, will persist a long time enough to correct the previous regional disparities which still exist. Obviously, even during the economic crisis, large cities with a diversified economic basis keep a great advantage. But the now depressed regions of the old industrial cores in north and north-east of France may solve their problems and perhaps will soon rediscover the benefits of their excellent geographical position with respect to the hearths of economic activity of north-western Europe. So the future of the shift in population from the "core" regions to the western and southern "periphery" which has been observed in France as in other western highly developed countries (Vining, Pallone [6.14]) remains uncertain: is it a durable reversal in the location of economic development or is it merely a temporary stage of deconcentration in a still highly concentrated development process at the national level? More detailed studies of migration flows for significant subgroups of population may allow to progress a little further in this kind of prediction.

6.7 Comparison with a Gravity Model

The same matrices of interregional migratory flows have been treated by more classic and static migration models (Pumain [6.15]). Several kinds of models were experimented, to adjust the migration matrix of each period, using various functions of the interregional distances (the models were developed by Poulain [6.16]). We will use here the results for the model which gave the better fits, with a Pareto function and an expression of the average physical distance between regions. We will further refer to this model as to "the gravity model"

$$w_{ji} = k n_i n_j d_{ij}^{-a} \tag{6.1}$$

where w_{ji} is the number of migrant from region i to region j, n_i and n_j are respectively populations of regions i at time t and j at time $t+1$, d_{ij} is the measure of the distance between the regions, a is a parameter. So the philosophy underlying the gravity model is quite different: one assumes that the number of migrants between i and j is proportional to the probability of finding the same person in region i at the beginning of the period and in region j at the end of the period (the probability itself is proportional to the product $n_i \cdot n_j$) and that the deterrence effect of distance is homogeneous over space and independent from the spatial distribution of populations. One assumes then that only the residuals between observed and computed flows should have to be explained by socio-economic variables.

We thought that a comparison of the results given by the two models could bring interesting insights on the specific contribution to the study of migrations of the new approach presented in this book. We will compare them on three points: i) the values that they use for the deterrence function between regions; ii) the global measure that they give in order to characterize the attractiveness of each region for each period; iii) the spatial distribution of their residuals.

i) The *deterrence function* in our model is estimated after the actual flows of migrants, whereas it is exogeneously given in the gravity model, using a physical distance between regions. But both expressions are symmetrical for every couple of regions and can then be compared. We plotted on Figure 6.10 the values of the term f_{ij} (which is an average over time of the mobility factors v_{ij} weighted by the global mobility v_0), against the distance d_{ij} used in the gravity model. As both scales are logarithmic, a strict correspondence between the two measures of the deterrence functions does exist if all points are situated along a straight line with a negative slope. It is far from being the case. Thus, as compared with physical distance, f_{ij} is a more phenomenological measure of the separation between regions. When mapped, the discrepancies between the two measures reveal the greater frequency of exchanges between remote regions (mainly between Ile-de-France and the four southern regions and Provence-Côte d'Azur and the four north-western regions, but also between other distant regions situated at the periphery of the country) whereas the flows between the regions surrounding Paris and between them and Limousin, Auvergne, Nord, Lorraine, Alsace and Franche-Comté would generally be less according to f_{ij} than one could expect from the physical distance d_{ij}.

This comparison shows then that the physical distance is only a rough surrogate for an expression of the symmetric effect of the separation between each couple of regions on migration flows. The measure f_{ij} better incorporates other factors which combine themselves to the physical distance in order to facilitate or to reduce the migrations, as for instance the differences in regional mobility levels, or the effects of nation-wide attraction centers (as the capital, or an area for retirement), or the short-circuiting of flows between the regions surrounding the largest population center.

ii) The *attractiveness of regions* and its variations over time is an interesting output of the two models. In the gravity model à "push-pull" index is computed as a weighted average of the residuals between the observed and adjusted flows, ranking from positive values, when the region is more attractive, to negative values

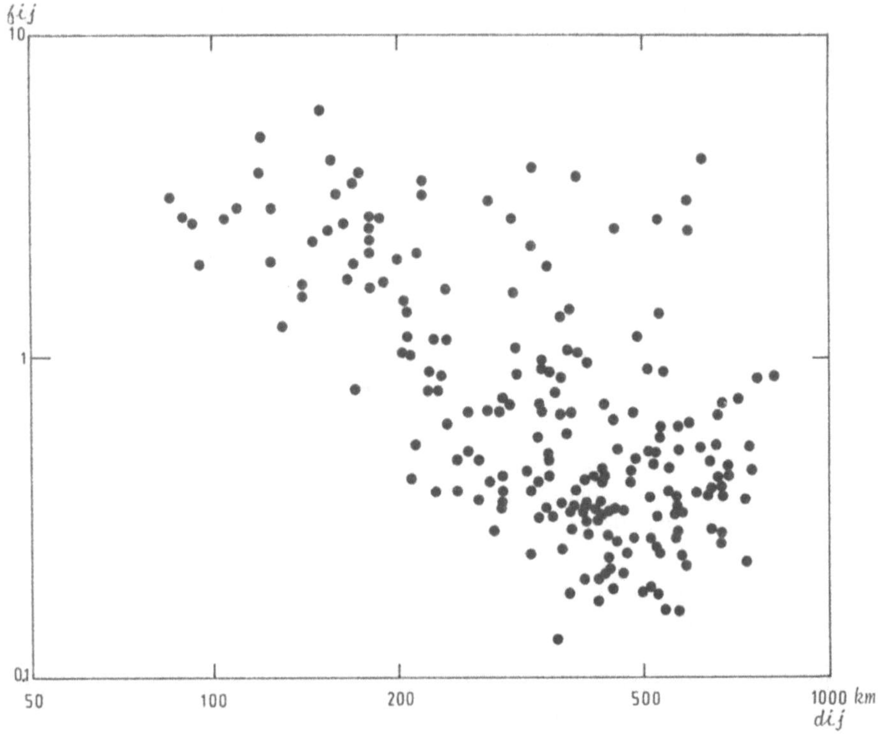

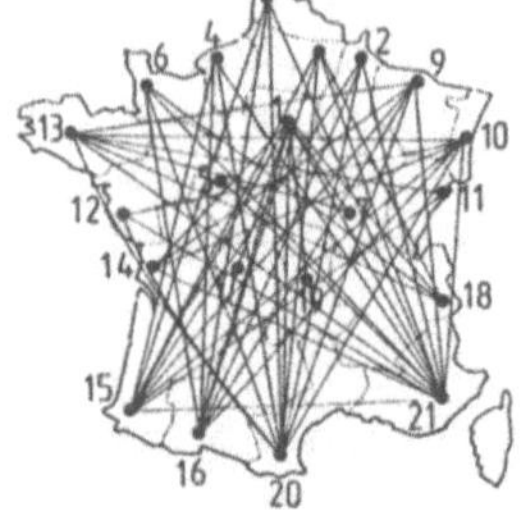

Fig. 6.10. Comparison of the effect of distance between the model of this book and the gravity model

when it is more repulsive. As the gravity model already incorporates the effect of the size of the regions of destination, this index is better to compare with the values that our model gives for "preferences" than with the "utilities" (Figure 6.11).

The ranking of the regions by the two indices do not differ too much. The similarity is higher for the first period (1954–1962) than for the last one (1975–1982).

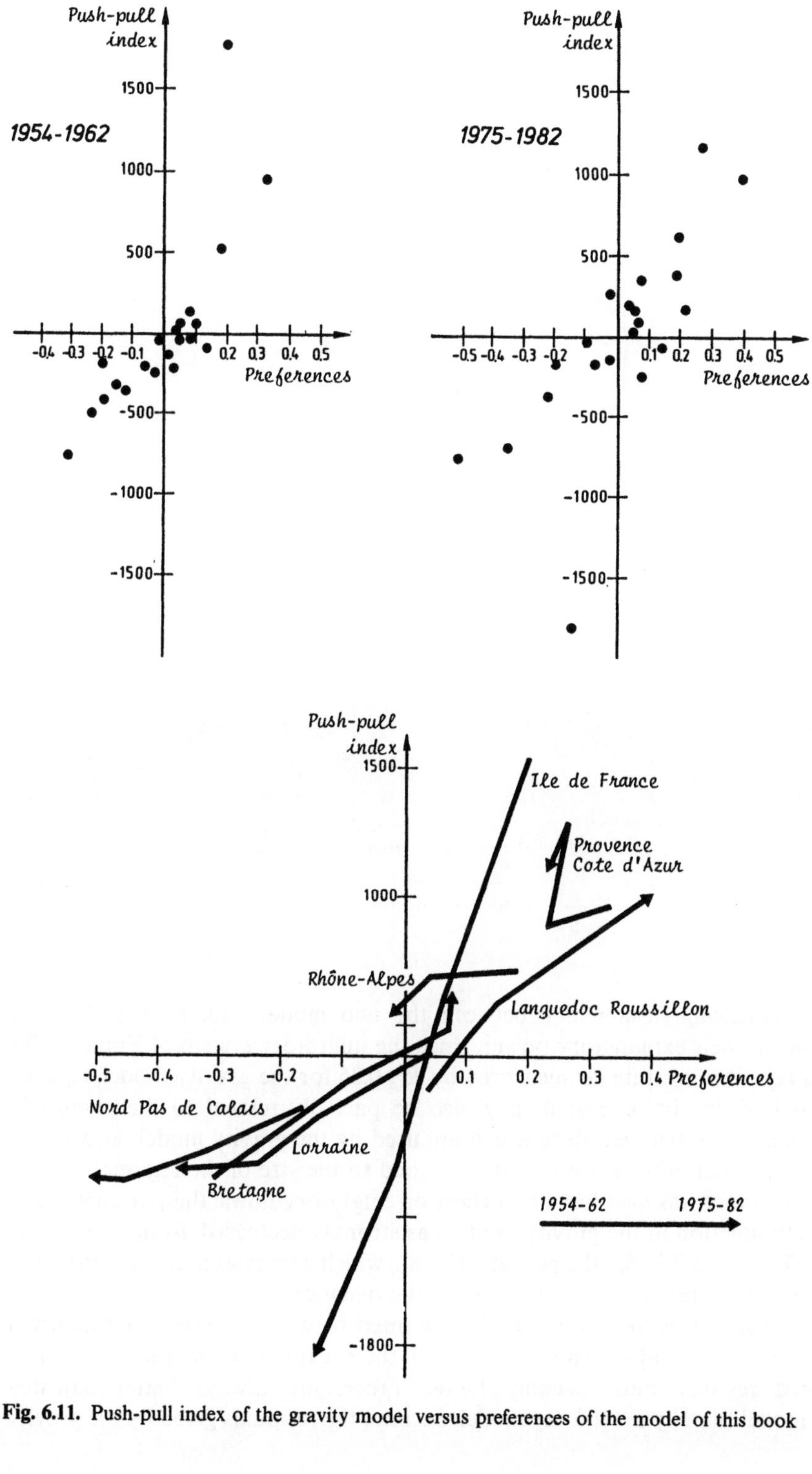

Fig. 6.11. Push-pull index of the gravity model versus preferences of the model of this book

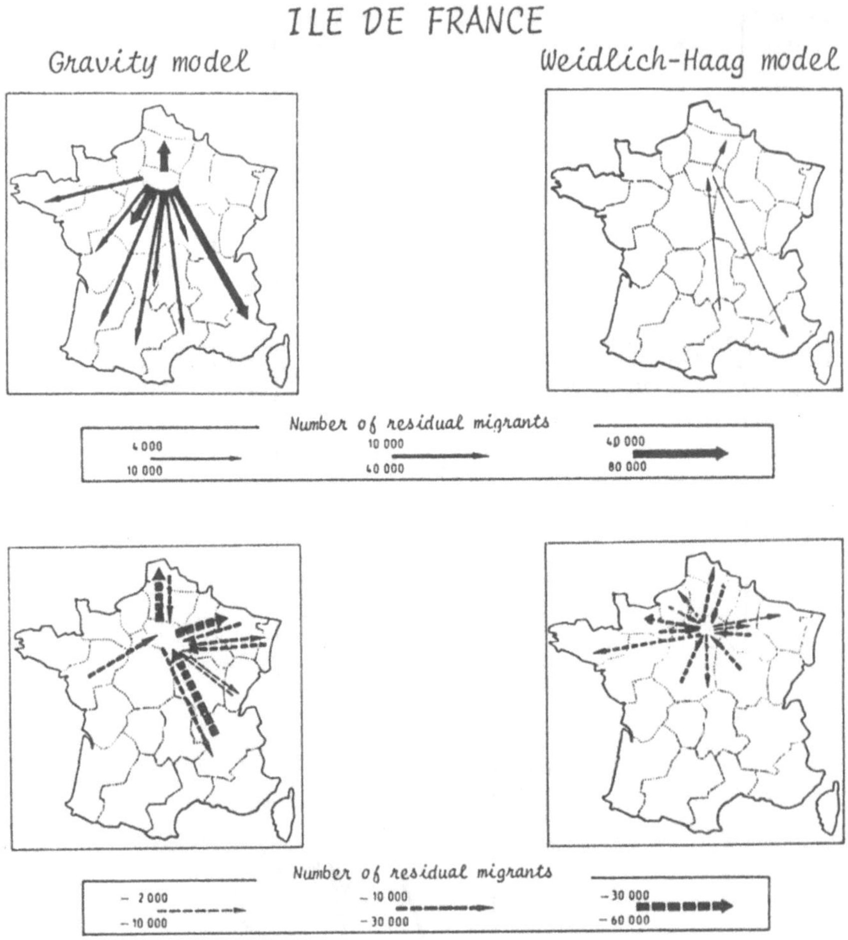

This increasing discrepancy between the two models can be explained by variations in their explanatory power, since the migratory system of France is far from its equilibrium state (as measured by R^2, 0.85 for the gravity model, 0.96 for the model of this book. But it may also be partially related to the change in significance of the two variables which are used by the gravity model: as we have seen, the regional utilities are less and less tied to the size of the regions, and the physical distance has also lost of its effect on migratory flows: the parameter a of the Pareto function in the gravity model has strongly decreased, from 0.77 for the period 1954–62 to 0.34 for the period 1975–82, which expresses a larger freedom of interregional migrants in consideration of the distance.

iii) The *spatial pattern of residuals* computed between observed and adjusted flows from both models tends to confirm these explanations. First, the most populated regions, and specially Ile-de-France, are always better adjusted by our model than by the gravity model. Second, the residuals given by the gravity

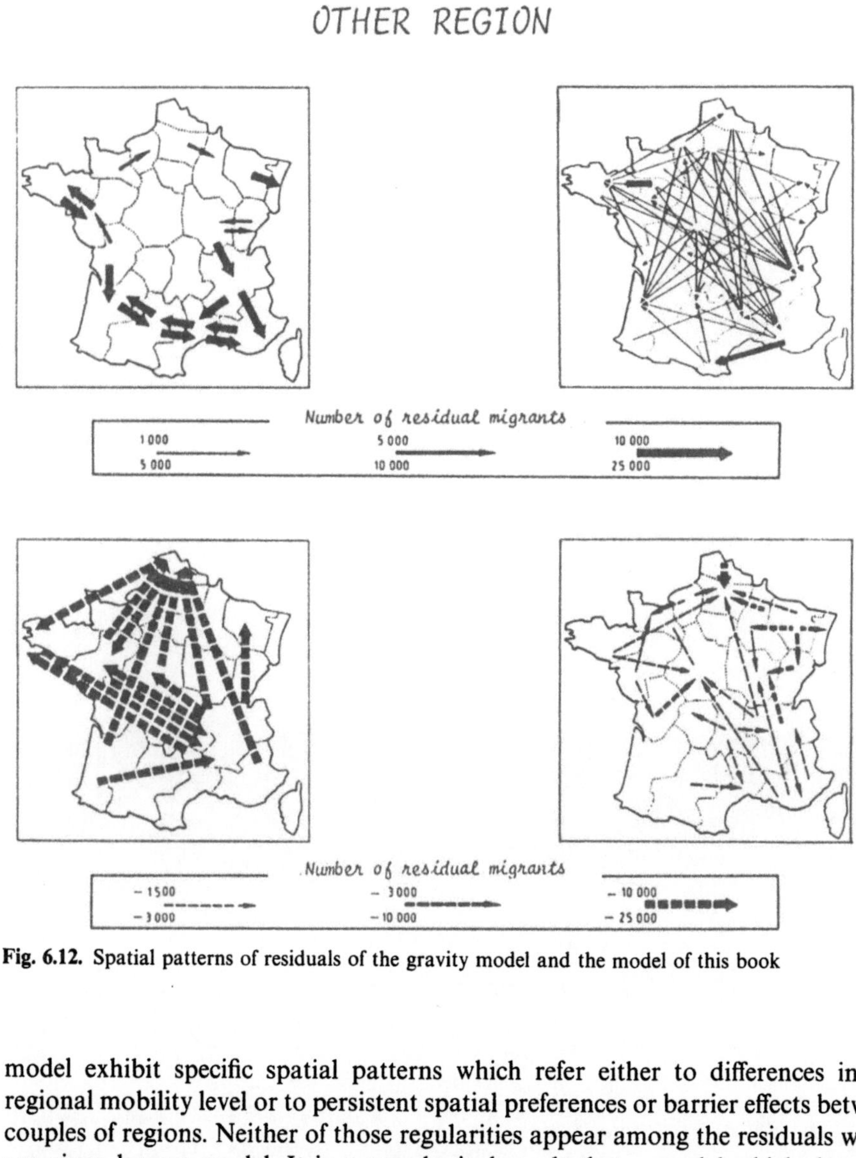

Fig. 6.12. Spatial patterns of residuals of the gravity model and the model of this book

model exhibit specific spatial patterns which refer either to differences in the regional mobility level or to persistent spatial preferences or barrier effects between couples of regions. Neither of those regularities appear among the residuals which are given by our model. It is a paradoxical result that a model which does not include explicitly the spatial properties of the regions finally eliminates better spatial bias from the residuals to be explained by socio-economic variables than a gravity-type model (Figure 6.12).

7 Israel

Michael Sonis

7.1 Introduction

The first steps of demographic, economic, social and political development of the State of Israel (since reaching political independence in 1948) were affected by dramatic "catastrophic" changes caused by huge waves of immigration. These waves doubled the Jewish population in the first three years. During the next three decades the volume of massive immigration was substantial though not such as in the period 1948–1951 [7.1].

The streams of immigration influenced essentially the process of metropolization of the three Israeli Metropolitan areas – Tel Aviv, Haifa and Jerusalem. The "cohorts" of the first immigrants shaped the main patterns of Jewish population distribution. The main feature of this population distribution is a higher population density in the coastal plain (Tel Aviv and Haifa) and in Jerusalem compared with the secondary population concentration in the Beer Sheva area and with the lower population density in the intercore and peripheral areas.

Over the years multiple changes have taken place in the demographic development of the Israeli Jewish population. These changes are characterized by

Table 7.1. Internal migration of Jewish population of Israel, 1965–1980 (in thousands)

Year	Total Jewish population	Internal migration		Interregional migration	
		absolute size	% of total population	absolute size	% of total population
1965	2239.2	237.1	10.6	94.0	4.2
1966	2299.1	191.4	8.3	62.0	2.7
1967	2344.9	143.3	6.1	52.3	2.2
1968	2383.6	165.7	7.0	58.0	2.4
1969	2434.8	242.5	10.0	75.1	3.1
1970	2582.0	196.9	7.6	65.6	2.5
1971	2662.0	203.9	7.7	70.1	2.6
1972	2752.7	184.0	6.7	68.1	2.5
1973	2845.0	188.0	6.6	68.2	2.4
1974	2906.9	221.8	7.6	84.4	2.9
1975	2959.4	202.9	6.9	73.3	2.5
1976	3020.4	236.0	7.8	81.4	2.7
1977	3077.3	278.2	9.0	86.0	2.8
1978	3141.2	252.8	8.0	74.6	2.4
1979	3218.4	232.7	7.3	76.7	2.4
1980	3282.7	244.9	7.5	78.8	2.4

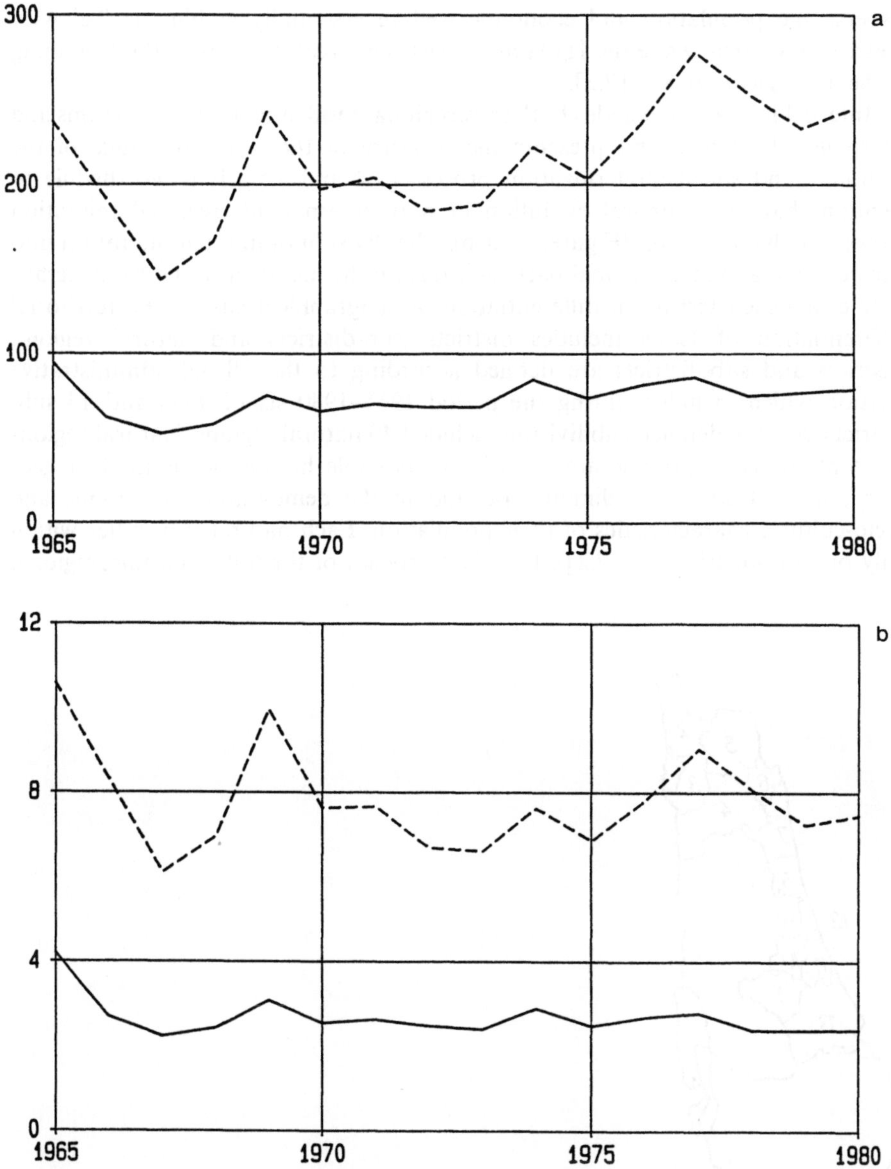

Fig. 7.1. (a): Total volume of internal (dashed) and interregional (solid) migration of the Jewish population in Israel from 1965 to 1980 (in thousands). (b): Internal (dashed) and interregional (solid) migration of the Jewish population in Israel from 1965 to 1980 (in per cent of total population)

transition from the "catastrophic" to "natural" population transformations, showing themselves in the development of the well-pronounced multiregional internal migration. The internal migration of the Jewish population in the last decade includes on average 7.5 per cent of the total Jewish population (see Table 7.1 and Figure 7.1b). During the 60's and 70's the internal migration tended to disperse the population and economic welfare essentially within the Tel Aviv Metropolitan area, while the Haifa and Jerusalem areas are only at the beginning of the analogical process [7.3].

Internal migration includes both intraregional residential mobility streams and interregional streams and presents the measure of the level and stage of the economic and social modernization process [7.4 and 7.5]. It is worthwhile to mention that the temporal evolution of both internal interregional migration shows a high correlation (Figures 7.1a, b). The division of internal migration into intraregional and interregional parts depends on the historically and bureaucratically established territorial differentiation of geographical space. The territorial differentiation of Israel includes districts, sub-districts and natural regions. Districts and sub-districts are defined according to the official administrative division which included during the period 1965–1980 six districts and 14 sub-districts. A more detailed subdivision included 40 natural regions. Natural regions are continuous geographic areas, as far as possible homogeneous in their geomorphological structure, climate, soil and in the demographic, economic and socio-cultural characteristics of their population. Each natural region lies within only one sub-district (see [7.2]). For the purposes of the following interregional

Fig. 7.2. Regional differentiation of Israel

migration analysis the territorial differentiation into 14 Israeli sub-districts was chosen (see Figure 7.2). According to this choice of regionalization the internal migration of Jewish population is divided into intraregional and interregional migration in the proportion 2:1, i.e., the migration of the Jewish population between 14 sub-districts includes on average about 2.5 per cent of the total Jewish population (see Table 7.1 and Figure 7.1b).

Table 7.2. Symbols used in the graphical representation of regional utilities and preferences of Israel. Division of sub-districts into two groups

1. Regions with higher population density		
symbol	region	name
+	1	Jerusalem
o	6	Haifa
Δ	12	Tel Aviv
□	14	Beer Sheva
2. Regions with lower population density		
symbol	region	name
+	2	Zefat
o	3	Kinneret
Δ	4	Yizreel
□	5	Akko
◊	7	Hadera
·	8	Sharon
*	9	Petah Tiqva
x	10	Ramla
-	11	Rehovot
>	13	Ashqelon

7.2. Sources and Limitations of Data

The legal grounds for the accumulation of the Israeli migration data are the Population Registration Ordinance, 1949, and the Register of Population Law, 1965. According to these laws every Israeli citizen must inform the Local Population Register Bureau of every change in his (or members of his family's) status and of a change of residence. This information is reported to the Central Population Register of the Ministry of the Interior. The computer disks for the internal migration data can be received from the Population Registration Section of the Ministry of the Interior through the Office Mechanization Centre. The internal migration data are also available in aggregated form from the publications of the Israel Central Bureau of Statistics in "Internal migration of Jews in Israel" [7.2] and in "Supplements to the Monthly Bulletin of Statistics" [7.2] (Hebrew only).

The internal migration data are systematically distorted because some citizens do not immediately report changes of address and the recorded information includes the time of report but not the time of an actual residential change. As a result, only about 87 per cent of annual reports are related to the residential change in the same year and 13 per cent on average are related to residential changes in the previous years. Besides, in an election year there is a bigger notification of the changes in address which occurred in previous years but were not reported. This phenomenon is clearly seen in Table 7.1 and more illustrative in Figure 7.1 where the election years 1965, 1969, 1974, 1977 respectively included a higher number of reports than in non-election years. Moreover, a person who changed his address more than once during the same year was registered more than once. New immigrants, for whom only incomplete addresses were registered on their first settling and who came to one of the Bureaux of the Ministry of the Interior to complete the details of their address, were registered as changing their address (see the explanations in "Statistical Abstracts of Israel"). It is important to underline the fact that the population and migration data from the Census of Population and Housing, 1972, are also the source for the systematically distorted description of the migration process (see Table 7.1 and Figures 7.3–7.6). The symbols used in the graphical representation of regional utilities and preferences are depicted in Table 7.2.

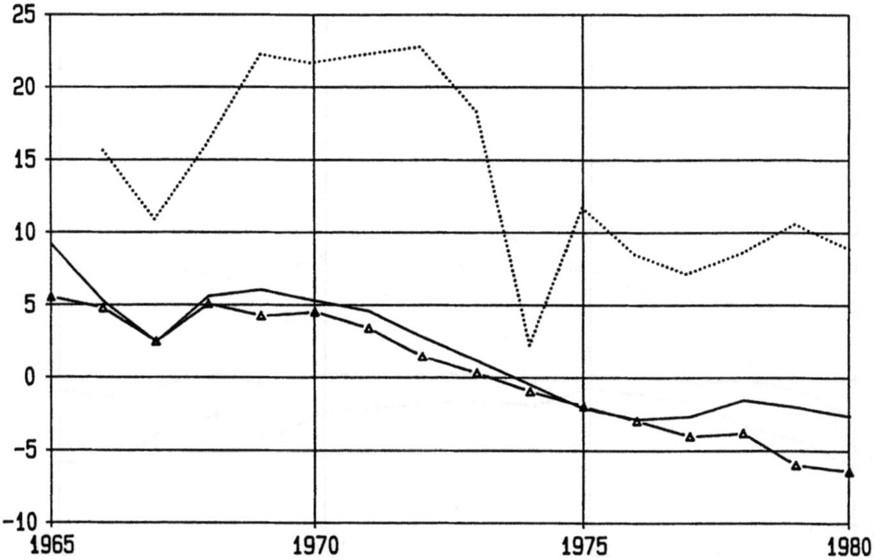

Fig. 7.3. Total annual population growth (. . .), internal net-migration (solid line) and preference (△) of Tel Aviv district, 1965 – 1980

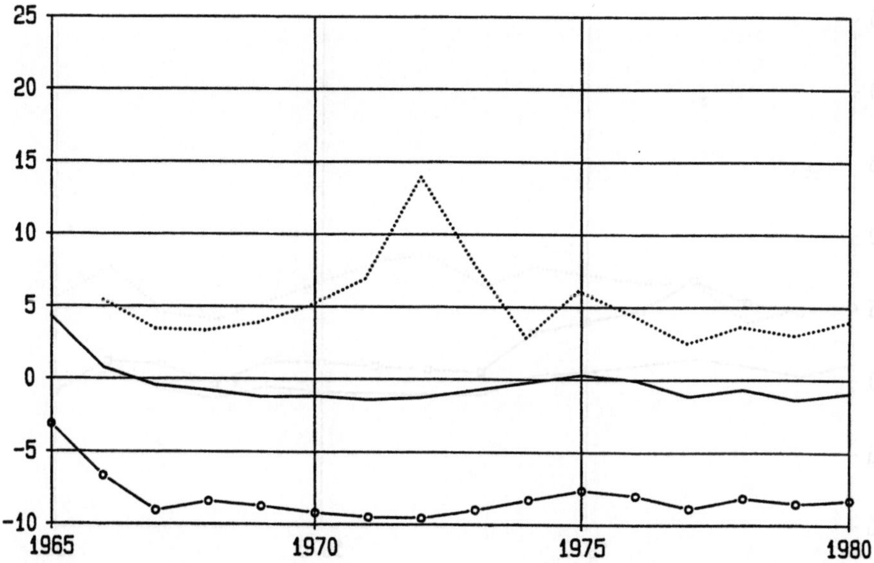

Fig. 7.4. Total annual population growth (. . .), internal net-migration (solid line) and preference (○) of Haifa district, 1965 – 1980

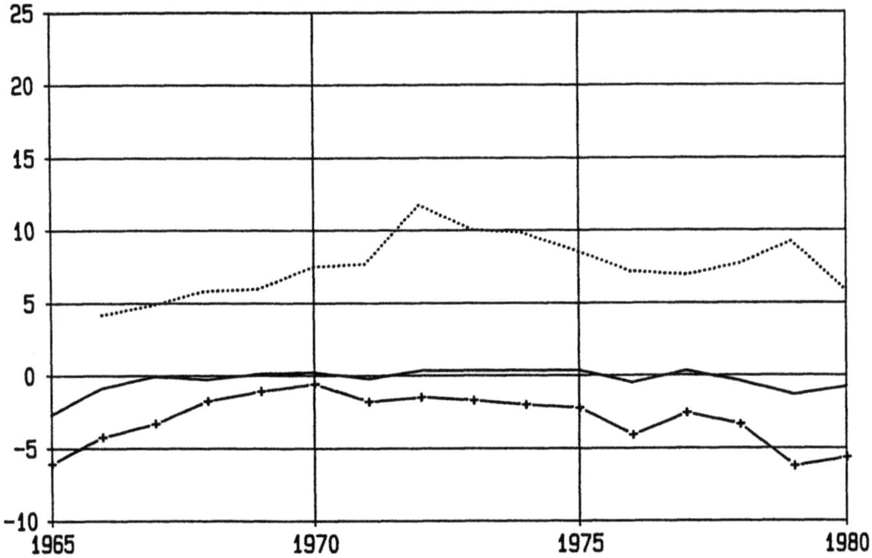

Fig. 7.5. Total annual population growth (. . .), internal net-migration (solid line) and preference (+) of Jerusalem district, 1965–1980

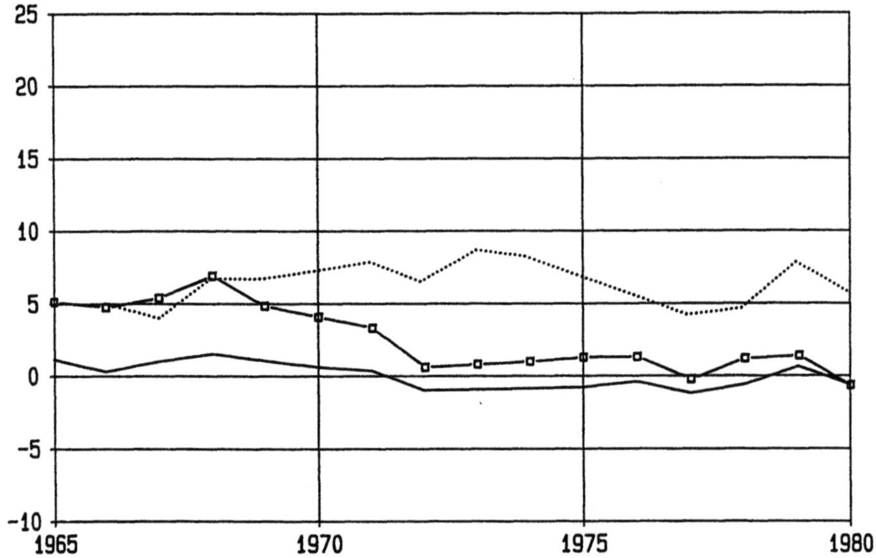

Fig. 7.6. Total annual population growth (. . .), internal net-migration (solid line) and preference (□) of Beer Sheva sub-district, 1965–1980

7.3 Impact of Internal Migration on Israeli Population Redistribution

The Israeli population change is characterized by the gradual decrease of the relative population rate of the Tel Aviv, Haifa and Northern districts and by the gradual relative increase of the population of Jerusalem, Central and Southern districts (see Table 7.3). There are four different sources for the relative population redistribution: 1) natural increase (the difference between births and deaths); 2) the reclassification (the change of the regional boundaries); 3) the external net-migration (the difference between in-migration from and out-migration to the country abroad); and 4) internal net-migration (the difference between in-migration from and out-migration to other regions of the same country).

The numerical description of the internal net-migration can be done by constructing the difference $M-M^T$ between the migrational origin-destination matrix M and its transposed matrix M^T. If there exists a clear hierarchy of net-migration, a simultaneous rearrangement of origins and destinations can be accomplished in such a way that the first chosen region will obtain the positive net-migration impulses from each other region; the second chosen region will obtain the positive net-migration from all other regions except the first one, and so on (see [7.2], no. 262). As a result, the reorganized $M-M^T$ matrix will include the positive

Table 7.3. Jewish population distribution between sub-districts and districts, 1965–1980

Districts	Sub-districts	Population distribution								
		1965		1970		1975		1980		
		1000's	%	1000's	%	1000's	%	1000's	%	
Jerusalem	1 Jerusalem	216.6	9.4	245.1	9.6	293.0	9.9	329.8	10.1	↑
Northern	2 Zefat	50.1	2.2	51.2	2.0	55.3	1.9	60.2	1.8	↓
	3 Kinneret	37.8	1.6	38.3	1.5	41.8	1.4	46.5	1.4	—
	4 Yizreel	84.2	3.7	89.8	3.5	102.0	3.5	113.8	3.5	—
	5 Akko	63.1	2.7	69.5	2.7	81.9	2.8	96.4	3.0	↑
Haifa	6 Haifa	295.9	12.9	317.1	12.4	354.8	12.0	372.2	11.4	↓
	7 Hadera	72.9	3.2	74.8	2.9	84.3	2.9	91.8	2.8	↓
Central	8 Sharon	99.5	4.3	109.1	4.3	129.4	4.4	147.9	4.5	—
	9 Petah Tiqva	153.8	6.7	176.6	6.9	224.4	7.6	266.8	8.2	↑
	10 Ramla	69.9	3.0	76.6	3.0	88.6	3.0	96.8	3.0	—
	11 Rehovot	116.8	5.1	133.9	5.2	172.3	5.8	216.4	6.6	↑
Tel Aviv	12 Tel Aviv	787.7	34.3	874.1	34.1	951.4	32.2	994.7	30.5	↓
Southern	13 Ashqelon	120.0	5.2	143.8	5.6	175.9	6.0	199.7	6.1	↑
	14 Beer Sheva	130.8	5.7	160.6	6.3	198.6	6.7	226.6	7.0	↑
Total Jewish population		2299.1		2560.5		2953.7		3259.7		

impulses over its zero diagonal. The deflexion from this form gives additional important information about the migration process. The arrangement of the regions according to the descending number of positive net-migration impulses represents the net-migration hierarchy of regions and measures accurately the influence of the internal net-migration on the regional population changes.

Tables 7.4a–d represent the measurement of the internal net-migration and net-migration hierarchy of Israeli sub-districts during the period 1965–1980. The impacts of the net-migration on the Israeli population redistribution can be described as follows.

The relative portion of the internal net-migration within the sub-districts population growth during the period 1965–1980 was subject to considerable changes. The biggest (by population size) and the oldest metropolitan district of Tel Aviv crossed the limits of the intensive urbanization stage and entered a stage of suburbanization. This crossover was accompanied by the strong population aging process which transformed the Tel Aviv district into the oldest urban area in Israel. The most intensive metropolitan development was transferred from the Tel Aviv district to the neighbouring Central district within the inner ring of the Tel Aviv Metropolitan conurbation. Accordingly, the relative portion of the internal net-migration was gradually changed from a positive balance of 50 per cent of the Tel Aviv annual population growth in 1965 to zero in 1973 and to a negative balance since then (see Figure 7.3). This change was caused by the strong streams of the out-migration from Tel Aviv to the neighbouring sub-districts of Petah Tiqva and Rehovot which started in 1970. The Six Days War (1967) and the Yom Kippur War (1973) added significantly to this process. The strong aging process, influencing the decline in natural growth, the cessation of external in-migration in about 1975, the rise of the external out-migration at the end of the 70's and the negative balance of the internal net-migration eventually brought to an end the population growth of the Tel Aviv district: from 1981 the Tel Aviv district shows an absolute population decline.

The Central district includes the inner and outer rings of the Tel Aviv conurbation. The Petah Tiqva and Rehovot sub-districts are at the top of the net-migration hierarchy. They have been attracting the population from almost all other sub-districts while the Sharon and Ramla sub-districts are in the middle of net-migration hierarchy and are losing population through out-migration.

The second Israeli metropolitan area – the Haifa district, entered the stage of the negative net-migration balance before the Tel Aviv district despite the fact that the Tel Aviv district is the oldest and most completely developed urbanized area. This phenomenon can be explained by the fact that the Haifa district started to lose its population in the middle of the 60's to the Tel Aviv district, and during the last decade the Central and Jerusalem metropolitan areas, the Beer Sheva sub-district and the neighbouring Akko sub-district have joined the Tel Aviv district (see Table 7.4).

The third Israeli metropolitan area – the Jerusalem district, entered the stage of the positive net-migration balance only after the governmental decision to expand the municipal area of the city of Jerusalem after the Six Days War (1967). The positive net-migration impulse lasted only for a short period till (1975). Compar-

Table 7.4. Interregional Israeli net-migration. (a) 1965, (b) 1970, (c) 1975, (d) 1980

1965	12	6	9	13	14	11	10	8	7	5	1	4	3	2	positive net-streams	negative net-streams	total net-migration
12 Tel Aviv	0	913	68	390	307	472	1052	651	992	623	1378	1231	434	690	9201	0	9201
6 Haifa	-913	0	299	570	328	277	145	267	618	633	302	1058	233	510	5240	913	4327
9 Petah Tigva	-68	-299	0	163	-9	4	153	181	287	108	280	230	122	163	1691	376	1315
13 Ashqelon	-390	-570	-163	0	70	122	-38	92	205	198	307	306	116	209	1625	1161	464
14 Beer Sheva	-307	-328	9	-70	0	150	17	208	230	309	286	295	115	243	1862	705	1157
11 Rehovot	-472	-277	-4	-122	-150	0	34	-13	60	51	119	126	67	106	563	1038	-475
10 Ramla	-1052	-145	-153	38	-17	-34	0	74	85	81	148	89	60	69	644	1401	-757
8 Sharon	-651	-267	-181	-92	-208	13	-74	0	167	68	31	75	74	94	522	1473	-951
7 Hadera	-992	-618	-287	-205	-230	-60	-85	-167	0	29	-25	19	17	56	121	2669	-2548
5 Akko	-623	-633	-108	-198	-309	-51	-81	-68	-29	0	13	95	30	76	214	2100	-1886
1 Jerusalem	-1378	-302	-280	-307	-286	-119	-148	-31	25	-13	0	58	45	60	188	2864	-2676
4 Yizreel	-1231	-1058	-230	-306	-295	-126	-89	-75	-19	-95	-58	0	22	65	87	3582	-3495
3 Kinneret	-434	-233	-122	-116	-115	-67	-60	-74	-17	-30	-45	-22	0	42	42	1335	-1293
2 Zefat	-690	-510	-163	-209	-243	-106	-69	-94	-56	-76	-60	-65	-42	0	0	2383	-2383

Table 7.4. continued

1970	11	9	12	14	1	8	13	10	5	6	7	4	3	2	positive net-streams	negative net-streams	total net-migration
11 Rehovot	0	82	20	57	83	41	105	194	52	47	60	132	32	119	1024	0	1024
9 Petah Tigva	-82	0	197	-3	34	97	86	119	68	75	60	171	26	133	1066	85	981
12 Tel Aviv	-20	-197	0	231	322	400	572	565	332	1055	403	1006	219	384	5489	217	5272
14 Beer Sheva	-57	3	-231	0	104	-46	46	28	8	308	124	91	46	182	940	334	606
1 Jerusalem	-83	-34	-322	-104	0	30	118	18	38	213	43	113	67	147	787	543	244
8 Sharon	-41	-97	-400	46	-30	0	20	54	32	-42	93	151	46	97	539	610	-71
13 Ashqelon	-105	-86	-572	-46	-118	-20	0	3	88	42	20	94	54	143	444	947	-503
10 Ramla	-194	-119	-565	-28	-18	-54	-3	0	6	48	33	76	15	56	234	981	-747
5 Akko	-52	-68	-332	-8	-38	-32	-88	-6	0	91	33	67	44	85	320	624	-304
6 Haifa	-47	-75	-1055	-308	-213	42	-42	-48	-91	0	107	278	89	190	706	1879	-1173
7 Hadera	-60	-60	-403	-124	-43	-93	-20	-33	-33	-107	0	33	3	32	68	976	-908
4 Yizreel	-132	-171	-1006	-91	-113	-151	-94	-76	-67	-278	-33	0	66	59	125	2212	-2087
3 Kinneret	-32	-26	-219	-46	-67	-46	-54	-15	-44	-89	-3	-66	0	88	88	707	-619
2 Zefat	-119	-133	-384	-182	-147	-97	-143	-56	-85	-190	-32	-59	-88	0	0	1715	-1715

Table 7.4. continued

1975	11	9	12	1	13	6	8	10	7	5	14	3	4	2	positive net-streams	negative net-streams	total net-migration
11 Rehovot	0	199	2437	232	133	251	179	435	105	120	433	18	169	55	4766	0	4766
9 Petah Tigva	-199	0	2744	131	95	179	98	14	130	107	99	7	92	55	3751	199	3552
12 Tel Aviv	-2437	-2744	0	143	190	600	255	182	465	291	155	91	451	206	3029	5181	-2152
1 Jerusalem	-232	-131	-143	0	53	257	76	21	154	91	10	4	158	55	879	506	373
13 Ashqelon	-133	-95	-190	-53	0	31	-15	43	36	48	38	11	69	39	315	486	-171
6 Haifa	-251	-179	-600	-257	-31	0	98	33	376	411	22	67	353	271	1631	1318	313
8 Sharon	-179	-98	-255	-76	15	-98	0	0	211	25	52	10	95	34	442	706	-264
10 Ramla	-435	-14	-182	-21	-43	-33	0	0	17	21	17	9	40	-1	104	729	-625
7 Hadera	-105	-130	-465	-154	-36	-376	-211	-17	0	8	-35	27	40	61	136	1529	-1393
5 Akko	-120	-107	-291	-91	-48	-411	-25	-21	-8	0	35	32	53	92	212	1122	-910
14 Beer Sheva	-433	-99	-155	-10	-38	-22	-52	-17	35	-35	0	7	48	-4	90	865	-775
3 Kinneret	-18	-7	-91	-4	-11	-67	-10	-9	-27	-32	-7	0	23	36	59	283	-224
4 Yizreel	-169	-92	-451	-158	-69	-353	-95	-40	-40	-53	-48	-23	0	44	44	1591	-1547
2 Zefat	-55	-55	-206	-55	-39	-271	-34	1	-61	-92	4	-36	-44	0	5	948	-943

Table 7.4. continued

1980	5	11	2	9	10	8	12	1	4	6	14	7	13	3	positive net-streams	negative net-streams	total net-migration
5 Akko	0	34	61	50	-9	57	118	48	49	405	69	54	30	54	1029	9	1020
11 Rehovot	-34	0	46	20	333	39	1441	205	17	74	221	53	249	-11	2698	45	2653
2 Zefat	-61	-46	0	31	44	5	65	104	14	-31	8	-6	12	52	335	144	191
9 Petah Tigva	-50	-20	-31	0	18	59	2800	106	52	123	49	62	65	49	3383	101	3282
10 Ramla	9	-333	-44	-18	0	37	0	10	22	1	94	61	46	4	284	395	-111
8 Sharon	-57	-39	-5	-59	-37	0	18	16	14	35	79	71	71	-17	304	214	90
12 Tel Aviv	-118	-1441	-65	-2800	0	-18	0	275	107	583	55	246	479	35	1780	4442	-2662
1 Jerusalem	-48	-205	-104	-106	-10	-16	-275	0	45	119	-12	75	-247	17	256	1023	-767
4 Yizreel	-49	-17	-14	-52	-22	-14	-107	-45	0	98	51	54	1	64	268	320	-52
6 Haifa	-405	-74	31	-123	-1	-35	-583	-119	-98	0	81	190	58	156	516	1438	-922
14 Beer Sheva	-69	-221	-8	-49	-94	-79	-55	12	-51	-81	0	44	-7	12	68	714	-646
7 Hadera	-54	-53	6	-62	-61	-71	-246	-75	-54	-190	-44	0	25	16	47	910	-863
13 Ashqelon	-30	-249	-12	-65	-46	-71	-479	247	-1	-58	7	-25	0	2	256	1036	-780
3 Kinneret	-54	11	-52	-49	-4	17	-35	-17	-64	-156	-12	-16	-2	0	28	461	-433

able with the Tel Aviv district the rate of net-migration for the Jerusalem district was considerably smaller and represented not more than 10 per cent of the population growth (see Table 7.4).

The net-migration features of the secondary Israeli population core-the Beer Sheva sub-district, resembles those of the Tel Aviv district (see Tables 7.5 and 7.6). This resemblance is only superficial, because the Beer Sheva sub-district completed its initial developmental stage in the middle of the 70's and transfer to the suburbanization stage took place.

The Northern district, which represents the national periphery, during the last 15 years lost its population to all other sub-districts. Only at the end of 70's did the Akko sub-district obtain a strong positive net-migration impulse, which includes about 20 per cent of the population change. In 1980 the Akko sub-district ascended to the top of the net-migration hierarchy which can be explained by the strong attraction of the population from the Tel Aviv and Haifa districts.

This brief and insufficiently detailed description of the impacts of net-migration in the Israeli regional population growth supports the general picture for other countries.

7.4 Spatial Organization of Temporarily Stable Migration Streams

Because there is no such a person as a net-migrant, the substantial description and understanding of the essence of the migration process must be based on the consideration of the origin-destination streams themselves. One can start such a consideration with the derivation of the spatial organization of temporarily stable migration streams.

As a first step, an annual rank-size sequence of the individual migration flows can be considered in such a way that the biggest migration flow will be on the top of the rank-size sequence. It is possible to consider only the qualitative rank-size sequence, which includes, instead of the absolute volume of each flow, its origin and destination (i, j).

Table 7.5 represents the qualitative rank-size flow sequences for the years 1965, 1970, 1975 and 1980 (only the first 30 migration flows are included). The shift over the time of the particular flows along the sequences represents the changes in spatial arrangement of migration streams. For example, the migration flow (11, 12) between Rehovot and Tel Aviv started in 1965 from the rank 18, in 1970 jumped to the rank 9, and in 1975 and 1980 occupied the rank 2. The migration flow (12, 10) from Tel Aviv to Ramla gradually changed its rank in sequence 11, 13, 16, 19. One can claim that such a movement corresponds to the deep changes in the multi-regional urban hierarchy.

Despite the existence of changes, the essential stability of the ranking of streams by size exists. This stability represents the spatial arrangement of the biggest migration flows which is temporarily stable. As the simplest measure of such a spatio-temporal stability, the average rank of stream and average absolute devi-

Table 7.5. Annual qualitative rank-size sequences of migrational streams. (j,i) represents the migrational stream from location i to j

	Years			
Rank	1965	1970	1975	1980
1	(12,9)	(9,12)	(9,12)	(9,12)
2	(9,12)	(12,9)	(11,12)	(11,12)
3	(12,14)	(12,6)	(12,9)	(12,9)
4	(12,6)	(12,14)	(12,14)	(12,1)
5	(12,1)	(12,1)	(12,1)	(12,14)
6	(12,13)	(14,12)	(14,12)	(14,12)
7	(14,12)	(12,13)	(1,12)	(12,11)
8	(13,12)	(1,12)	(12,6)	(1,12)
9	(12,4)	(11,12)	(12,11)	(12,6)
10	(12,11)	(12,4)	(12,8)	(12,13)
11	(12,10)	(12,11)	(6,5)	(5,6)
12	(6,12)	(12,8)	(6,12)	(13,12)
13	(6,4)	(12,10)	(8,12)	(8,12)
14	(12,7)	(14,6)	(12,13)	(12,8)
15	(6,14)	(6,12)	(6,4)	(6,12)
16	(12,8)	(6,4)	(12,10)	(4,6)
17	(13,14)	(13,12)	(11,14)	(6,5)
18	(11,12)	(12,7)	(12,4)	(10,12)
19	(14,13)	(14,13)	(13,12)	(12,10)
20	(6,13)	(5,6)	(12,7)	(13,14)
21	(6,5)	(6,14)	(1,14)	(14,13)
22	(1,12)	(8,12)	(14,1)	(11,13)
23	(14,6)	(13,14)	(5,6)	(6,4)
24	(6,7)	(14,1)	(13,14)	(11,14)
25	(12,2)	(4,6)	(10,12)	(14,1)
26	(12,5)	(6,5)	(1,6)	(1,14)
27	(6,1)	(12,5)	(6,14)	(6,14)
28	(14,1)	(12,2)	(14,13)	(13,1)
29	(8,12)	(1,6)	(14,6)	(11,10)
30	(12,3)	(1,14)	(4,6)	(14,6)

Table 7.6. Spatio-temporal stable flows of the Israeli internal migration 1965–1980. (j,i) represents the migrational stream from location i to j

	Migration streams	Average rank	Average deviation
1	(9,12)	1.250	0.375
2	(12,9)	2.438	0.883
3	(12,14)	4.438	0.688
4	(12,6)	5.188	1.961
5	(12,1)	5.375	1.047
6	(11,12)	5.500	3.813
7	(14,12)	5.875	1.047
8	(12,11)	8.875	1.172
9	(12,13)	9.250	1.531
10	(1,12)	9.625	2.688
11	(12,10)	12.563	2.203
12	(12,8)	12.813	1.461
13	(6,12)	13.375	1.094
14	(13,12)	15.813	3.484
15	(6,4)	16.063	2.227
16	(12,4)	17.813	5.391
17	(8,12)	18.938	4.945
18	(6,5)	19.125	3.109
19	(12,7)	20.813	4.094
20	(14,6)	20.938	3.938
21	(13,14)	21.625	2.422
22	(5,6)	23.063	7.180

ation can be used. Table 7.6 includes average ranks and deviations for a number of the biggest Israeli interregional migration flows for the period 1965–1980. Such new rank-size ordering represents a temporarily stable spatial organization of the biggest migration flows (see Figure 7.7). One can claim that only such spatio-temporal stable parts of the overall migration stream allow meaningful prediction.

Figure 7.7 underlines the important role of the Tel Aviv district in the spatio-temporal structure of the Israeli internal migration: the spatio-temporal stable part of the overall migration stream includes merely the in- and out-migration flows exchanged with the Tel Aviv district. The Haifa sub-district exchanges migration flows with Tel Aviv and Akko sub-districts and pushes migration to the Beer Sheva sub-district. The Beer Sheva sub-district exchanges the

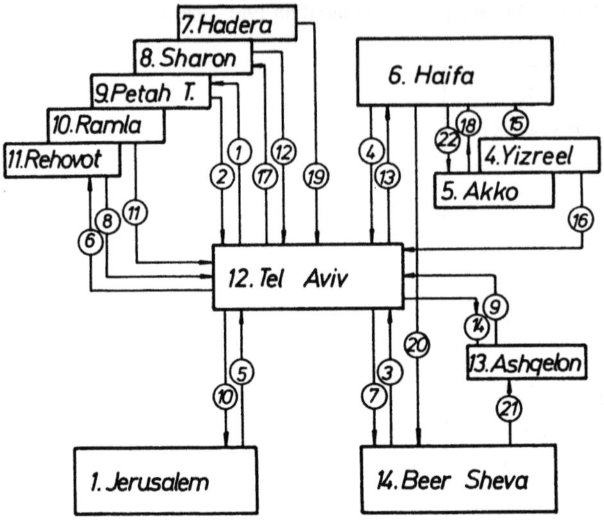

Fig. 7.7. Graphical presentation of spatio-temporal stability of Israeli internal migration streams, 1965–1980

migration flows with Tel Aviv, attracts migrants from Haifa sub-district and pushes the population into the Ashqelon sub-district.

It is expected that this spatial organization of the internal migration streams will be preserved in the next future.

7.5 Global Mobility

The level of residential mobility of Israeli Jewish population is much less than mobility rates of the majority of European countries: about 7.5 per cent of total population (see Table 7.1). The total volume of interregional migration does not follow the growth tendencies of the total population: from the peak point of 94 thousands of migrants in 1965, the volume of interregional migration declined to a minimum of 52.3 thousands in 1967 and increased with strong variations to the level of 78.8 thousands in 1980.

It is very interesting to see that the evolution of the time-dependent global mobility parameter $v_0(t)$ (Figure 7.8) calculated by the model, followed almost accurately the variations of the share of Israeli interregional migration (especially between 1965 and 1973, see Figure 7.9). The temporal variations in the share of interregional migration show a strong correlation with the number of beginnings of housing constructions with a two-year delay. The building activities reflect the attitude among Israeli population to become owners of dwelling. (It is important to underline that Israeli housing rent market is practically closed.) The building activities reflect also the economic, political and geo-political attitudes of the Jewish populations and Israeli Government. The Government is deeply involved in

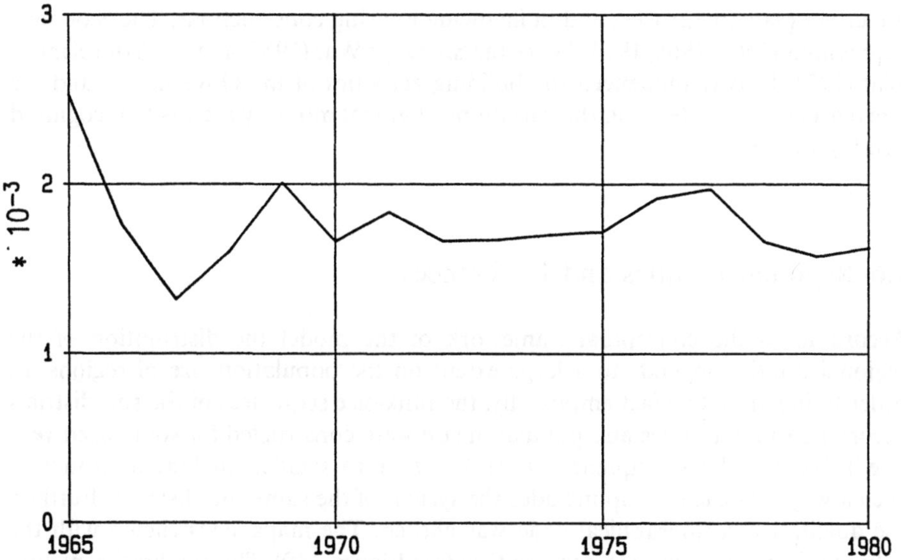

Fig. 7.8. The global mobility $v_0(t)$ of the Jewish population in Israel

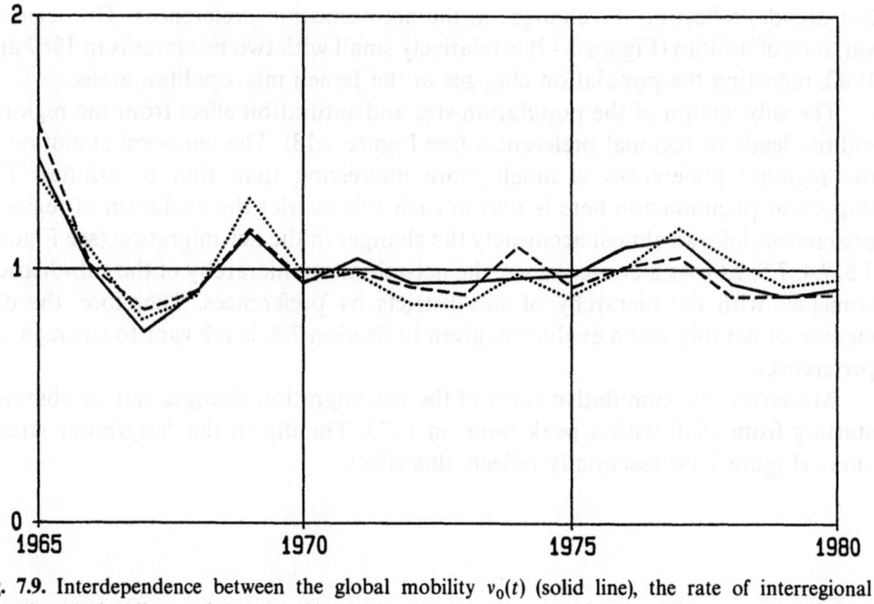

Fig. 7.9. Interdependence between the global mobility $v_0(t)$ (solid line), the rate of interregional migration (dashed line) and the rate of internal migration (dotted line). All variables are normalized to their temporal mean value

the urban planning process and in financing housing constructions. The economic depressions (1965–1968, 1974–1977), the Six Days War (1967) and the Yom Kippur War (1973) heavily influenced the building activities of the Government and the free market. Their effects on the variations of global mobility are easily recognized (see Figure 7.8).

7.6 Regional Utilities and Preferences

According to the conceptual framework of the model the distribution of the regional utilities depends to a large extent on the population size of regions. In order to illustrate this fact empirically, the rank-size sequences of the sub-districts according to the utilities and populaton size were constructed for some fixed year (see Table 7.7). These sequences were divided into smaller qualitative groups in such a way that each group includes the system of the same sub-districts. Further, the qualitative cartographical code was chosen. The maps, constructed with the help of this qualitative code, are identical (see Figure 7.10). This qualitative picture remains stable throughout the period 1965–1980, while the utility of the biggest by population size Tel Aviv district is decreasing gradually, and the utilities for the Jerusalem and Haifa metropolitan areas remain stable (see Figures 7.11a,b). The utilities of the Petah Tiqva, Rehovot and Akko sub-districts have increased in the last decade, reflecting the changes in the net-migration preferences. The regional variance of utilities (Figure 7.12) is relatively small with two minimums in 1967 and 1980, reflecting the population changes in the Israeli metropolitan areas.

The subtraction of the population size and saturation effect from the regional utilities leads to regional preferences (see Figure 7.13). The temporal evolution of the regional preferences is much more interesting than that of utilities. The important phenomenon here is that in each sub-district the evolution of regional preferences follows almost accurately the changes in the net-migration (see Figures 7.3, 7.4, 7.5, 7.6). As a consequence, the net-migration hierarchy of the sub-districts coincides with the hierarchy of sub-districts by preferences. Therefore, the discussion of net-migration evolution, given in Section 7.3, is relevant to the regional preferences.

Moreover, the cumulative effect of the net-migration changes can be observed starting from 1970 with a peak point in 1975. The dip in the "migratory stress" curve (Figure 7.14) essentially reflects this effect.

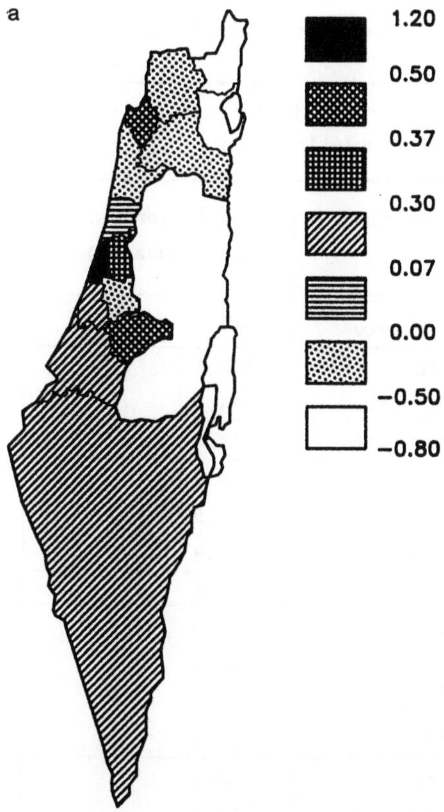

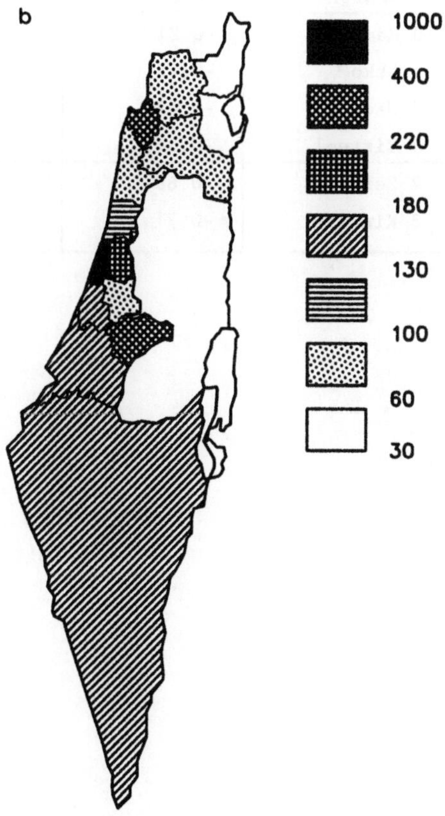

Fig. 7.10. Cartographical representation of the interconnections between regional utilities (a) and population size (b)

Table 7.7. Qualitative rank-size analysis of interconnections between regional utilities and population size, 1972

Rank-size sequence for utilities		Rank-size sequence for population size	
Sub-districts	Regional utilities	Sub-districts	Population (in thousands)
12 Tel Aviv	1.15	12 Tel Aviv	919.2
1 Jerusalem	0.43	6 Haifa	338.1
6 Haifa	0.39	1 Jerusalem	264.7
9 Petah Tiqva	0.35	9 Petah Tiqva	201.4
11 Rehovot	0.29	14 Beer Sheva	174.9
14 Beer Sheva	0.12	13 Ashqelon	160.5
13 Ashqelon	0.09	11 Rehovot	141.9
8 Sharon	0.055	8 Sharon	119.1
10 Ramla	-0.21	4 Yizreel	96.0
5 Akko	-0.42	10 Ramla	83.3
7 Hadera	-0.42	7 Hadera	80.1
4 Yizreel	-0.44	5 Akko	75.6
2 Zefat	-0.68	2 Zefat	53.7
3 Kinneret	-0.71	3 Kinneret	34.9

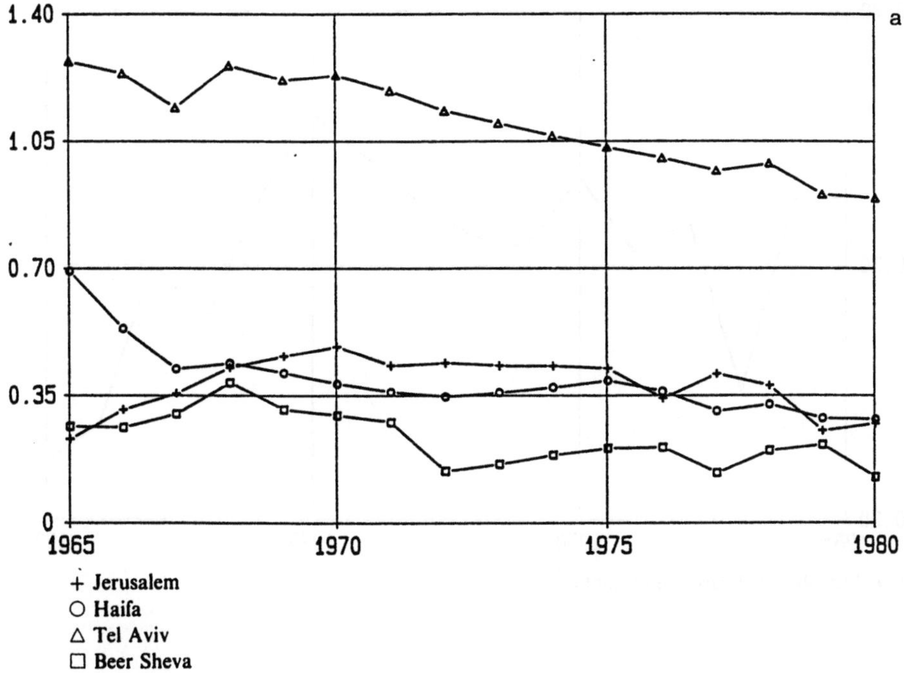

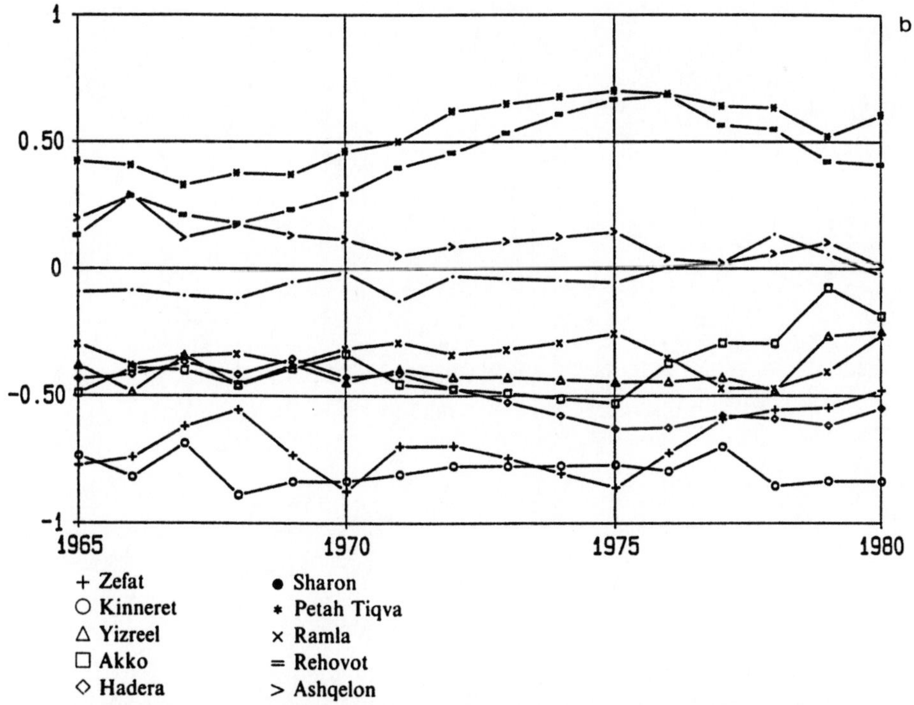

Fig. 7.11. Evolution of the Israeli regional utilities. (a): regions 1, 6, 12, 14, (b): regions 2–5, 7–11, 13

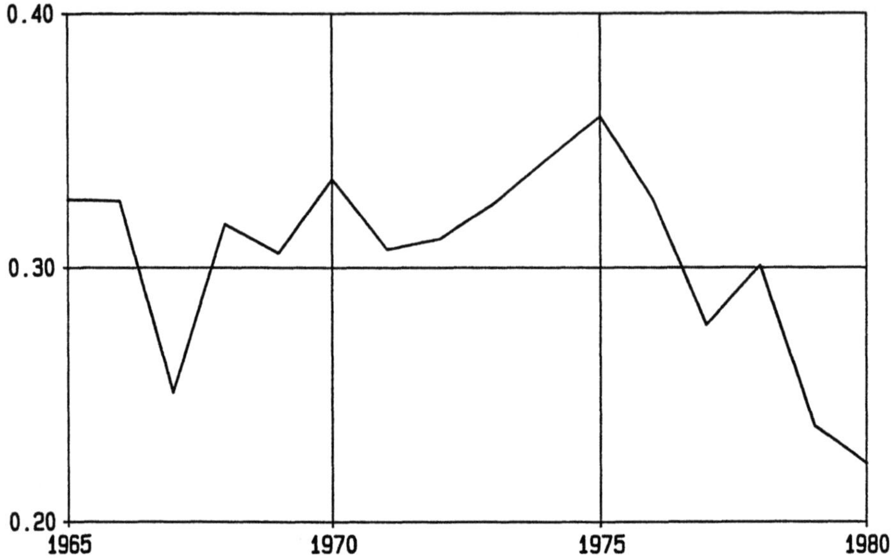

Fig. 7.12. Regional variance of utilities

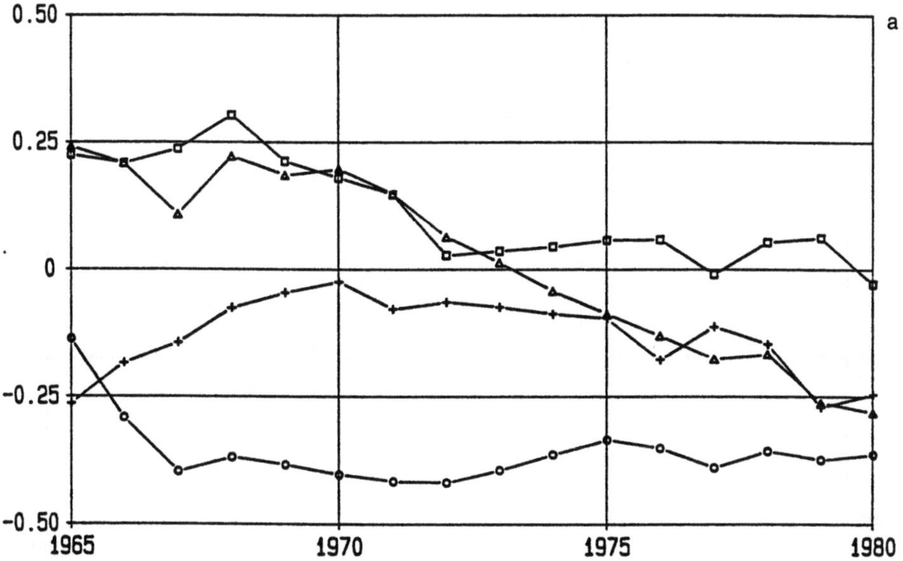

+ Jerusalem
○ Haifa
△ Tel Aviv
□ Beer Sheva

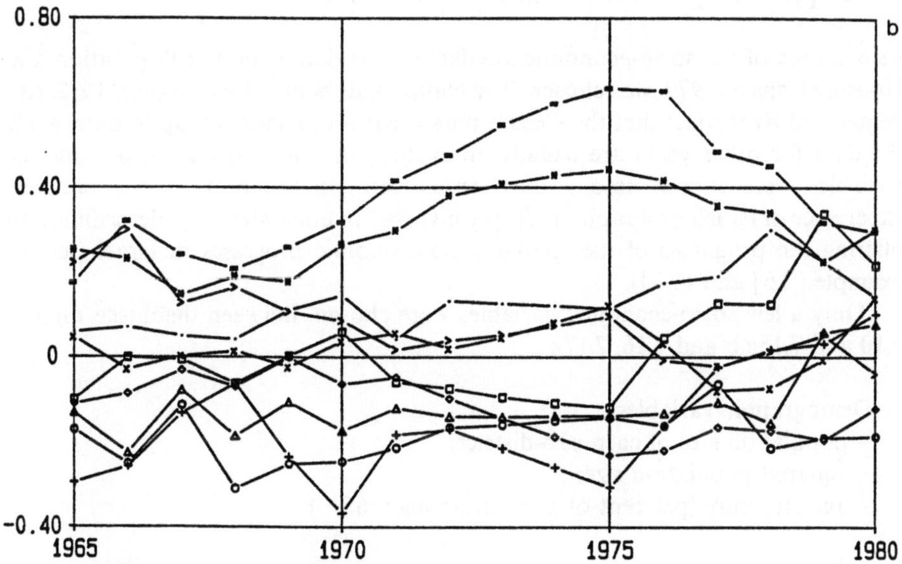

+ Zefat
○ Kinneret
△ Yizreel
□ Akko
◇ Hadera
● Sharon
* Petah Tiqva
× Ramla
= Rehovot
> Ashqelon

Fig. 7.13. Evolution of the Israeli regional preferences. (a): regions 1, 6, 12, 14, (b): regions 2–5, 7–11, 13

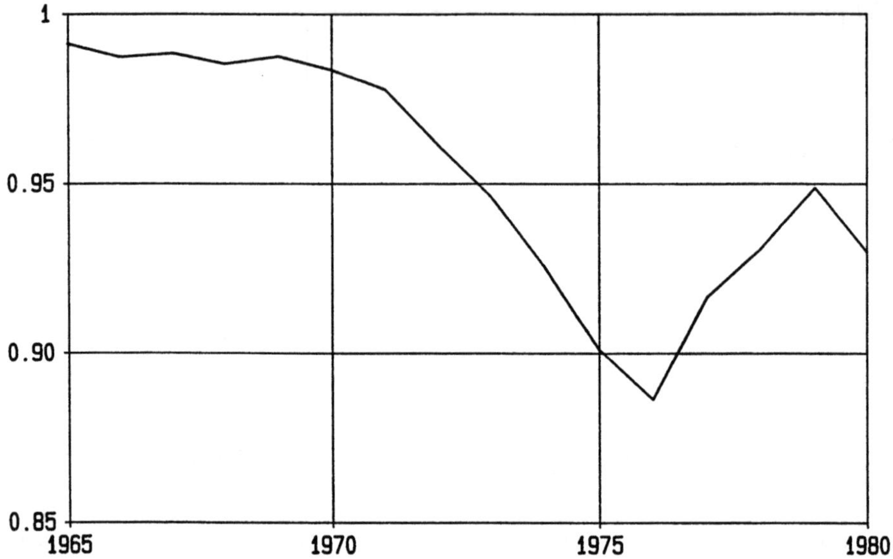

Fig. 7.14. Correlation between actual population and the stationary solution

7.7 Explanatory Socio-Economic Key-Factors

As a source of the socio-economic key-factors the data from the Population and Housing Census, 1972, was chosen. The choice of data only for one year, 1972, can be justified by the fact that the Census data are most precise and up to date, while the data for other years are usually an evaluation. The specific socio-economic variables were chosen among the Census data on the basis of the previous experience of Israeli geographers, deeply involved in understanding, describing and offering the prognosis of the spatial socio-economic processes in Israel (see for example [7.6] and [7.7]).

Only a few socio-economic variables were chosen. Between them (see Figures 7.10 and 7.15a,b and 7.16, 7.17):

1. Demographic variables:
   - population size of each sub-district,
   - squared population size;
   - age structure (per cent of population aged 65+).

2. Labour market:
   - weekly civilian labour force participation.

3. Income distribution:
   - average number of persons per room;
   - auto ownership (per cent of households with car ownership).

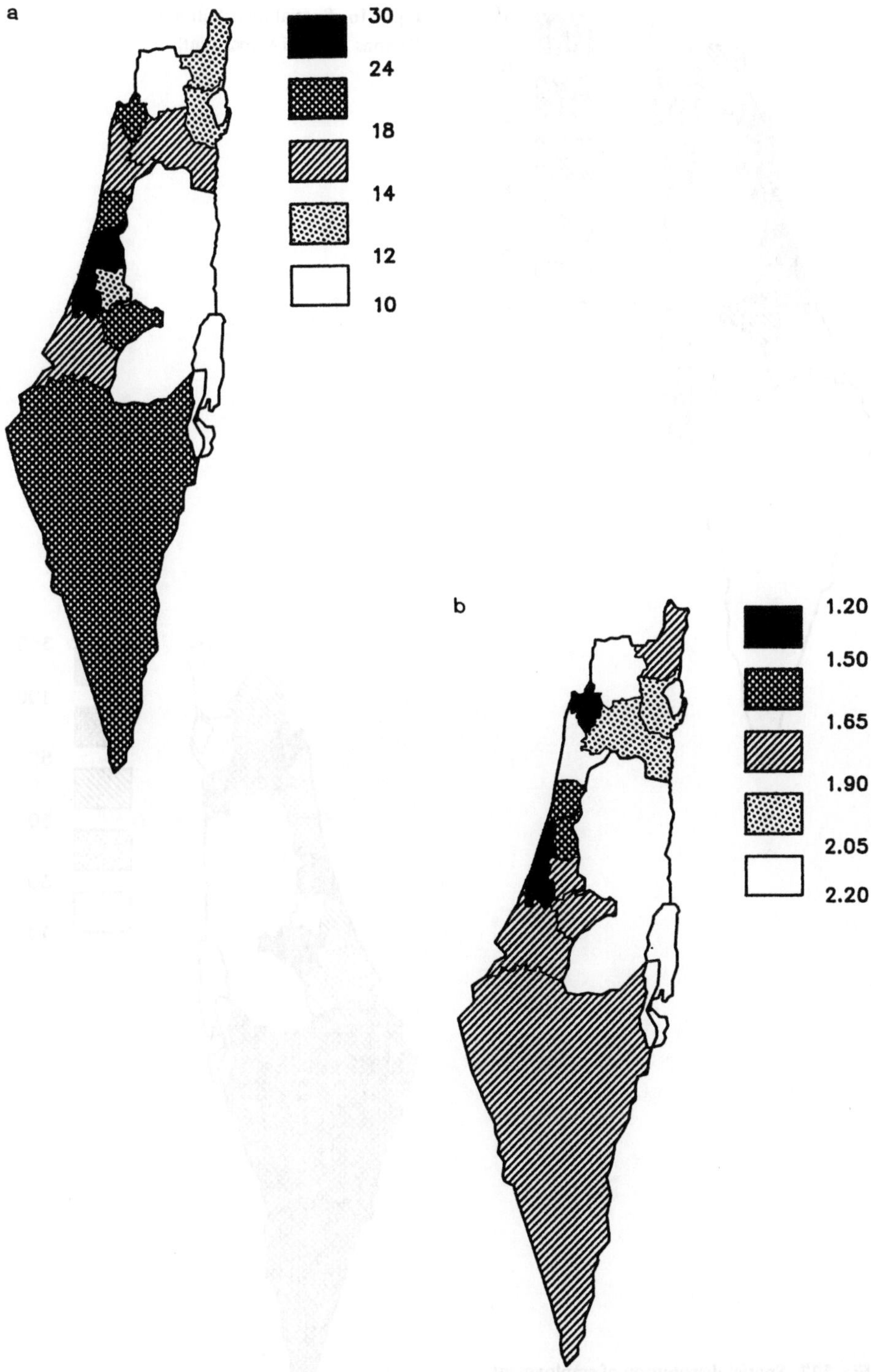

Fig. 7.15. Spatial distribution of income. (a): Households car ownership (per cent), (b): Average number of persons per room

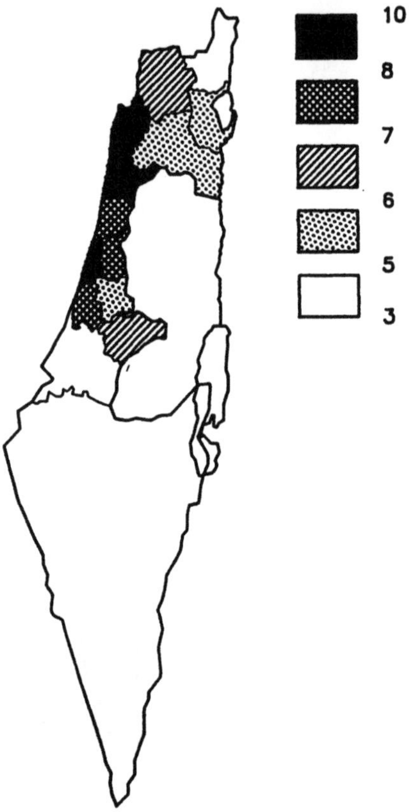

Fig. 7.16. Spatial distribution of age structures: Persons aged 65+ (per cent)

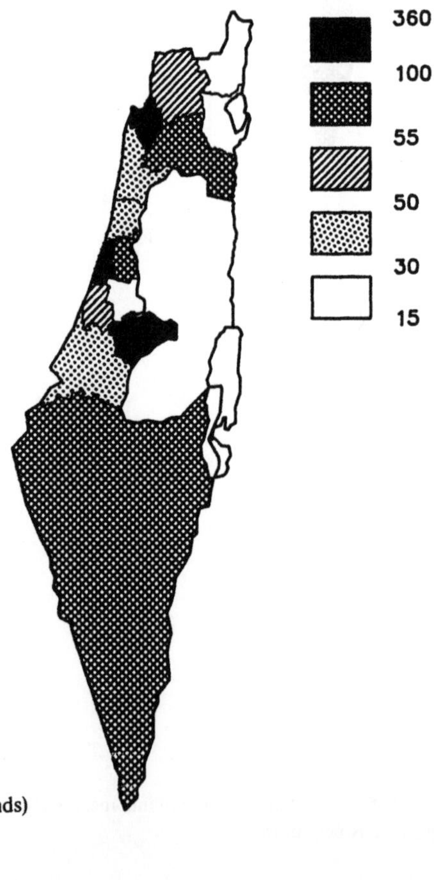

Fig. 7.17. Spatial distribution of employment: weekly civilian labour force participation (in thousands)

Surprisingly, only five key-factors are necessary to explain with some significance the complicated socio-economic structure of the regional utilities (see Table 7.8). The signs of the first four key-factors are intuitively justified: the population size and the car ownership tend to increase the regional utilities, whereas squared population leads to saturation, and the age structure has negative effects. The number of persons per room has a positive effect; one may claim that this variable presents not only the distribution of income, but also the population density – this explains the positiveness of the coefficient.

The factor of the weekly labour participation was automatically excluded by the model because its spatial rank-size distribution resembles the distribution of population between the sub-districts, and is therefore already taken into account by the population variable.

Table 7.8. Regression analysis of regional utilities

Socio-economic key factors	Timelag	coefficient	t - value
Population size	0	0.596	21.5
squared population size	0	-0.079	-3.5
car ownership	0	1.755	9.1
age structure	0	-0.423	-3.2
persons per room	0	1.172	2.7

Correlation $R^2 = 0.974$
Corrected $R^2 = 0.958$

Fisher's test value: 60.4

7.8 Concluding Remarks

The use of the model gives new insights for the deeper understanding of the socio-economic essence of Israeli migratory dynamics. Moreover, the Israeli case study reveals the interconnections between the global mobility parameter and total volume of internal migration, and between the regional preferences and net-migration hierarchy of regions. This points at the possibility to improve the model.

Other directions of the possible extensions of the model lie in the incorporation of predictions of the intraregional residential mobility, and the possibility of derivation of the hierarchical structure of the migration streams, resembling the graph-theoretical Push-Pull approach (cf. [7.8] and [7.9]).

8 Italy

Martin Munz and Giovanni Rabino

8.1 The Study Area

8.1.1 Introduction

The aim of this chapter is threefold:

Its first purpose is, to extend the experience of stochastic modelling of inter-regional migration by adding to the list of case studies of this book. In this context a specific attention is paid to the problem of the evaluation of uncertainties in population and migration counts.

Secondly, we want to go on with a longstanding tradition of quantitative analysis of Italian demographic facts. Vitali [8.1] 1978 performed the earliest stable population analysis for the whole of Italy (the evolution of population into its equilibrium) and Campisi, La Bella and Rabino [8.2] applied the Rogers [8.3] multiregional model to the five main regions of Italy. After that Rabino [8.4] in 1984 applied to the 20 Italian regions a multiregional model with exogeneously variable rates. In this context the present contribution is not only a continuation of this tradition but due to the linking of economic and demographic facts it represents a considerable improvement in the modelling of migratory systems.

Thirdly, it should be possible to improve the capacity of forecasting of the evolution of the population as a consequence of a better understanding of economic and demographic relationships in a period of fast transformations of the economy and of social characteristics of population.

After introducing the regional system of Italy (8.1.2) and after presenting the main demographic evolutions of the last century (8.1.3), we briefly describe the spatial and socio-economic pattern of migration flows in Italy (8.1.4). Section 8.2 is devoted to the conditions of the applicability of the model of this book. In particular in Section 8.2.1 we focus on the experimental conditions (i.e. zoning system, time period, source of data etc.) and in Section 8.2.2 we treat the problem of uncertainties in input data. The purely demographic results of the study are presented in Section 8.3. As in the other case studies of Part II we consider in turn the global mobility (8.3.1), the structure of the regional utilities (8.3.2), the regional preferences (8.3.3) and finally the migration stress (8.3.4). The key determinants of migration are analysed in Section 8.4. After some consideration of the availability of economic data (8.4.1), the key-factor analysis of the global mobility is carried out in Section 8.4.2 and of the regional utilities in Section 8.4.3. A general comment (8.4.4) on the interrelation between economic and demographic factors concludes the Section 8.4.

8.1.2 The Country and its Subdivision into Regions

Located in the south of Europe with approximately 57 millions of inhabitants, Italy after the U.S.S.R (191 millions) and the FRG (62 millions) is the third largest country of Europe with respect to population, having a higher population than the U.K. (56 millions) and France (53 millions). The population of Italy is rather inhomogeneously distributed over the national territory, with a density of population varying from very few people per square kilometres like in Basilicata (62) or Valle d'Aosta (35) to more than 350 inhabitants per square kilometres in regions like Lombardia (365) and Campania (381). The geomorphical configuration of Italy (being a mountainous country), together with the urbanization process, have strongly affected this distribution of population, much more than the north-south difference.

Despite of its long social and cultural history, the Italian Republic as a nation is established only since 1861. After the decline of the Roman Empire and until 1861, Italy always consisted of a multitude of small countries with different government systems changing with time. The actual regional subdivision (see Figure 8.1) still reflects this history, and despite of the strong process of homogenization having occurred in the most recent years (related to strong interregional migration) considerable differences in regional population are still persisting. These differences appear in dialects, style of living and to some extent also in physical characteristics of the people. The 20 regions have administrative purposes and the regional governments have significant independence in implementing socio-economic and territorial planning policies. The regions are divided into about 100 *provincie* (roughly corresponding to urban systems) which in turn are divided into more than 8000 *comuni* (corresponding to cities and towns). For statistical purposes the 20 regions are aggregated in 5 clusters known as *grandi ripartizioni.*

I. Northwest: Valle d'Aosta, Piemonte, Lombardia, Liguria
II. Northeast: Trentino-Alto Adige, Veneto, Friuli-Venezia Giulia, Emilia Romagna
III. Central: Toscana, Umbria, Marche, Lazio
IV. South: Abruzzi, Molise, Campania, Puglia, Basilicata, Calabria
V. Islands: Sicilia, Sardegna

In the following the 20 regions of Italy used in this analysis are numbered according to Table 8.1. The numbering is the usual one and for sake of clarity in the figures we plot separately regions of *ripartizioni* I (Northwest), II (Northeast) and III (Central) and of *ripartizioni* IV (South) and V (Islands).

8.1.3 Brief Overview of the Italian Demographic Situation and of its Roots

The present demographic situation of Italy can be clearly explained in terms of the two major transformations having occurred in the country in the last 80 years: the so-called demographic transition (from high fertility and high mortality to low fertility and low mortality) and the industrialisation process (from an agricultural society to a highly industrialized society). A central role in this explanation plays the difference in space and time of the occurrence of these two facts. Following the analysis carried out by Golini [8.5] it is convenient to start the consideration with the decline of mortality appearing since the beginning of the century as a consequence of the improvement in medical science and the quality of life. Therefore the acceleration of the growth of population has given rise to strong

Fig. 8.1. Map of Italy and the subdivision into 20 regions

emigration at first directed towards the most industrialized countries abroad and later to the most developed Italian regions, when industrial development started to rise also in some parts of Italy. To give an idea of this reduction of the rate of mortality we can remember that in 1900 the crude death rate was 22.4 per thousand inhabitants, which declined to 9.9 in 1950 with an approximately linear trend (with a small positive deviation due to the first and second world war) and remained approximately constant after 1950 until now. Concerning the emigration to abroad we can observe that since the decade 1951–1960 – where we have approximately 3 million emigrants with the balance of emigration and re-immigration being positive of approximately 1.6 million persons – it progressively reduces to 2.65 million in the decade 1961–1970 with a balance of 800 thousands, and to 1.1 million in the decade 1971–1980 with approximately a zero balance. In fact during the sixties we have the most massive transfers of population from the southern regions to the northern ones, in particular to the big cities of Piemonte and Lombardia.

Among the consequences of these transformations from our point of view we have to stress the decline of the natality rate as a result of the new urban style of life

Table 8.1. Symbols and numbers used for the graphical representation of the regional utilities and preferences

Part I: North and Central		
symbol	region	name
+	1	Piemonte
o	2	Valle d'Aosta
Δ	3	Lombardia
□	4	Trentino-Alto Adige
◊	5	Veneto
·	6	Friuli-Venezia Giulia
*	7	Liguria
x	8	Emilia-Romagna
=	9	Toscana
>	10	Umbria
<	11	Marche
#	12	Lazio
Part II: South and Islands		
symbol	region	name
+	13	Abruzzi
o	14	Molise
Δ	15	Campania
□	16	Puglia
◊	17	Basilicata
·	18	Calabria
*	19	Sicilia
x	20	Sardegna

and in particular because of the larger attention to birth control. In fact from a rate of 32.5 per thousand people in 1900 we observe a reduction to 18.3 in 1950 and to 11.0 in 1981 (with a trend which does not seem to stop!). The correlation of the decline of the birth rate with the new style of life is clearly evidenced by the fact that, for instance, even if in 1900 Lombardia and Basilicata had approximately the same birth rate, 34.3 and 35.8, respectively, in 1950 Lombardia already reached a rate of 15.2 while Basilicata was still remaining at 26.5. Today Lombardia has arrived at a rate of 10.3 and Basilicata was reached 16.7, showing that now there exists the trend of equalizing birth rates, despite of relevant differences still remaining. We discuss the consequence of these differences later.

The effect of all above mentioned transformations having occurred in the Italian population can be summarized by two aspects of the present Italian population structure: the structure by age and the structure of location. To illustrate the second point we represent in Table 8.2 the share of population over the size of the cities at the census dates 1951, 1961, 1971 and 1981. In general we observe that small cities have lost population in favour of larger ones, especially of cities with more than 100 thousand inhabitants. Only in the most recent census we see a reverted trend (counter-urbanization) related to two factors: a larger diffusion over space of the new style of life made possible by the modern means of communication, and the declining quality of life in the bigger cities as a consequence of problems like pollution, transport congestions, violence etc. These last considerations introduce to the complexity of the migration flow structures to be analysed in the following and remember of the necessary caution in analysing the results of our model treating flows of homogeneous populations only.

In Table 8.3 we give the age-structure of the population at the census dates 1951, 1961 1971 and 1981, both in absolute values and in percentages. The most stringent characteristic of this table is summarized by the following index: the ratio of the population over 65 years to the population of an age less than 15 years. This index is raised from 0.32 in 1951 to 0.65 in 1981 showing that in 30 years there is a shift from one elder people per two young peoples to a ratio of two elder peoples per one young people. It is worthwhile to mention that this transformation must also affect the structure of migration flows, thus introducing another argument for caution in the analysis of the results of our model. More significantly this change allows to forecast that in a situation of increasing demand for man power (which can be supposed if the technological innovation occurring in the industrialized north will be able to generate a strong economic development) a new immigration will be necessary to cover the needs of the industry. We have already mentioned that the process of aging of population does not occur at the same rate in all regions, and that the birth rate declines more slowly in the south than in the north. This means that in the next years in the southern regions we shall have a supply of labour higher than in the northern regions, so that migration from the south to the north may occur to cover the demand of man power. However, it seems certain that this kind of migration will be not enough to cover all the demand of man power. Mainly, the demand for unskilled work will not be covered by the Italian younger people since they do not accept this kind of jobs and also do no more accept the emigration as a solution to the employment problem. In this

Table 8.2. Share of the total population over the size of cities at the census dates.

size of the city	1951	1961	1971	1981
till to 500	0.3	0.4	0.4	0.4
501 - 1000	1.3	1.5	1.6	1.5
1001 - 3000	12.3	11.3	9.9	9.0
3001 - 10000	30.6	26.8	23.2	22.2
10001 - 20000	14.2	13.0	12.5	13.5
20001 - 30000	6.4	6.5	6.4	6.6
30001 - 50000	6.8	7.1	8.8	9.1
50001 -100000	7.6	8.7	8.1	9.5
100001 -250000	4.2	5.4	8.4	8.7
250001 -500000	5.0	4.8	5.3	4.9
over 500000	11.3	14.5	15.4	14.6
total	100.0	100.0	100.0	100.0

Table 8.3. Age-structure of the population at the census years: absolute values (in thousand) as well as percentages

age	1951		1961		1971		1981	
	abs.	%	abs.	%	abs.	%	abs.	%
5-	4332	9.1	4197	8.3	4428	8.2	3362	6.0
5 - 9	3874	8.2	3979	7.9	4618	8.5	4215	7.5
10 -14	4216	8.9	4229	8.4	4182	7.7	4550	8.0
15 -24	8142	17.1	7851	15.5	7944	14.7	8833	15.6
25 -44	13653	28.7	14194	28.0	14737	27.2	15152	26.8
45 -64	9404	19.8	11346	22.4	12127	22.4	12960	22.9
65 -74	2653	5.6	3162	6.2	4009	7.4	4801	8.5
75+	1242	2.6	1666	3.3	2092	3.9	2684	4.7
total	47516	100.0	50624	100.0	54137	100.0	56557	100.0

context a considerable immigration probably will take place from underdeveloped countries, facing Italy for the first time of its recent history with the problem of a multiracial society and the demand for guest workers. The first symptoms of these problems have already appeared.

As a general conclusion we can observe that the two decades of migration movements analysed in this chapter are occurring in a period of fast change both in the spatial and in the socio-economic structure of Italy. An analysis capable of connecting the migration to the economic and social key determinants as the work done in the following therefore is not only useful but to some extent necessary.

8.1.4 Migration Pattern

As a typical example of the structure of migration flows the migration matrix for the last year of the present analysis is listed in Table 8.4.

The most simple observation emerging from this table is that the highest migration flows consist of internal migration within the regions: The diagonal of the matrix amounts to around 70 per cent of the total migration of the region; the lowest value is found for Umbria (54 per cent) and in general there seems to exists a correlation between the degree of closure of the region and its population size. A more detailed spatial analysis could reveal that these intraregional migrations are mainly consisting of intraprovincial migration. In other words a high percentage of the migration flows are related to changes of residence within cities and towns of the same urban system.

With respect to the interregional migration we can see that 6 regions show the strongest interrelation. They are:

i. Piemonte interacting mainly with Sicilia (6725), Lombardia (5446), Puglia (4880) and Campania (4014).
ii. Lombardia interacting mainly with Sicilia (7375), Puglia (6725), Emilia Romagna (5747) and Campania (5169).
iii. Lazio interacting mainly with Campania (4779), Lombardia (3234), Toscana (3007) and Abruzzi (2729).
iv. Campania interacting mainly with Lombardia (7815), Lazio (7806), Toscana (3868) and Piemonte (3860).
v. Puglia interacting mainly with Lombardia (6889), Piemonte (3626), Lazio (2842) and Emilia Romagna (2361).
vi. Sicilia interacting mainly with Lombardia (9398), Piemonte (5524), Lazio (3183) and Toscana (3052).

From these data two main features appear: The first is a clear size-effect, in the sense that a couple of regions with high populations have higher migration flows. The second is the coexistence of a gravitational effect linking neighbouring regions (like Lombardia and Emilia Romagna, Lazio and Campania) with a long range effect linking northern and southern regions. In this context there seem to appear some kind of preferential relationships between specific pair of regions (like

Table 8.4. Migration matrix for the year 1982

	1	2	3	4	5	6	7	8	9	10	11	12	13	14	15	16	17	18	19	20
1	94783	629	5446	176	1707	526	3821	1629	1551	348	551	2110	819	191	4014	4880	909	4265	6725	2328
2	525	2619	82	17	31	5	55	32	39	12	11	42	23	4	48	34	6	117	67	30
3	5108	103	179999	835	4118	1217	3014	5747	2703	683	1241	3147	1305	336	5169	5947	936	4257	7375	1992
4	168	17	876	13659	1317	198	85	386	233	59	99	297	70	12	188	214	15	116	196	70
5	955	31	3734	831	70209	2400	406	2018	723	196	336	1222	321	44	823	830	88	450	1007	396
6	305	10	748	136	2260	20259	144	376	302	65	154	656	115	30	484	522	24	201	508	168
7	3658	85	2586	105	420	218	22618	976	1906	134	227	1031	296	23	843	846	107	840	1548	1015
8	829	26	3880	207	1646	293	785	62880	1424	278	1189	1218	527	144	2308	1343	279	778	1681	537
9	759	42	1643	127	569	199	1428	1489	53321	1004	346	2600	267	57	1946	926	263	580	1831	636
10	119	7	419	52	116	62	70	274	746	8266	419	1882	152	13	178	116	8	58	123	101
11	255	4	856	49	312	113	144	1281	374	414	20474	1191	960	40	264	384	37	89	218	107
12	1517	74	3234	331	1223	812	1106	1596	3007	2458	1705	57863	2729	470	4779	2013	311	1404	2503	1454
13	468	6	1053	57	298	113	187	707	321	187	879	2664	20399	464	406	545	52	176	207	74
14	202	2	378	13	97	25	34	262	133	34	92	698	727	4923	609	350	24	41	42	23
15	3860	58	7815	333	1513	970	1290	3805	3868	317	518	7806	722	801	102658	2139	869	1320	1218	485
16	3626	51	6899	251	1155	726	939	2361	1591	206	644	2842	796	431	1629	48030	1156	796	1308	276
17	937	7	1482	30	135	68	160	736	639	47	87	543	106	36	908	1554	6242	473	112	28
18	4123	233	6427	139	674	258	1001	1670	1188	177	157	2306	134	34	1021	926	388	31255	1543	103
19	5524	63	9398	317	1545	741	1566	2732	3052	224	344	3183	270	86	1092	1247	142	1473	88857	470
20	1464	50	2034	112	366	219	917	867	1025	184	175	1736	115	39	371	286	26	129	419	32329

Piemonte with Sicilia and Lombardia with Puglia). The roots of these interrelations may be found in the micro-behaviour of migrants, where parentage, friendship and socio-cultural similarities play an important role in the choice of destination.

The flows of migration (see Table 8.5) from the southern regions – despite of the fact that they are partially compensated by immigration from the northern or central regions – are so high that the net-balance of these regions is negative (for instance Campania (−12000), Puglia (−2500), Basilicata (−2500), Calabria (−5000) and Sicilia (−5000)). In the central regions this balance is usually positive (for instance Toscana (8000), Umbria (2000), Marche (2000) and Lazio (4500)) and this fact will be explained by our key-factor analysis of regional utilities. Finally, the northern regions show a variety of different situations. In contrast to the high negative balance of Piemonte (−8000), Lombardia and Veneto show small positive balances. The difference depends on whether or not out-migration from the northern regions is compensated by in-migrations from the south. It also depends from the redistribution of population between the northern regions.

Table 8.5. Net-balance of the migration flows

region	net-migration
Piemonte	-8223
Valle d'Aosta	318
Lombardia	3757
Trentino	-498
Veneto	2691
Friuli	1955
Liguria	288
Emilia Romagna	9572
Toscana	8113
Umbria	2112
Marche	2082
Lazio	4448
Abruzzi	1590
Molise	-531
Campania	-12627
Puglia	-2581
Basilicata	-2448
Calabria	-4939
Sicilia	-4838
Sardegna	-241

At the end of this analysis it should be mentioned that the migratory behaviour of the population depends on sex, age-structures and socio-economic conditions. For instance, if we consider the share of sexes in the total migration we find in the year considered a value of 0.5, but if we go into spatial detail of this share we observe that females are prevailing over males in the intraregional migration whereas males prevail females in interregional migration. This fact confirms again that intraregional migrations are mainly intraurban changes of residence involving both sexes. On the other hand interregional migrations seem to be connected to moves of employed persons (where males have a higher share than females).

Concerning the age-structure of the migration flows, Italy shows the typical pattern of industrialized countries [8.6] with the highest part 48.9 per cent of the flows coming from the age class from between 16 and 35 years. We have to stress that in the age pattern is not yet appearing the so-called "return migration" peak at the age of approximately 65 years. This fact could be explained in terms of the observation that interregional migration of employed peoples is not only related to the first occupation and to retirement but also to the transfer of the working place during the professional career. This phenomenon is particularly important for the employment within the public tertiary sectors (such as transport and telecommunication services), where hiring in the northern regions and the following transfer to a southern region is a common practice. This last point opens a discussion of the complex socio-economic characteristics of migrants (such as educational degree, sector of activity, professional positions etc.). A detailed analysis of this kind have been carried out by Rabino [8.7] for the whole of Italy and for the region of Piemonte in particular. This type of study allows to verify for instance that Piemonte is at the same time losing people with high educational degree (like diploma or Ph.D.) towards Lombardia and the central regions and gaining people with the same degree from the southern regions. But the second group of people is younger than the first one.

In conclusion a detailed analysis of migration of subpopulations could be very profitable in understanding the relation between the moves and their micro-motives, but at the same time such a detailed study is at risk to lose the capability of connecting the demographic factors with the macro-economic trends. In this respect the analysis carried out in the following, despite of the restrictive assumption of homogeneity of population, is still maintaining its specific interest.

8.2 Design of the Application

8.2.1 Choice of the Zoning System

In this sub-section we briefly analyse the choice of the zoning system used for the application of the model as well as the choice of the time period considered. Finally we list the sources of migration and economic data.

Concerning the zoning system, we have already mentioned in Section 8.1 that in the present analysis we use the 20 administrative regions of Italy. There are, however, two weaknesses of this choice:

i. A relative degree of inhomogeneity of the population within the regions. As we have discussed in Section 8.1.4 it seems that in the migration flows two components are present with different characteristics: The intraurban change of residence and the interurbran interactions. From this point of view the use of another zoning system as for instance the *provincie* could be better, even if also the *provincie* not always correspond very well to the urban system mentioned above. We have to add that recent studies show that in some cases in the last years the economic differences *between* the regions are becoming smaller than differences *within* regions. This is particularly true for some southern regions where the economic development takes place mainly in the cities, so that today, for instance, the difference between Bari (the main city of Puglia) and Turin (in Piemonte) is less than the difference between Bari and other parts of its region. This is the problem of the under-development of the so-called "aree interne".

ii. The difference in population size of the regions. In fact they are ranking from a maximum of 15.7 per cent of the total population in Lombardia to a minimum of 0.2 per cent in Valle d'Aosta. These differences of size are not a problem from the point of view of the pure model application, but they are weaknesses with respect to the degree of statistical validity of the migration data of the different regions. Moreover the difference of size makes different the degree of inhomogeneity of the populations within the regions. So for instance in Piemonte, which has 7.9 per cent of the total population of Italy, are internal differences between the *provincia* of Turin (4.1 per cent) and the other *provincie* of Piemonte (3.8 per cent). That could be avoided if the regions would be disaggregated in two parts like Abruzzi and Molise, which are quite similar, having only 2.2 per cent and 0.6 per cent of the total population, respectively.

On the other hand there are three strong practical arguments in favour of the zoning system used:

i. This zoning is the one commonly used in almost all studies both demographical and socio-economical concerning Italy. In this way we guarantee the comparability of the results with other investigations mentioned. This also guarantees an easier interpretation of the outcome of the model, without expressing the results in an unusual way.

ii. Even if demographic data are available on a more disaggregated level as we will see below, the availability of economic data necessary for the key-factor analysis is just limited to the zoning system with 20 regions. We can add that on the other hand the use of a more aggregated zoning system (for instance clusters of regions) is not appropriate because the loss of homogeneity of regions largely compensates the statistical benefits of the aggregation of the smaller regions. Also the arguments of point i. disfavours this kind of aggregation.

iii. In any case, even with other zoning systems it seems to be impossible to find lower relative differences in the size of the zones. For instance if we consider the system of *provincie* we have variations from 7.1 per cent of the total population of the *provincia* of Milano in Lombardia to 0.16 per cent of the *provincia* of Isernia in Molise. In fact these differences are intrinsic because of the uneven distribution of the population over the country.

Concerning the zoning system a final observation is necessary with respect to the hypothesis of the migratory independence of Italy from other countries. We have stressed in Section 8.1.4 that emigrations from Italy to abroad and internal migration flows from less developed regions to more industrialized regions were competing, the first one prevailing in the first part of the century and the second one becoming more and more important in the last decades. On the other hand the analysis carried out by Rabino [8.7] shows that at least in the recent years the characteristics of the two types of migrant populations are quite different. As a consequence of this reasoning both the choice of treating the system as closed or the choice of considering an external zone implies some degree of approximation of the reality. We choose the first alternative of treating Italy as a closed system, also in view of the technical difficulties of introducing external zones, for instance in defining the population and the economic variables of an external zone. It is worthwhile to mention that if the net-migration rate from abroad would affect all regions homogeneously then it would be possible to separate this external migration from the interregional migration flows (compare Section 2.2.2).

The time period considered in the present analysis extends from 1965 to 1982. This is the longest period compatible with the availability of data. In fact interregional migration data, despite of having been collected since a quite long time, have started to be published and stored on magnetic media since 1965 only. The data of the year 1982 are the most recent ones available at present.

The source of all data used for the Italian application of the model are informations collected and published (or distributed) by the ISTAT, the National Bureau of Statistics of Italy. Regional population numbers and migration flows are reported in [8.8]. The time series of economic data have been extracted from [8.9, 8.10, 8.11]. These time series allow the key-factor analysis for the whole period under consideration.

8.2.2 Uncertainties of Demographic Input Data

The migration process is an intrinsically stochastic process, which should be treated by means of the master equation approach, as done in Chapter 2. To apply the model to a real case, however, one must go over to the meanvalue equations (see Section 2.2) treating the available input data in a deterministic way both in the parameter estimation and in the key-factor analysis. It is a peculiarity of the Italian case, related to the procedure of data collection, that we are able to compute an estimation of the uncertainties of the demographic input data. Although this fact does not allow to use the full stochastic treatment, it is at least possible to compute the effect of the uncertainties of input data on the estimated trendparameters.

The rationale for the evaluation of uncertainties is explained with the help of Figure 8.2. We have to stress that in Italy the population registers are still very far

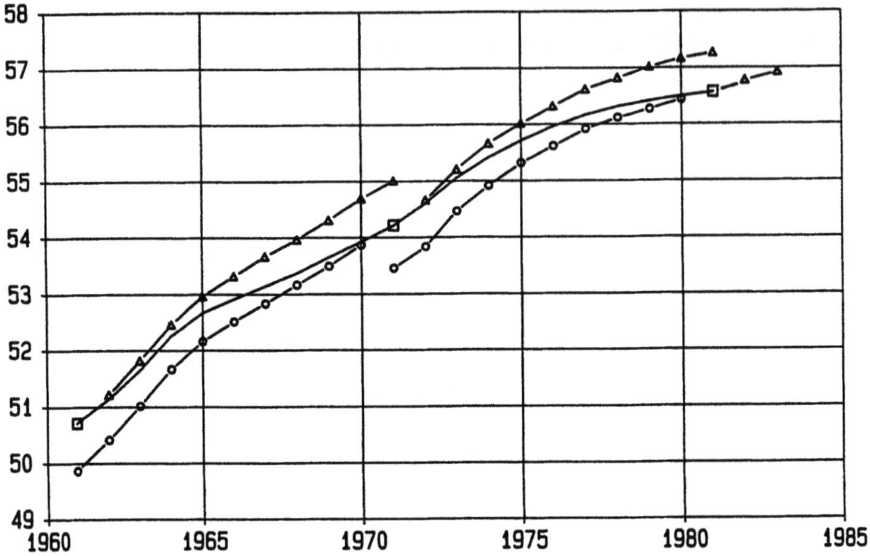

Fig. 8.2. Estimation of the population number for Italy (forward (△) and backward (○) estimation)

from being completely computerized (for instance even very big cities like Palermo and Bari are still using "hand-written" registers!). As a consequence population counts are based on censuses taken every ten years. For the intercensal years the numbers are updated by accounting of the demographic movements (births minus deaths, in-migrations minus out-migrations). If we compare, for a census year, the data computed in this way and the census values, it is possible to find far from negligible discrepancies (see Figure 8.2, the data of the years 1961, 1971 and 1981). These discrepancies are present at any level: national, regional, urban etc. They are varying from case to case. Some significant examples are listed in Table 8.6 for the census years 1971 and 1981. The computed values systematically exceed the census values (with a small number of negative exceptions). The reasons for this are not yet fully understood. Two causes surely are involved:

i. The census, because of the procedure adopted, tend to lose units of population as compared to the real one.

ii. Registers tend to overestimate the population because, unlike births, deaths and in-migrations, out-migrations may be not registered. In fact migrations are recorded in the in-migration zone and then the information is communicated to the emigration zone. Loss of information is possible in this transfer, especially for people migrating abroad.

These hypotheses are to some extent confirmed by the fact that the uncertainties in general are larger (see Table 8.6) in southern regions with a negative net-migration balance.

Table 8.6. Discrepancies in population numbers at the census years 1971 and 1981

1971	n_i^{calc}	n_i^{obs}	per cent
Valle d'Aosta	111033	109210	1.67
Piemonte	4479506	4432971	1.05
Sicilia	4874069	4682345	4.09
Sardegna	1509888	1475803	2.31
Italy	54989052	54177297	1.50
Maximum: Sicilia 4.09 per cent		Minimum: Lombardia 0.07 per cent	
1981	n_i^{calc}	n_i^{obs}	per cent
Valle d'Aosta	114300	112329	1.75
Piemonte	4496407	4473195	0.52
Sicilia	5052504	4907514	2.95
Sardegna	1618740	1594627	1.51
Italy	57250706	56536507	1.26
Maximum: Sicilia 2.95 per cent		Minimum: Piemonte 0.52 per cent	

As a consequence of the previous considerations we have to estimate population numbers for each year (see Figure 8.2): A value computed forward from the previous census and a value computed backwards. We assume the difference to be a measure of the uncertainty of population counts. On the basis of the hypothetic causes, we attribute all differences to the uncertainties of the migration flows. Figure 8.2 shows with a continuous line an interpolated population assumed to be the best approximation for the real one and therefore used in the model application.

8.3 Demographic Trendparameters of Migration

8.3.1 Global Mobility

The global mobility $\nu_0(t)$ is estimated according to eq. (3.60) and is plotted in Figure 8.3. This quantity is a measure for the individual frequency to change location within the unit of time. From this point of view it is not surprising that this curve looks very similar to the time evolution of the total rate of migration (number of migrants per 1000 persons of population). The values of Figure 8.3 are about one fourth of the corresponding rate for interurban migrations (see for instance [8.12]), and this fact gives a rough idea of the difference in mobility between the two spatial scales of analysis. But care must be taken in this evaluation because of the influence of the zoning system on the absolute numerical value of the result.

Uncertainties of the value of the global mobility as a consequence of the uncertainties of population and migrations count (see Section 8.2.2) are not recorded in the plot, because they are very small (approximately $0.4 \cdot 10^{-6}$).

In 17 years the global mobility fell by approximately 29 per cent. The reasons are on the one side those also responsible for the reduction of interregional migration in other European countries, but on the other hand they are specific to the Italian situation. To the first set of reasons belongs the influence of economic stagnation, triggered off by the "oil crises" (in fact the global mobility increases slowly from 1965 to 1972 and then decreases rapidly beginning in 1973). The second set of reasons includes both labour market and housing market conditions. A rigid rent control and high taxation made houses for rent unprofitable for the owner, thus reducing the supply of new residencs. At the same time a limited public housing supply has made more dramatic the problem of housing availability, especially in metropolitan and urban areas of immigration. Concerning the labour market, we observe the first appearance of the structural change in industrial production related to technological innovations, together with the consequences of the international economic stagnation. This implies a further reduction of the demand for unskilled man power which previously was furnished by migration of unskilled people from the southern to the northern regions. Both factors (housing and labour market conditions) have led to a reduction of the global mobility.

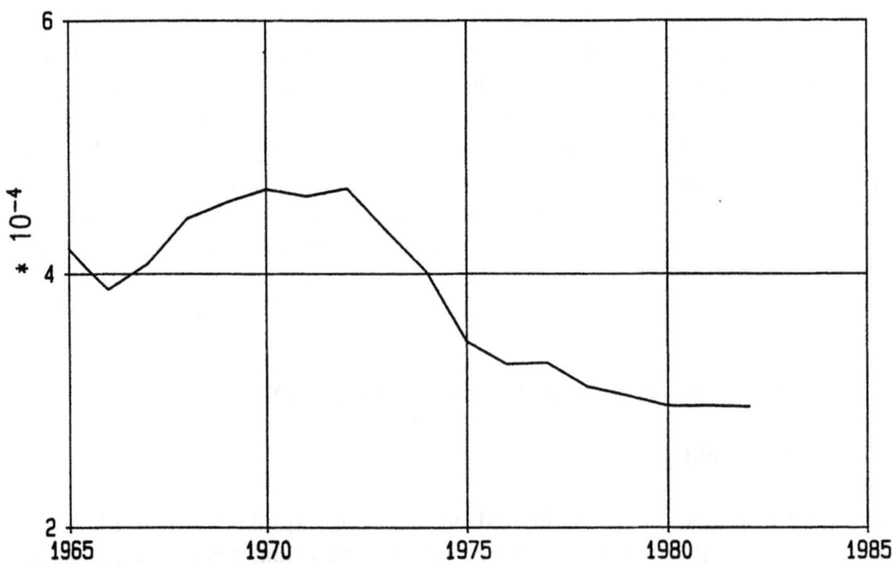

Fig. 8.3. The global mobility of Italy

8.3.2 Regional Utilities and their Variance

The regional utilities are estimated according to eq. (3.58) and their time evolution is shown in Figure 8.4. In Figure 8.4a the utilities for the northern regions are plotted, and in Figure 8.4b the utilities for the southern ones. By definition the utilities reflect the overall attractiveness of a region for a migrant.

Two major facts can immediately be read off from these figures: First of all the clear correlation of regional utilities with the population size of the regions. Secondly, that northern and central regions, which are attractors of migration flows, have in general a positive utility (see Figure 8.4a), while southern regions, which are donors of migration flows, have in general a negative utility. The regional utilities have in general a stable long term trend. This is certainly due to the fact that the utilities scale with the regional population size, which does not change significantly in the time period under consideration. An extra comment is required for the utility of Valle d'Aosta (see Figure 8.4a). The stronger negative value is in part due to the small population (Valle d'Aosta is with 0.2 per cent of the total population the "smallest" region), but also to the fact that Valle d'Aosta has an autonomous regional status which among other things implies some restrictions on the freedom of immigration.

With respect to the spatial variance of the regional utilities, plotted in Figure 8.5, we observe a broad similarity between the trends of the global mobility and of this spatial variance. Since the two variables are completely independent within the model, the similarity suggests that the economic factors determining the temporal change of the global mobility and of the variation of the spatial variance of utilities could be correlated. In fact from an intuitive point of view it seems probable that factors capable of producing a greater imbalance between regions, also affect the global mobility of the total system.

To conclude this section we analyse the influence of the above mentioned uncertainties of population and migration data on the estimation of the regional utilities. The effects are shown in Figure 8.6, where the relative uncertainty of utilities is plotted versus the relative uncertainty of population numbers. To distinguish the northern and southern regions, the symbols of the southern regions are marked with a circle. The results of Figure 8.6 are obtained running several times the estimation procedure for utilities assuming different values for migration and population counts within the range of their variance and then estimating the uncertainty (standard deviation) of the utilities over all runs [8.13]. It is obvious that there is some correlation between the uncertainty of population and the uncertainty of utilities. But it appears that the effect of the uncertainty of input data does not affect in a significant way the results of the model.

8.3.3 Regional Preferences and their Variance

The regional preferences are computed applying the regression analysis described in Chapter 3 to the regional utilities with the population numbers and their squares as explanatory variables. By definition the remaining residuals of this regression

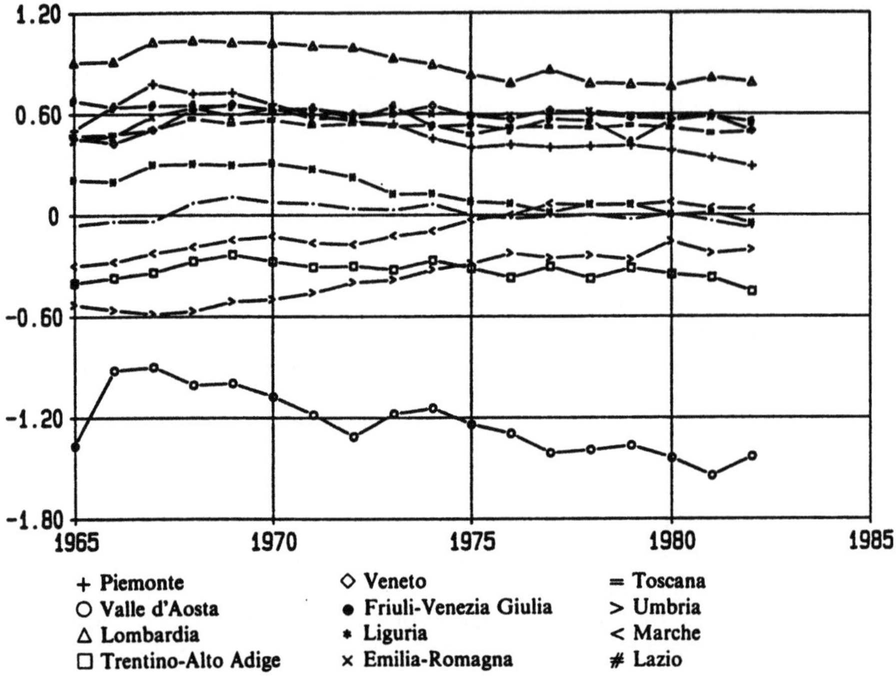

Fig. 8.4a. Regional utilities: northern and central regions

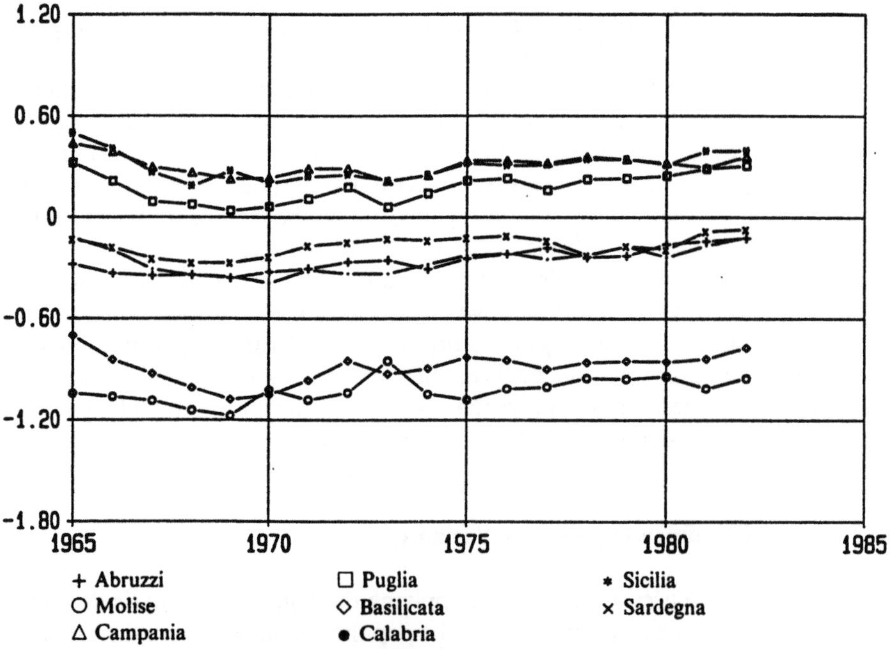

Fig. 8.4b. Regional utilities: southern regions

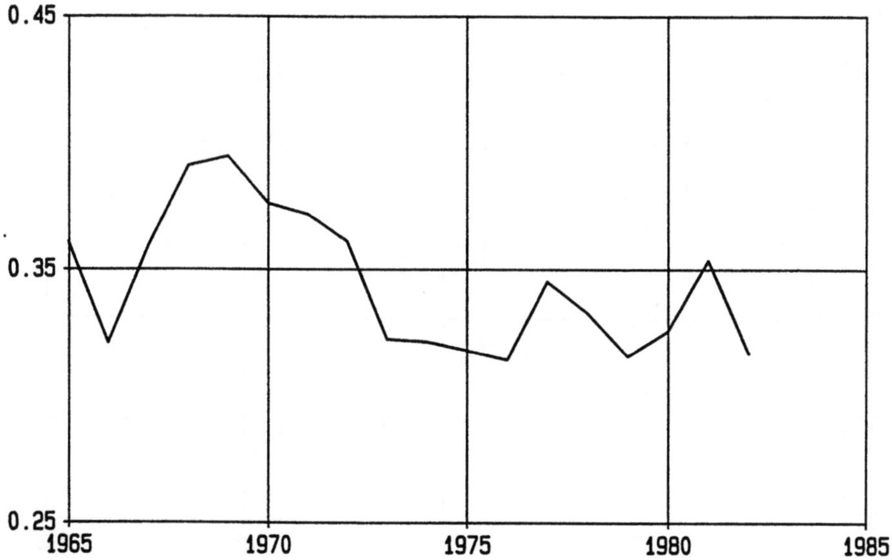

Fig. 8.5. Spatial variance of regional utilities

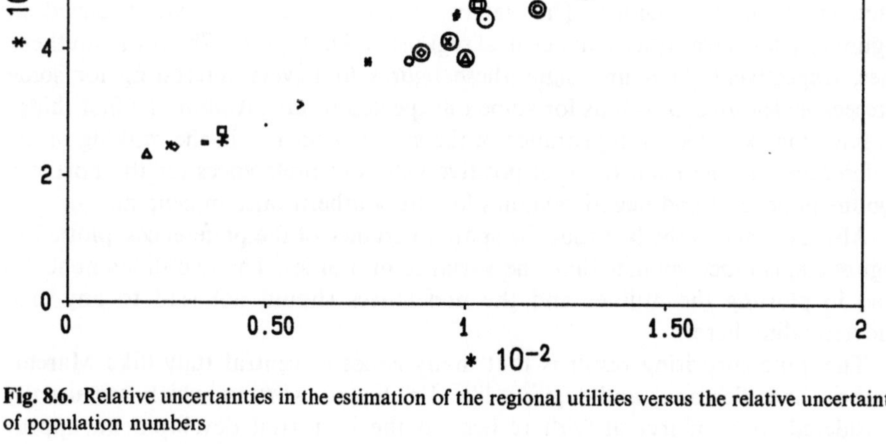

Fig. 8.6. Relative uncertainties in the estimation of the regional utilities versus the relative uncertainty of population numbers

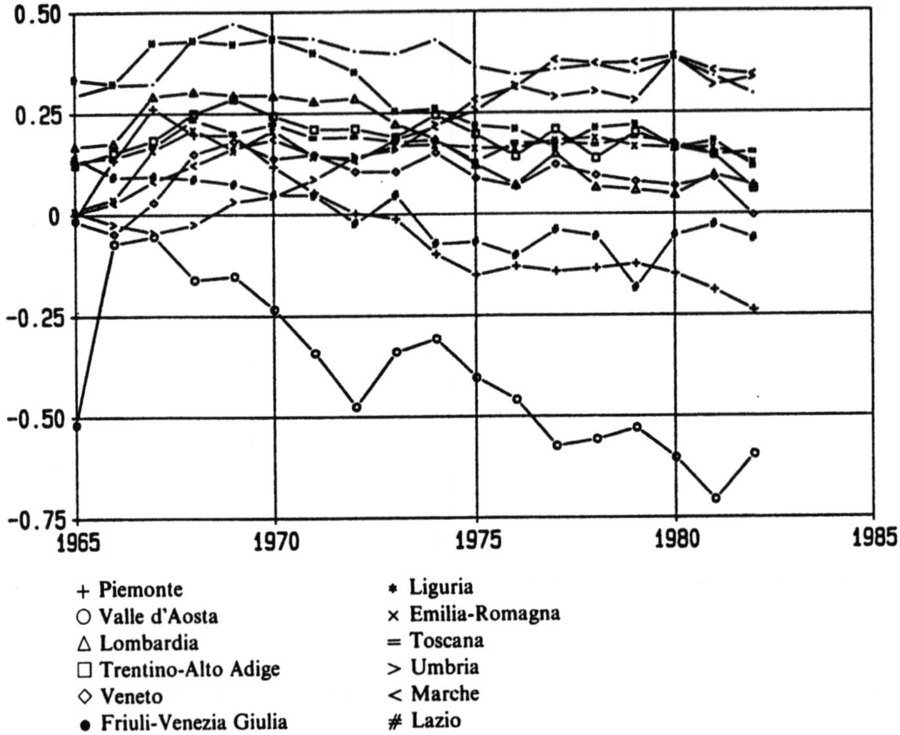

Fig. 8.7a. Regional preferences: northern and central regions

are the regional preferences. They are measures of the size-independent attractiveness of the regions. The values of these preferences are depicted in Figure 8.7a for the northern and central regions and in Figure 8.7b for the southern ones, respectively. At a first sight these figures look very interesting for some foreseeable features as well as for some unexpected results. Among the first things we have the expected disappearance of the size-dependence in the ranking of the preferences and the persistence of positive values of preferences for the northern regions in general and negative values for the southern ones in general.

Also expected is the fact that the spatial variance of the preferences, plotted in Figure 8.8, is much smaller than the variance of utilities. The two different scales used in plotting the utilities and the preferences should not lead to any misunderstanding here!

The quite surprising result is that many zones of central Italy (like Marche, Umbria and Toscana) and regions like Friuli and Abruzzi which are usually considered to be marginal with respect to the industrial development, appear among the regions with highest positive preference. On the other hand regions like Piemonte, Lombardia and Emilia Romagna, which are usually considered to be the most industrialized and favoured regions, appear in the middle or even (Piemonte) in the negative part of the ranking of preferences.

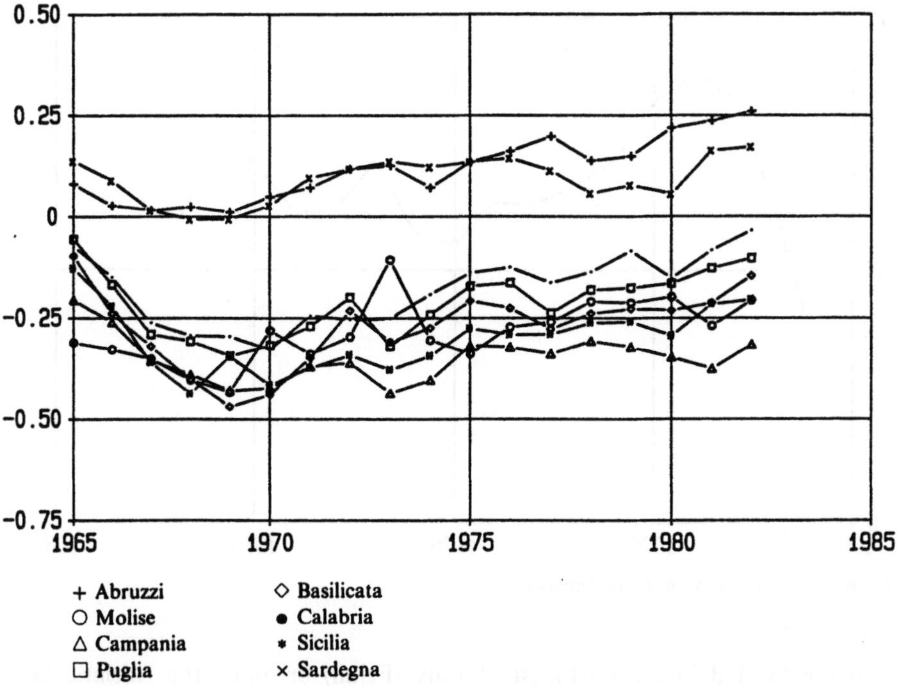

Fig. 8.7b. Regional preferences: southern regions

Going over to the analysis of trend evolutions of preferences of single regions in more detail, we can see that:

i. Piemonte, Liguria, Lazio and Valle d'Aosta show a consistent *decline* of their preference.
ii. Campamia, Sicilia, Sardegna, Abruzzi and all other regions of the south show a *slight increase* of their preference.
iii. Marche and Umbria show a *strong increase* of their preference.
iv. Finally the other regions show a rather *stable* trend.

All these trends have a sound economic interpretation. In fact they correspond to a new model of economic development appearing in the most recent years in Italy and known as "sviluppo della terza Italia" (development of the Third Italy) [8.14]. This model of economic development introduces a change of view with respect to the traditional dualistic interpretation of the development of Italian economy based on the coexistence of "two" Italies, the north (First Italy) and the south (Second Italy), respectively, each having its specific characteristics and growing with independent trends. This traditional model leads to an increasing gap between northern and southern regional development. On the contrary, the new model (the model of "three" Italies) assumes that the economic development spread from the older industrialized northern regions to neighbouring regions. These regions,

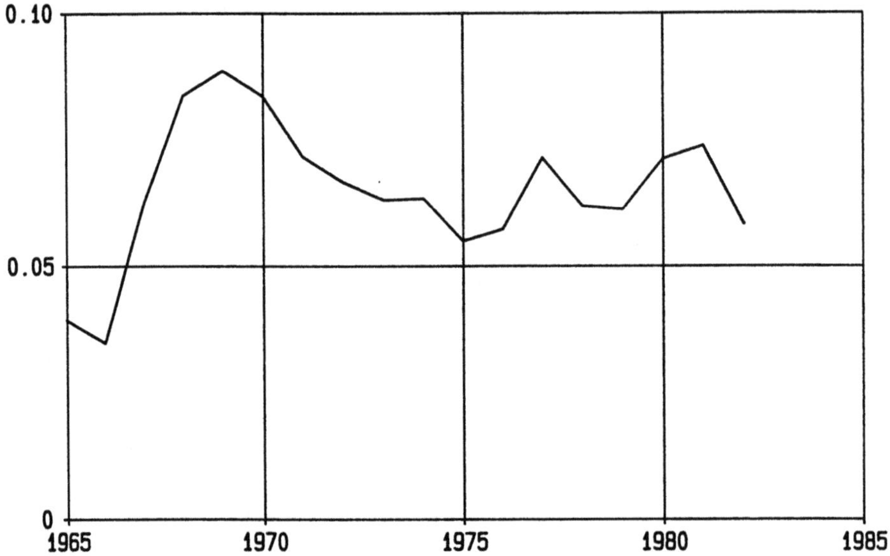

Fig. 8.8. Spatial variance of the preferences

which are located in the northeast of Italy (Friuli) or along the Adriatic Sea (Marche, Umbria and Abruzzi) constitute the Third Italy. The relative values and the trends of regional preferences show that migration movements follow the new spatial pattern of economic development. In this model also the decline of the preference of Piemonte, which much more than other regions (like Lombardia or Emilia Romagna) was related to the old model of economic development, is perfectly justified.

A final comment could be devoted to the preference of Valle d'Aosta. As already mentioned above, the low utility of this region is not only due to its small population size but to specific constraints with respect to immigration. The low value of the regional preference of Valle d'Aosta clearly confirms this fact.

8.3.4 Migratory Stress

Following Section 2.2.3 the migratory stress is evaluated. In Figure 8.9 the correlation between the actual population distribution and the virtual steady state population as implied by the observed migration rates (determined by the regional utilities) is plotted. In a more general sense the migratory stress is a measure of the distance from equilibrium of the migratory system.

Two comments should be made on this figure:

i. The migratory stress is higher in the period between 1966 and 1971 when the global mobility is increasing and the spatial variance of utilities has a maximum. It seems quite plausible that these last two features produce a spatial structure of

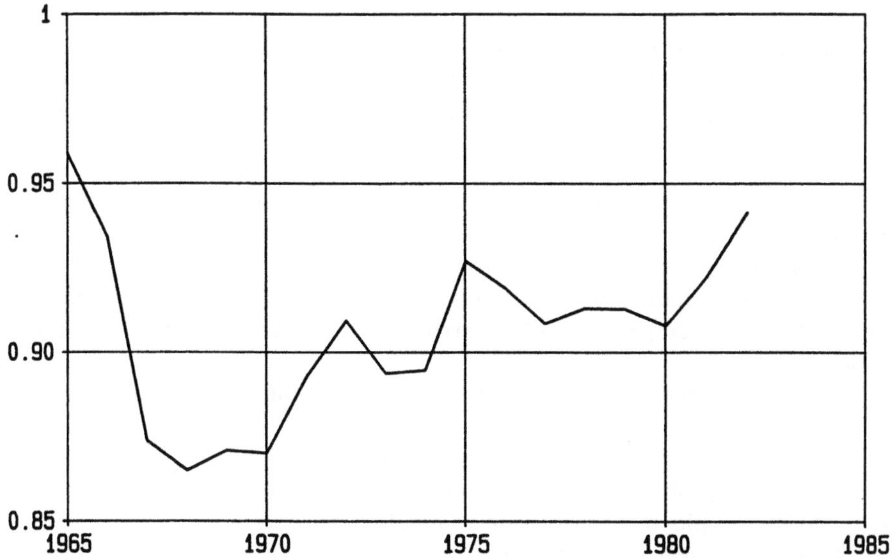

Fig. 8.9. Correlation of the actual population distribution with the virtual equilibrium distribution

migration flows much farther away from equilibrium than other situations, and thus lead to a higher migratory stress.

ii. In Section 8.1.4 we stressed the existence of preferential relationships between couples of regions (such as Piemonte and Sicilia, Lombardia and Puglia). These relationships imply some degree of discrepancy with an entropic spatial distribution of the migration flows (as implied by the stationary equilibrium solution). This could be the reason for the fact that the migratory stress of Italy is higher than in most other European countries.

8.4 The Key-Factors of Migration

8.4.1 Choice and Availability of Economic Data

Demographic data concerning both population and migration counts are largely available even to a high degree of detail. In this respect the Italian case is better, also compared with other European countries. Quite different is the situation concerning economic data. The national bureau of statistics, ISTAT, collects and publishes every year a large amount of informations, as for instance health care statistics, labour market statistics, statistics on prices, financial expenditures etc. These data, useful for specific micro-economic analysis and administrative purposes, unfortunately are quite useless to build up good and general macro-economic indicators. Because of the collection modality these data are often affected by a considerable bias.

In the following we list the economic variables which we were able to collect and which we used for the analysis of the trendparameters of migration. A short comment on the possible bias of these variables is added. Starting with directly available variables, we have for each of the 20 regions the time series of:

1. *Population numbers* $n_i(t)$: These demographic data do not need any further explanation. We can just remind of the evaluation of their uncertainties carried out in Section 8.2.2.

2. *Square of population numbers* $n_i^2(t)$

3. *Agricultural employment* $AE_i(t)$: This statistics is obtained by samples carried out in the families. Compared to other variables it can be considered to be a quite reliable indicator.

4. *Industrial employment* $IE_i(t)$: This variable needs the same comment as the former one.

5. *Employment in tertiary sector* $TE_i(t)$: The comment made for the variable 3 still holds; it must be stressed that a weakness of this statistics lies in the evaluation of the "hidden" employment in private commercial services.

6. *Total employment* $E_i(t)$: This variable sums up the variables 3, 4 and 5. As a general comment we should mention the problem of the evaluation of "black" labour and the problem of double jobs which affects differently the various regions.

7. *Unemployment* $UE_i(t)$: The estimation of this variable is rather difficult because of the presence of different forms of unemployment in a modern society. Specific of the Italian case is the so-called "cassa integrazione guadagni", a situation where workers receive a complete salary but they do a partial time job. In this case an "economic salary integration" is paid by public administrations.

8. *Labour force* $L_i(t)$: This variable sums up the variables 6 and 7.

9. *New constructions (rooms)* $C_i(t)$: To be precise this statistics reports the local governmental authorizations released each year to build up new rooms. It must be taken into account that especially in the central and southern regions a considerable amount of constructions occurred without any permission.

10. *Housing stock* $HS_i(t)$: This variable is computed adding each year the new constructions $C_i(t)$ to the census data of the stock. To the considerations made for $C_i(t)$, we have to add that we do not take into account the demolition of rooms (which is small but not negligible).

11. *Energy consumption* $EN_i(t)$: This statistics records the electric power consumption only. Due to the high dependence of the Italian economy on oil, the significance of this variable is doubtable.

12. *Gross regional product* $GRP_i(t)$: This indicator is computed according to the method SEC defined by the European Community [8.15]. The limited quality of the input data used by this method affects the degree of significance of the results.

From the direct variables listed above, it is possible to introduce another set of indirect variables:

Employment structure index $EI_i(t) = (IE_i(t) + TE_i(t) - AE_i(t))/E_i(t)$

Rate of employment in agricultural sectors $AR_i(t) = AE_i(t)/E_i(t)$

Rate of employment in industrial sectors $IR_i(t) = IE_i(t)/E_i(t)$

Rate of employment in tertiary sectors $TR_i(t) = TE_i(t)/E_i(t)$

Unemployment rate $UR_i(t) = UE_i(t)/LF_i(t)$

New constructions per capita $CPC_i(t) = C_i(t)/n_i(t)$

Employment per capita $EPC_i(t) = E_i(t)/n_i(t)$

Labour force per capita $LFPC_i(t) = LF_i(t)/n_i(t)$

Housing stock per capita $HSPC_i(t) = HS_i(t)/n_i(t)$

Energy consumption per capita $ENPC_i(t) = EN_i(t)/n_i(t)$

Gross regional product per capita $GRPPC_i(t) = GRP_i(t)/n_i(t)$

In the key-factor analysis of the global mobility we used in addition the *Density of population* $DN(t)$ of Italy computed as the ratio total population/total area.

8.4.2 Key-Factor Analysis of the Global Mobility

The key-factor analysis of the global mobility is carried out following the method introduced in Chapter 15. The detailed results of the regression analysis are presented in Table 8.7. Among the different economic variables introduced in

Table 8.7. Results of the ranking regression analysis of the global mobility

variable	timelag	coefficient	t-value
DN	1	$-1.73 \cdot 10^{-5}$	-86.3
AE	2	$-1.30 \cdot 10^{-5}$	-9.2
GNPPC	1	$5.12 \cdot 10^{-2}$	7.0
constant	-	$0.97 \cdot 10^{-2}$	-

correlation: $R^2 = 0.9763$
corrected : $\bar{R}^2 = 0.9712$
Durbin-Watson test: 2.20

abbreviations of the variables: DN: density of population, AE: agricultural employment, GNPPC: gross national product per capita

Section 8.4.1, the regression analysis reveals as key-factors of the global mobility: the density of population, agricultural employment and the gross national product per capita. Using these variables we obtain a fit of the global mobility which is quite impressive for the quality of the result (see Figure 8.10). In fact it is seen in Table 8.7, that the coefficient of correlation between the two curves is about 0.97. Discussing the observed time evolution of the global mobility, it is seen, that its main part is due to the first key-factor, i.e. the density of population. This is shown in Figure 8.11, where the result of the stepwise regression analysis is plotted: 1. step (+) using $DN(t)$, 2. step (○) using $DN(t)$ and $AE(t)$, 3. step (△) using $DN(t)$, $AE(t)$ and $GNPPC(t)$.

The increase with time of the density of population seems to act as a handicap to the global mobility (note the negative coefficient in the regression) and its temporal evolution describes the general trend of the mobility.

The second relevant key-factor is the agricultural employment plotted in Figure 8.12. In this figure we have to stress the concavity of the curve which implies a slowing down of the decline of the agricultural employment. In fact, this concavity is responsible for a modulation of the temporal evolution of the mobility as produced by the first key-factor, by reducing at first the rate of decline (even producing a slow increase of the mobility in the period 1965 to 1972) and later by accelerating the speed of decline.

Finally, the third relevant key-factor, the gross national product per capita, is plotted in Figure 8.13. The growth of this key-factor in the time period considered, acts as a corrective of the trend determined by the first two key-factors: in broad terms it adds an extra mobility quantity, in particular in the tail of the curve (from 1979 to 1982).

It is worthwhile to stress that all these results fit very well to the above mentioned so-called "sviluppo della terza Italia" model of economic development. First of all we have a general decline of the mobility, because the movements from less developed to industrialized regions are slowing down. The increase of the global density of population which is related to the urbanization process is clearly the most suitable key-factor to describe this relationship. Also the second key-factor is connected to this relationship. Migration from less developed to more industrialized regions is mainly composed of unskilled people, preferentially of agricultural workers. Indeed during the period of a strong decline of agricultural employment (1965–1972) we observe an increase of the mobility. Later on in the period 1972–1982 the slower decline of the agricultural employment leads to a decrease of the mobility. Finally the new type of mobility, related to the development of the Third Italy, is probably consisting of migration of more skilled people. The gross national product seems to be the most appropriate key factor to evaluate the level of this kind of mobility (note that living standard key-factors, such as the gross national product, are also appearing as relevant in explaining the mobility of more developed European countries).

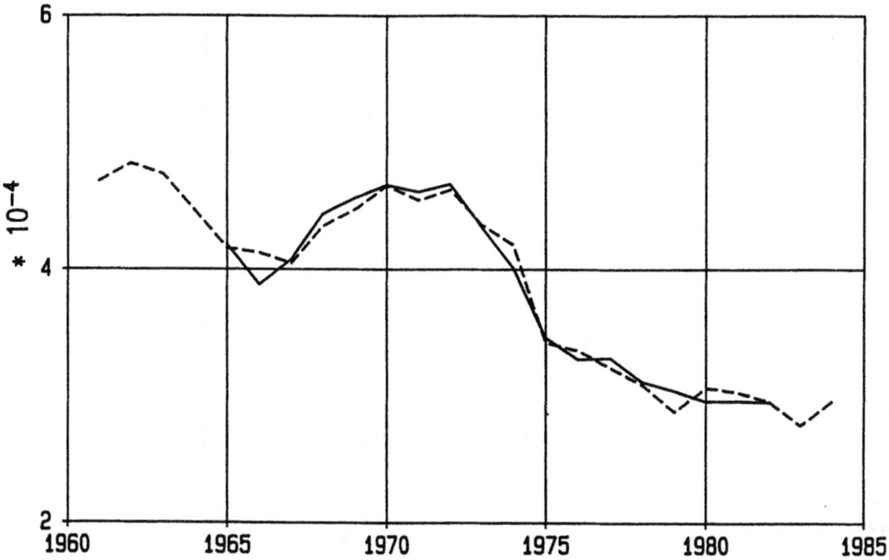

Fig. 8.10. Result of the ranking regression analysis of the global mobility: estimated mobility (solid line) and mobility represented by key-factors (dashed line)

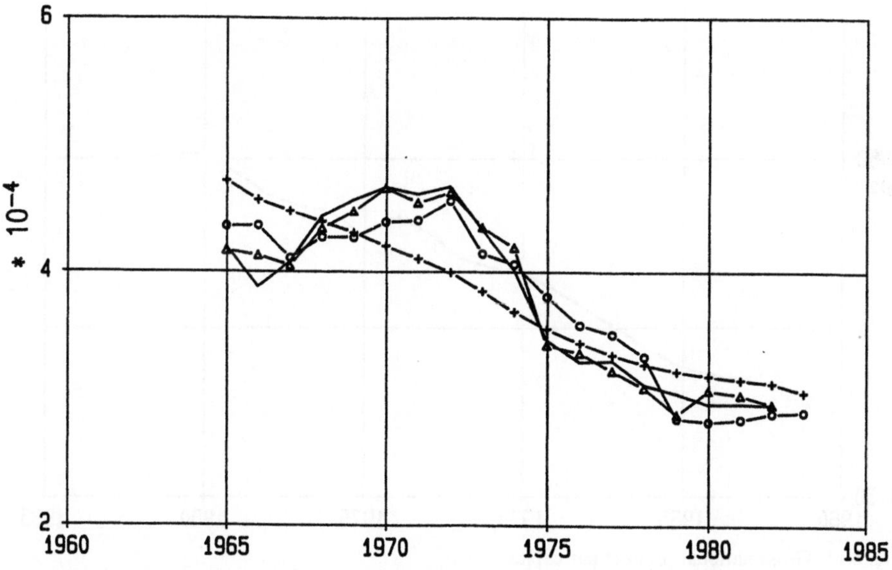

Fig. 8.11. Stepwise regression of the global mobility: 1. step (+), 2. step (○), 3. step (△)

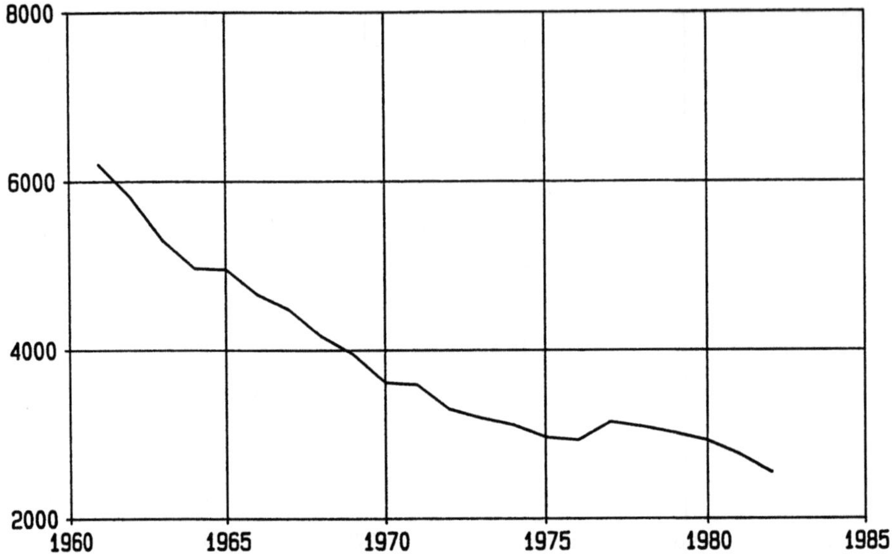

Fig. 8.12. Agricultural employment

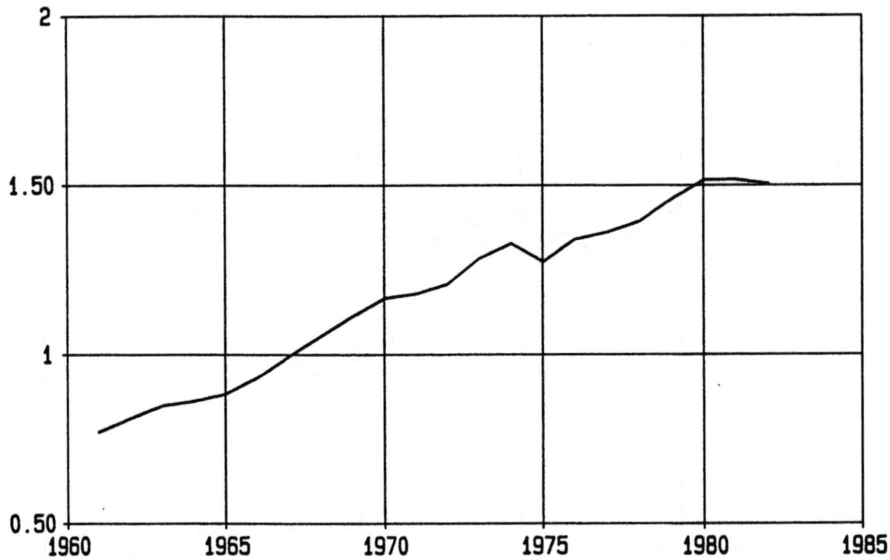

Fig. 8.13. Gross national product per capita

8.4.3 Key-Factor Analysis of the Regional Utilities

The key-factor analysis of the regional utilities is done according to the procedure presented in Chapter 3. The detailed results of the ranking regression analysis and some statistical tests are given in Table 8.8. As already mentioned in Section 8.3.2 the level of the regional utilities is strongly dependent on the size of the regions. It is satisfactory that from the regression analysis $n_i(t)$ and $n_i^2(t)$ appear as the most relevant key-factors. The regional population numbers are shown in Figure 8.14. Let us add that the appearance of a saturation effect (note the negative sign of $n_i^2(t)$) could be connected to the appearance of the population density as a first key-factor of the global mobility (with a negative coefficient). The reader will remember that, in analysing the variance of regional utilities in Section 8.3.2, we presented the hypothesis, of the existence of correlations between the global mobility and the spatial variance of utilities. The above finding seems to add some confirmation to this hypothesis and it suggests further analysis of this point.

The third and the fourth key-factors are the employment index and the number of new constructions, respectively. (The time evolutions of these variables, normalized according to Chapter 3, are plotted in Figure 8.15 and Figure 8.16, respectively.) The appearance of these variables as key-factors is very satisfactory, because we can again explain it in terms of the above mentioned model of economic development. Among the economic variables the employment index, which includes both the tertiary and the industrial sector of employment, seems the most suitable key-factor to characterize the new emergent regions of development. In fact, other economic variables, like unemployment or industrial employment, are

Table 8.8. Results of the ranking regression analysis of the regional utilities

variable	timelag	coefficient	t-value
n_i	0	1.107	88.4
n_i^2	0	-0.277	-21.7
EI_i	1	0.298	12.4
C_i	0	0.183	9.4
correlation: R^2 = 0.9041			
corrected : $\bar{R}^2$ = 0.9030			
Fishers F : F = 789.6			
abbreviations of the variables: EI_i: regional employment index, C_i: number of new constructions			

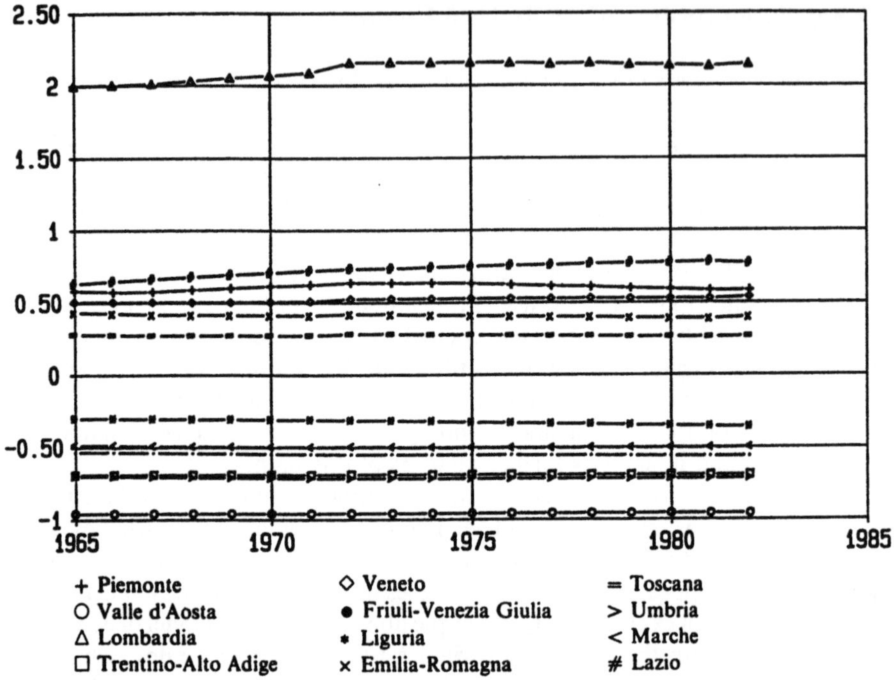

Fig. 8.14a. Normalized population numbers: northern and central regions

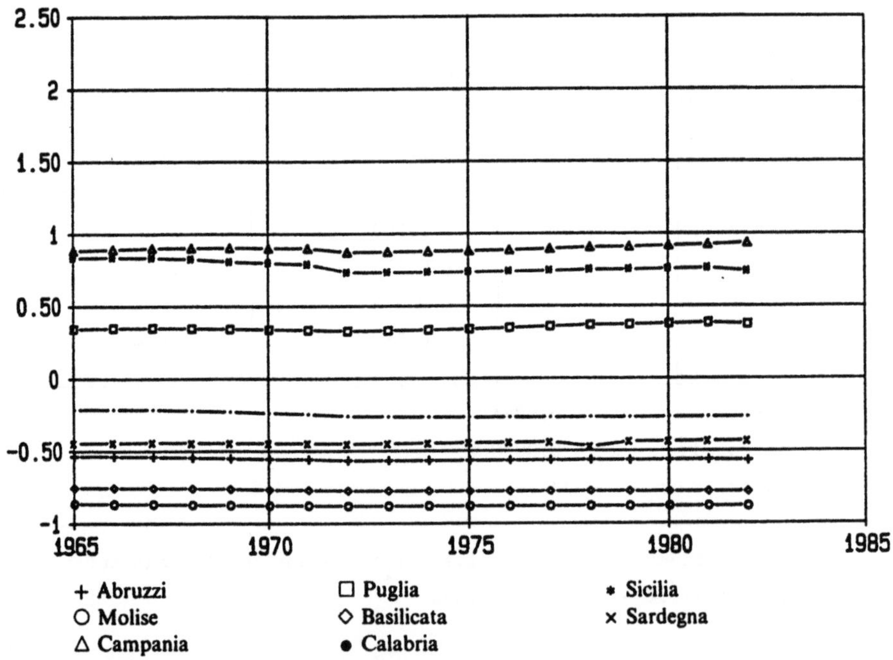

Fig. 8.14b. Normalized population numbers: southern regions

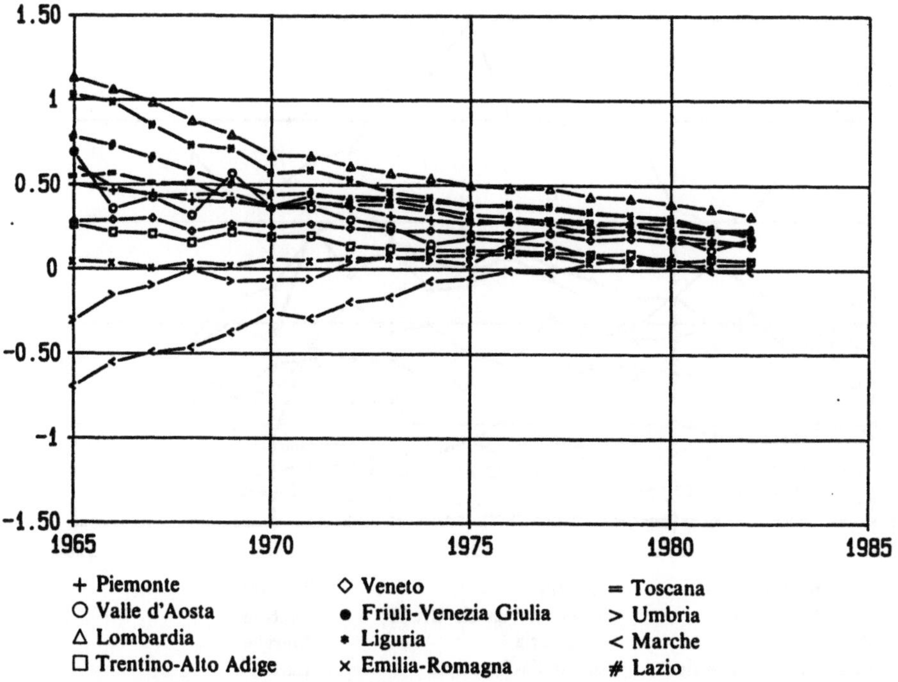

Fig. 8.15a. Normalized employment index: northern and central regions

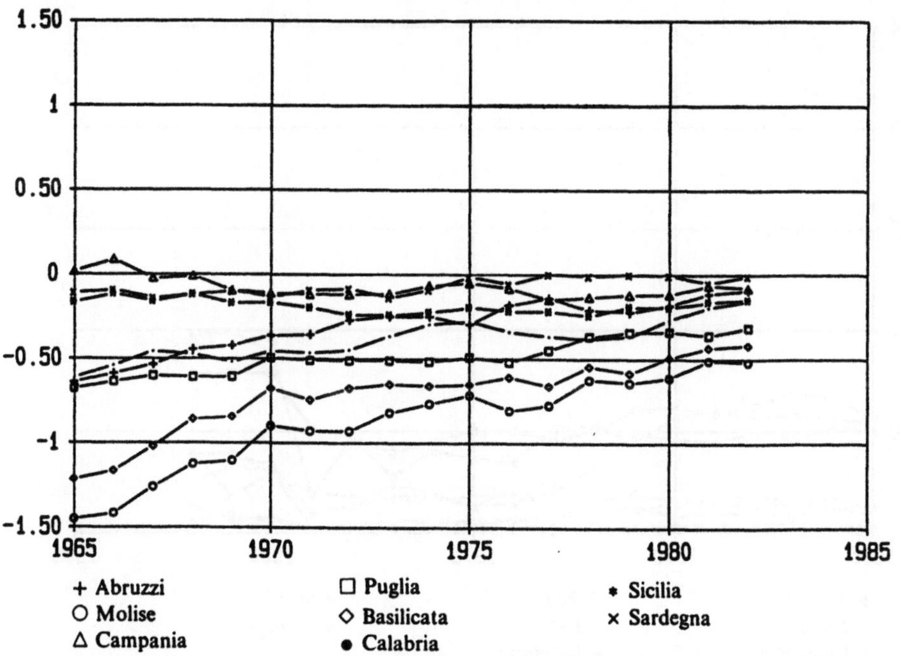

Fig. 8.15b. Normalized employment index: southern regions

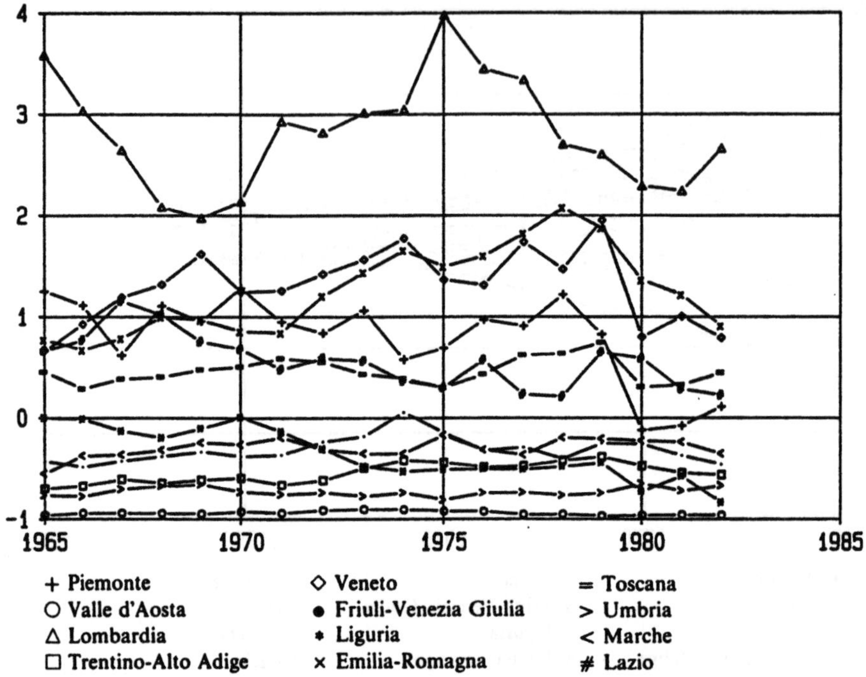

Fig. 8.16a. Normalized constructions of new rooms: northern and central regions

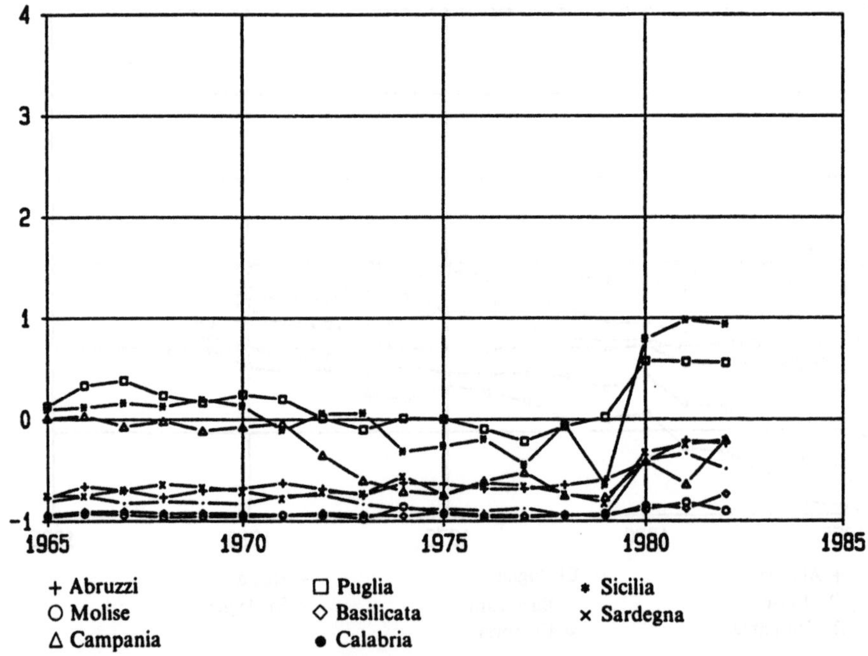

Fig. 8.16b. Normalized constructions of new rooms: southern regions

characterizing all industrialized regions or the regions of older industrialisations and not only the new ones of the Third Italy.

The variable of new constructions is then one of the most appropriate to characterize the dynamics of the residential stock. This dynamics obviously characterizes the developing of the Third Italy.

To finish this section we go to present the fit of the regional utilities using the key-factors listed in Table 8.8. To this end, according to the theory of Chapter 3, we have to introduce the natural preferences of the regions. These constant factors, which take into account socio-economic conditions not considered by the key-factors, have to be added to the latter ones. The natural preferences with the corresponding *t*-values are reported in Table 8.9. We can see that only four regions

Table 8.9. The natural preferences of Italy

region	natural preferences	t-value
1	0.000	-1.02
2	-0.515	-3.01
3	0.000	0.83
4	0.000	1.34
5	0.000	-1.85
6	0.178	3.55
7	0.000	1.82
8	0.000	-0.06
9	0.000	1.10
10	0.000	1.15
11	0.219	4.17
12	0.000	-1.08
13	0.182	5.41
14	0.000	-0.58
15	0.000	-0.54
16	0.000	-0.52
17	0.000	-1.05
18	0.000	-0.14
19	0.000	-0.90
20	0.000	1.87

correlation : 0.9638 adjusted : 0.9629
Fishers F : 1036.015

have a natural preference significantly ($t > 3$) different from zero. In three cases (Friuli, Marche and Abruzzi) the natural preferences are increasing a bit the utilities of already attractive regions. In one case (Valle d'Aosta) the natural preference is strongly negative. It is clearly related to the constraint on in-migration already mentioned in Section 8.3.2. This constraint obviously can not be explained by economic key-factors.

Both fittings of the utilities – namely without or with including the natural preferences are quite satisfactory. In fact, we have a correlation coefficient of 0.90 and 0.96, respectively. The fitting, using the four relevant natural preferences, is shown in Figure 8.17.

8.4.4 A General Concluding Comment

In the first part of this chapter we have stressed many cautions about an analysis of migration not considering the different types of inhomogeneity present in the migration pattern (such as age-structures, social composition of the migration flows etc.). On the other hand we also stressed that a too disaggregated analysis may be unable to capture the interrelationship between the micro-behaviour and the demographic and economic trends on the macro-scale.

The results obtained with the application of the model presented in this book show without any doubt that also an aggregated analysis can be very profitable. We were able to extract from the empirical data evidences for a theoretical model of Italian development and, within our model of migration, to connect in a quantitative way demographic and economic variables. The application of this approach described for forecasting purposes and the development of more refined versions of the model seem to be promising directions for future research.

Fig. 8.17. Results of the ranking regression analysis of the regional utilities: estimated utilities (represented by symbols) and utilities represented by key-factors (solid lines). a: Northeast, b: Northwest, c: Central, d: South, e: Islands ►

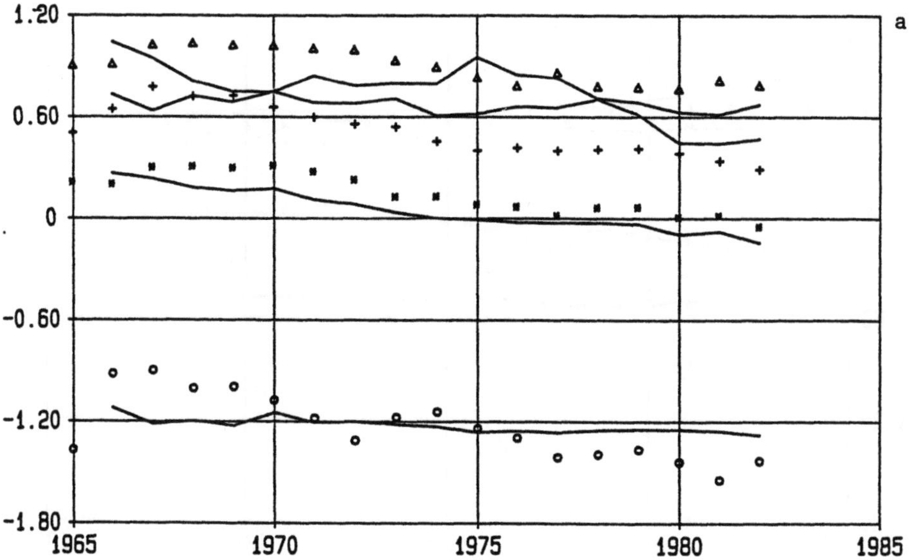

+ Piemonte
○ Valle d'Aosta
△ Lombardia
* Liguria

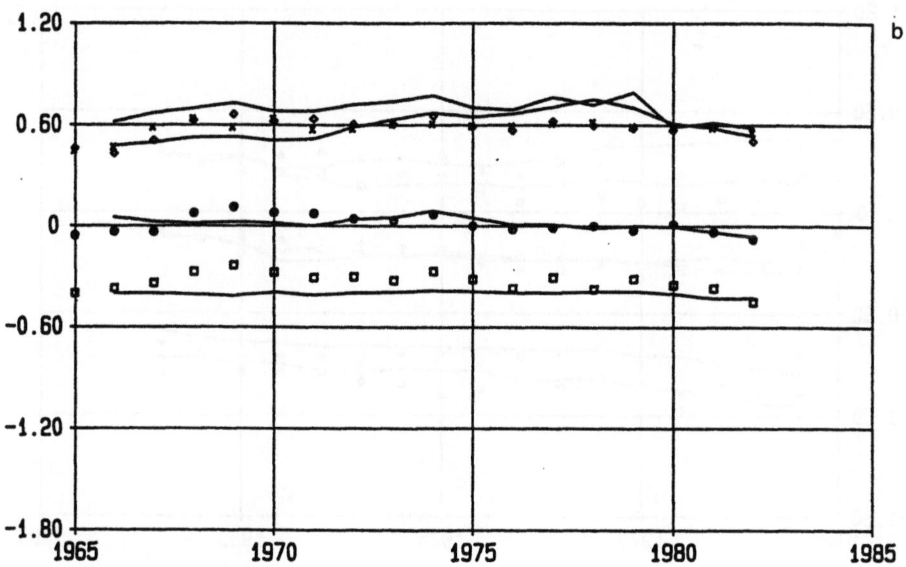

□ Trentino-Alto Adige
◇ Veneto
● Friuli-Venezia Giulia
× Emilia-Romagna

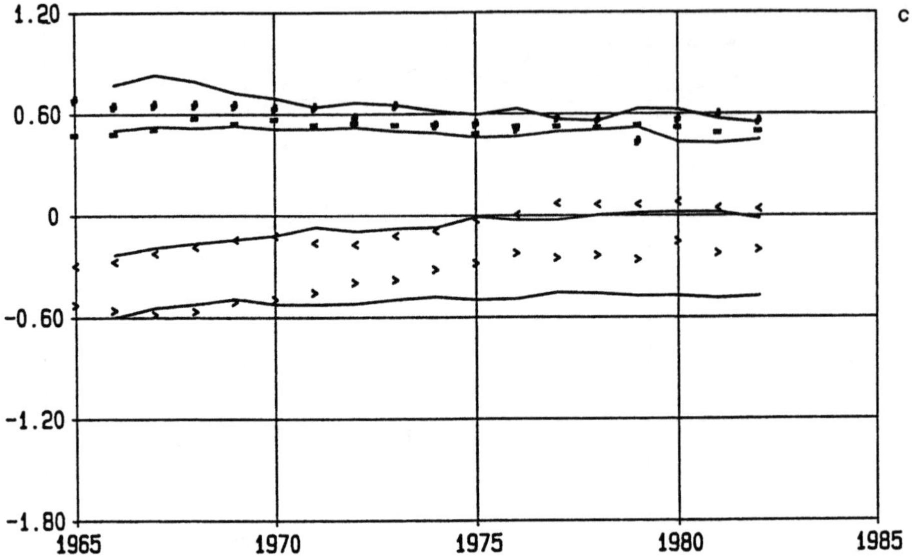

= Toscana
> Umbria
< Marche
Lazio

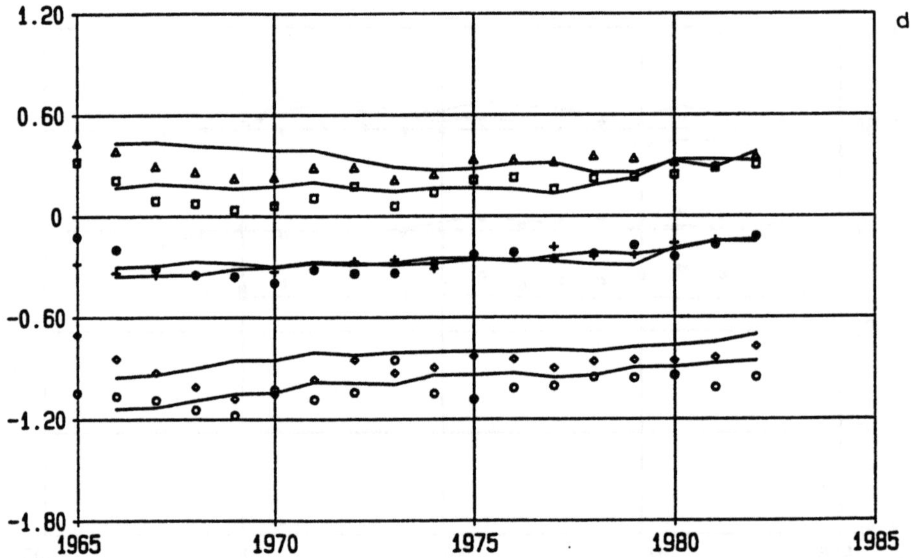

+ Abruzzi
○ Molise
△ Campania
□ Puglia
◇ Basilicata
● Calabria

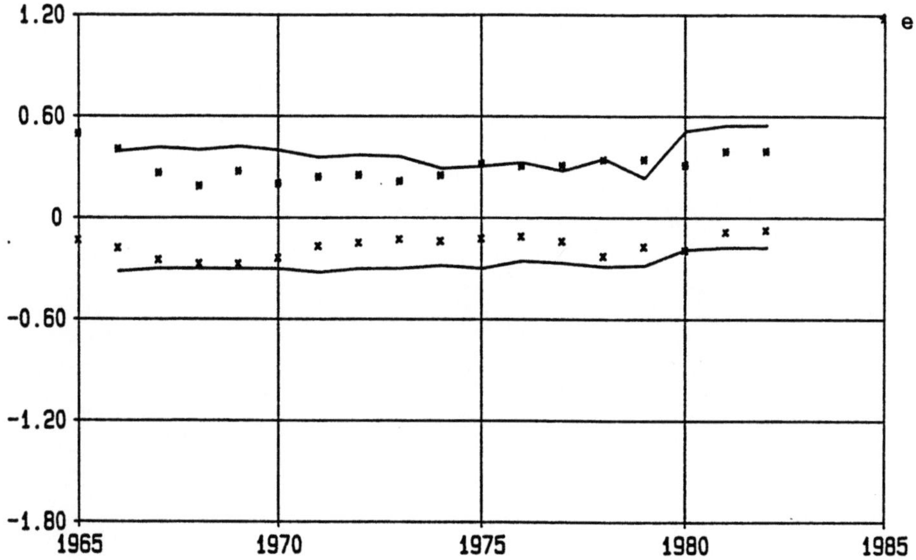

* Sicilia
× Sardegna

Fig. 8.17c–e. Legend see p. 218

9 Sweden

Ingvar Holmberg and Nicias Sarafoglou

9.1 Regional Subdivisions of Sweden

Sweden has one of the oldest and most comprehensive population statistics systems of the world dating back to the beginning of the mid-18th century. The basic unit in this statistical system is the parish, of which there are about 2,570. Usually, these are aggregated into much larger units for statistical purposes, such as the 280 municipalities, 70 labour market regions or to 24 counties. In a more cursory analysis the counties may be aggregated into 8 or 9 county regions. It is also possible to distinguish between only three major regions, viz. the metropolitan region (consisting of the Greater Cities of Stockholm, Gothenburg and Malmö), the forest region (consisting of Upper North and Lower North county regions) and the rest of Sweden. The boundaries of the counties, the 8 county regions (see Table 9.1) are all shown in Figure 9.1.

In the following discussion we will only use the last two classifications because they are more comprehensive and will therefore give a better overview of the demographic and economic development in Sweden. Moreover, we think that in this way we get a manageable set of data, without losing too much of interesting details.

The main purpose of this chapter is to apply the basic migration model, as described in part one, to the Swedish population development during the period 1968–1982.

In order to provide a deeper background to the study we will, however, start much further back in time, more precisely in the mid-18th century. The following Section 9.2 gives an overview of regional population development since the beginning of Swedish population statistics.

However Section 9.3 starts with a brief review of labour and housing market and reasons for migration in a more general and qualitative manner. In the present application we consider only one homogeneous population and we instead con-

Table 9.1. The county regions of Sweden. Numbers and symbols used in the graphical representation of regional quantities

symbol	region	name
+	1	Stockholm
o	2	East Middle
Δ	3	South Middle
□	4	South
◊	5	West
·	6	North Middle
*	7	Lower North
x	8	Upper North

Fig. 9.1. The map of Sweden and its subdivisions into eight county regions

centrate on the spatial differences of socio-economic factors. In Section 9.3 different reasons for migration are considered. In Section 9.4, the traces of the regional utility profiles, their variance, the regional preferences, and the migratory stress are exhibited. In Section 9.5 those socio-economic variables are identified which potentially could be relevant for the explanation of interregional migration of a homogeneous population (thus not taking into account its divisions into subgroups). In Section 9.6, finally, we present the result of the selection of key-factors in the representation of mobilities and utilities.

9.2 A General Outline of Regional Population Growth in Sweden

9.2.1 Historical Trends in Regional Population Development

The regional distribution of the Swedish population is available since the middle of the 18th century. Population data aggregated to the eight county regions are illustrated in Figure 9.2. The data suggest the following general remarks:

The long-term migration movement appears to be from inland to coastal areas, with some concentration to the metropolitan region of Stockholm. In a historical

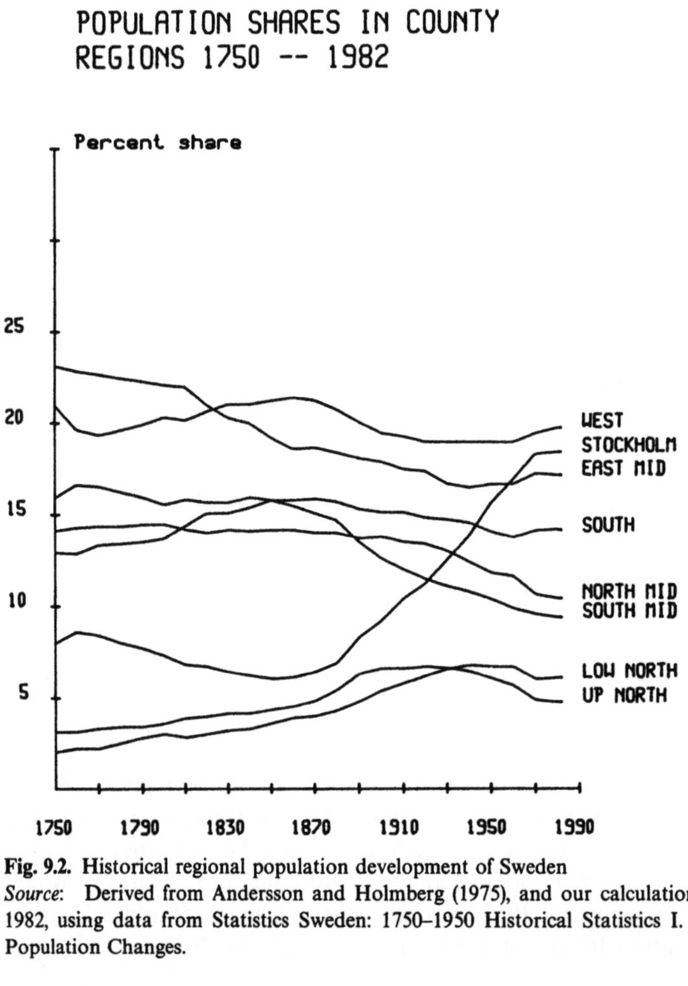

Fig. 9.2. Historical regional population development of Sweden
Source: Derived from Andersson and Holmberg (1975), and our calculations for the period 1975–1982, using data from Statistics Sweden: 1750–1950 Historical Statistics I. Population; 1961–1982 Population Changes.

perspective there has been no real increase in the relative population shares of the West and South county regions.

On a more detailed scale the following remarks can be made:

– the Stockholm region has been increasing its relative share of the population from 6 to 19 percent at a steady rate from 1850 to 1970. In the early 1970s there was a break in the general trend but after only a few years the population share increased again;

– the counties of the periphery (Upper and Lower North) have also shown a steady increase in their relative share of the total population. From a level around 5 percent in 1750, they increased to 8 percent in the 1850s and arrived at a peak level of approximately 13 percent in 1950. From then on the share of these two regions has been decreasing down to a level around 11 percent. In recent years there seems to have been a stabilization in the relative share close to this level;

– all the counties of the eastern and mid-inland parts of the country (South Middle, East Middle, and North Middle) have declined in relative economic and demographic importance. These areas together had around 53 percent of the population in 1750, a figure that had decreased to 44 percent by 1900 and to only 37 percent in 1970. In the last decade, however, the relative population share of the East Middle region seems to be increasing again, a fact that may be interpreted as an effect of the proximity to the Stockholm metropolitan region;

– the most stable parts of the country are the southern and western regions which have had approximately 14 and 19 percent of the population, respectively, for the time period studied here. When seen in this long run perspective, the relative increase in the population shares in the most recent years might be seen as just a reflection of the general wave-like movement in the shares during the whole period.

The general change in settlement patterns illustrated in Figure 9.2 is accompanied by changes both in population densities and in the regional income distribution.

The Stockholm region stands out as an exception in this context. Being mainly a totally urbanized region, its population density has shown a very rapid increase which lies far above the increase rates of all other regions. Also the two other metropolitan regions South (Malmö) and West (Gothenburg) show an increase in density, whereas all other regions have an almost constant population density for a very long time.

A characteristic problem observed in most industrialized countries in recent decades is the great migration of people from rural to urban regions. Although statistics are scarce for earlier periods it is possible to study the development of the urban population from the beginning of the nineteenth century, and some rough estimates can be made back to the mid-eighteenth century.

Even today large areas of Sweden are very sparsely populated, especially in the northern parts of the country. From 1880 and onwards the population increase became more and more concentrated around certain growth poles, where the population increased more rapidly than it did in Sweden as a whole.

The major reason for this locally concentrated growth is the emergence of cities. For the mid-eighteenth century the urban population has been estimated at approximately 9 percent of the total population. In 1880, when actual data are available for the first time, the urban share had not yet reached 10 percent.

From then on the urban population has been growing at an increasing rate. Today more than 80 percent of the total population lives in localities which comprise both cities as well as smaller urban agglomerations. The main reason for this large share is of course the fact that a very liberal definition of a "locality" is used in Sweden. By locality is meant an agglomeration with at least 200 people in a contiguous location.

9.2.2 Current Trends in Regional Population Growth

Detailed population statistics on all levels of aggregation with respect to spatial classifications are, in principle, available since 1960. However, the statistics published in official reports by Statistics Sweden (or formerly SCB) are much less specified. This means that we cannot make comprehensive studies of regional population development at any level of aggregation without making special tabulations from the original statistical sources. These are notifications of changes from the parishes that are sent via the County Boards to Statistics Sweden[1].

For the years prior to 1960 the availability of regional population data is very scarce. For the counties (and consequently for the county regions) we have information about births and deaths, while data on total migration can be obtained only as a residual when comparing natural increase with total population growth.

For the presentation in this chapter we will restrict the time-period to the years after 1968, but in a forthcoming study we will go as far back as to 1950. The following diagrams (Figure 9.3) show the contribution of various demographic components to total population growth.

The total domination of the metropolitan regions with respect to population growth was broken in the early 1970s. Then, suddenly, the foreign migration surplus for the country as a whole became negative for the first time since long before World War II. The surplus towards other parts of the country was considerably reduced and even negative in some of the years during the 1970s.

In the Stockholm Metropolitan region this reduction in the regional migration surplus was soon compensated for by an increase in the foreign migration surplus, and the population of the region started to increase again by the year 1975. For the other two metropolitan regions (South and West) this was not the case and for some years the population even declined.

In the last few years there seems to have occurred a new change of the trend: The migration streams are once again directed more towards the metropolitan regions at the expense of more remote areas. The evolutionary pattern of migration

[1] Because of problems with the computer tapes it is practically impossible to get migration data for the years 1961–1967 as well.

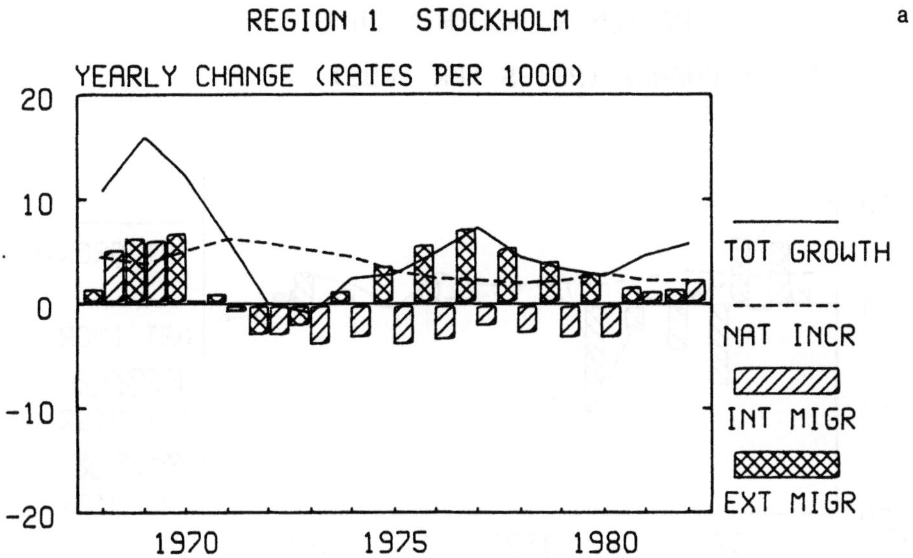

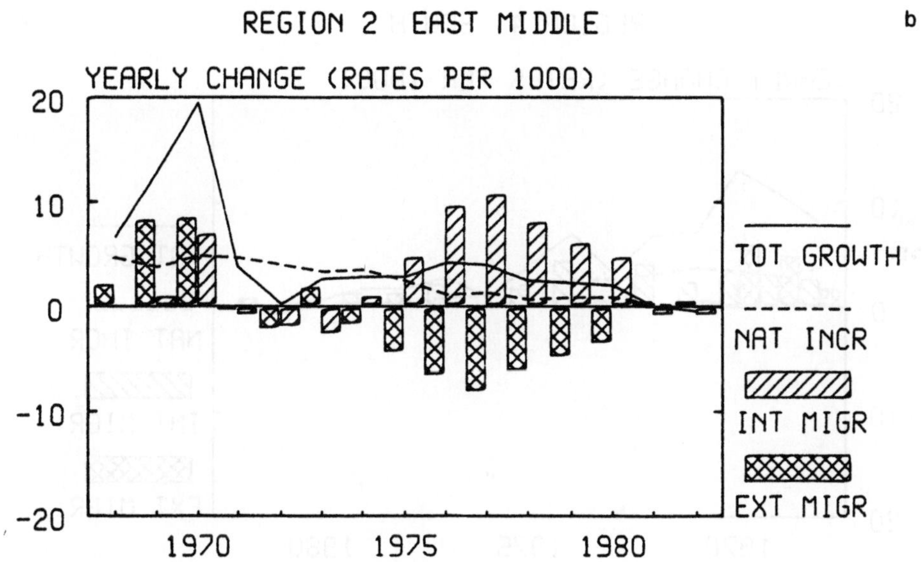

Fig. 9.3a–h. Population changes between 1968 and 1982
Source: Statistics Sweden, 1968–1982 Statistical Abstract of Sweden

thus seems to exhibit a cyclical trend of the type considered within the framework of the Core-Periphery approach. Three main stages can be identified: urbanization, deurbanization, and reurbanization. This type of evolution has been observed in many industrial countries (see e.g. R. E. Dickinson) [9.1].

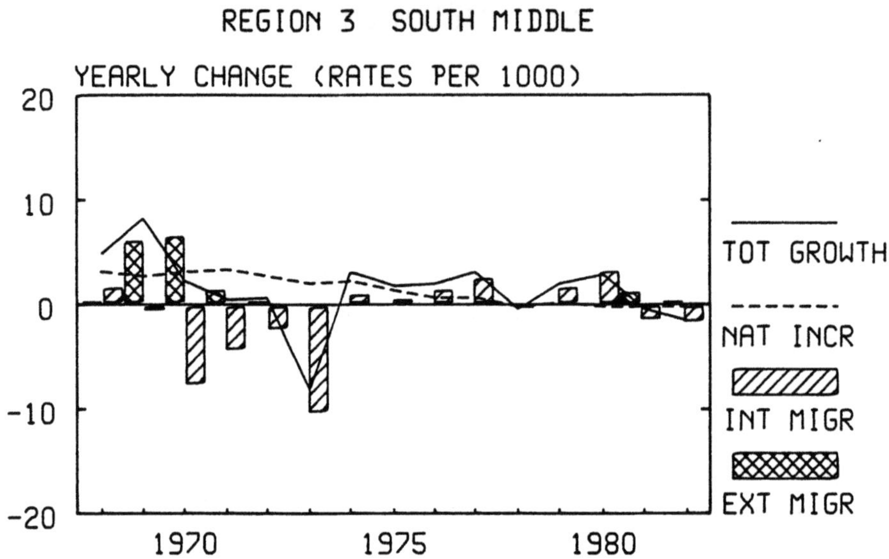
REGION 3 SOUTH MIDDLE
YEARLY CHANGE (RATES PER 1000)
20
10
0
-10
-20
1970
1975
1980
TOT GROWTH
NAT INCR
INT MIGR
EXT MIGR

c

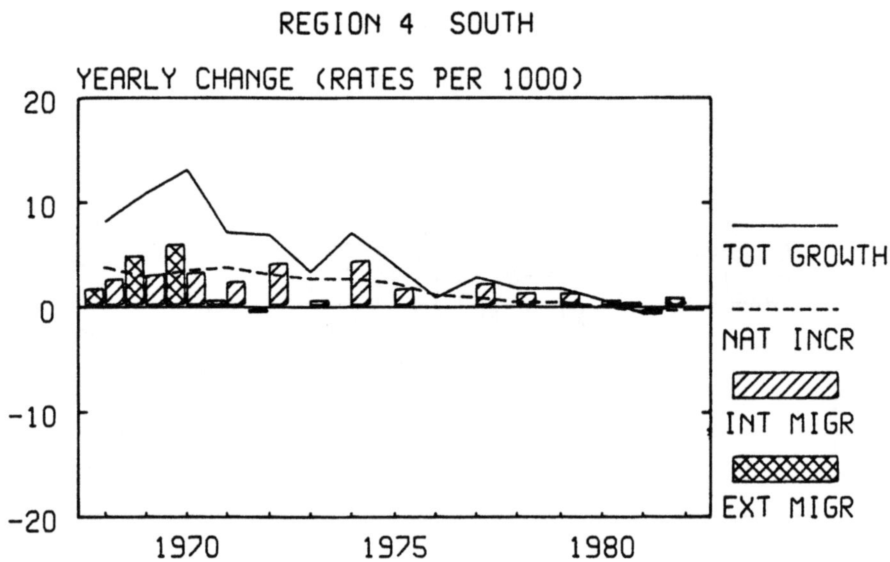
REGION 4 SOUTH
YEARLY CHANGE (RATES PER 1000)
20
10
0
-10
-20
1970
1975
1980
TOT GROWTH
NAT INCR
INT MIGR
EXT MIGR

d

e

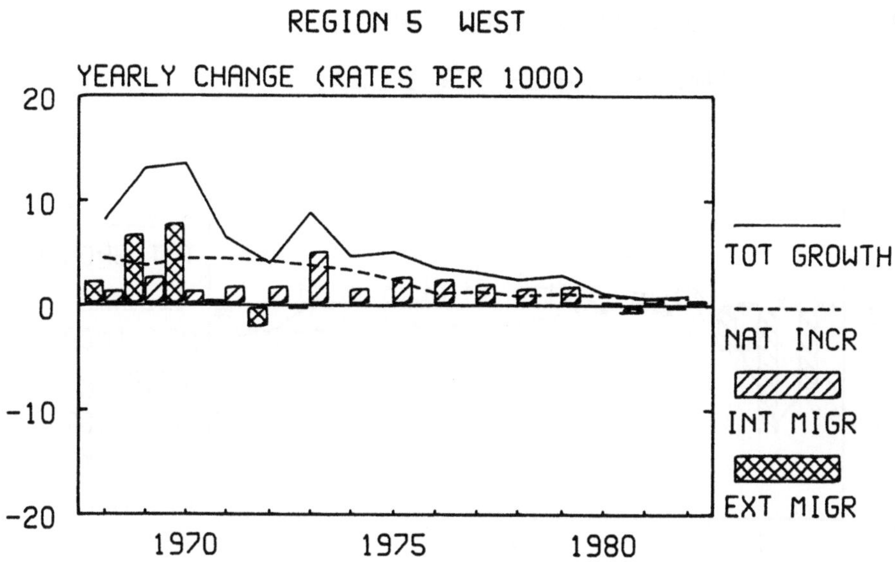

f

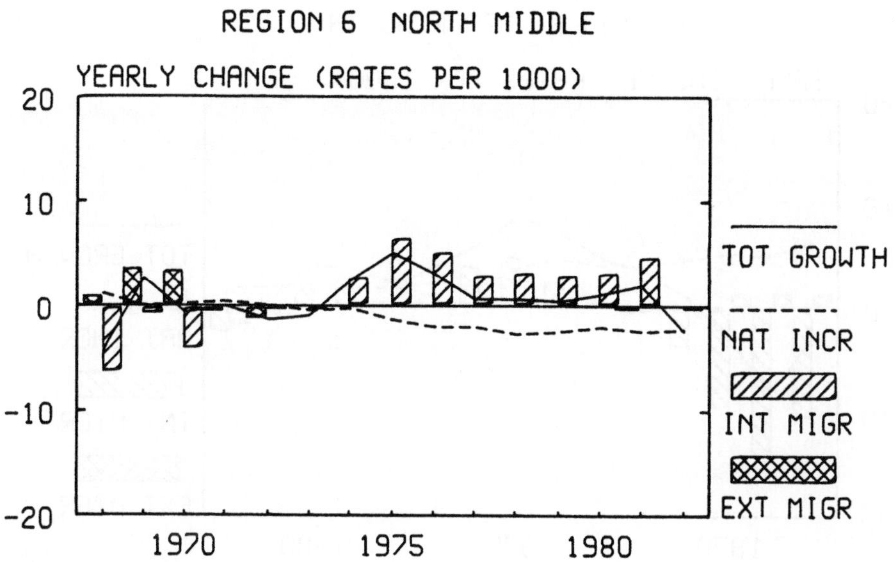

Fig. 9.3c–f. Legend see p. 229

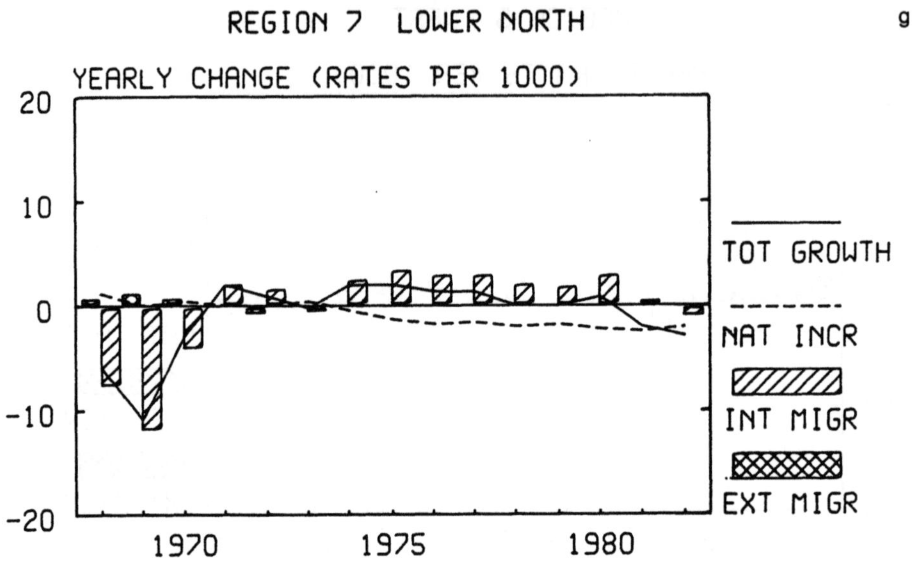

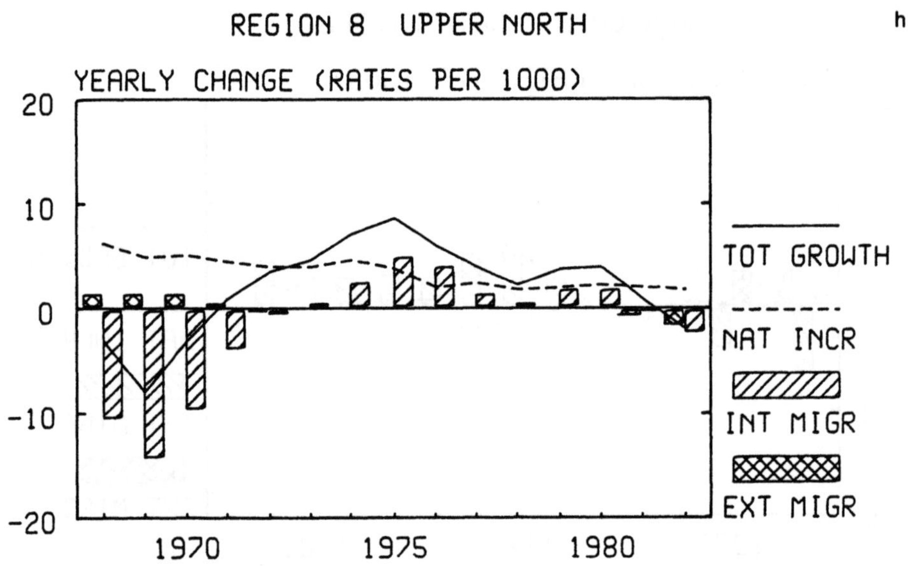

Fig. 9.3g, h. Legend see p. 229

9.3 Reasons for Migration

9.3.1 Labour Market

In analysing labour market performance it is common to consider labour supply and demand, labour disequilibria, labour policy, and, finally, the adjustment process through mobility.

Labour Supply

The labour supply can be measured either by the total number of workers or by the total number of working hours. It is necessary to use and decompose both indices in order to have a true picture of the labour supply growth.

Labour supply, as measured by the number of workers, has increased marginally during the whole period. This increase has two components: the economic one, changes in the labour force participation rates, and the demographic one, the population change.

As can be seen from the following diagram (Figure 9.4) the yearly changes in labour supply due to the demographic causes were on the average less than 1 percent, while the effects of changes in the participation rates were more significant.

Total labour supply, as measured by the number of working hours, shows a negative growth rate up to the year 1978. This is due to the fact that the absolute value of the decrease in the average working time and the time of absence was larger than the increased working time contributed by the number of new workers entering the labour market.

Later in the model application we will use the labour supply measured in terms of the number of persons, although, this might not be the best measure of labour

Table 9.2. The total labour force age 16–65, 1976–1982. Percent change from previous year
Source: Statistics Sweden, 1976–1982 Labour Force Surveys, Basic Tables, Yearly, Averages

year	1976	1977	1978	1979	1980	1981	1982
1. Labour force measured by number of persons	0.4	0.6	0.8	1.3	1.1	0.7	0.4
2. Average working hours per week	-0.6	-1.2	-1.9	-0.1	-0.3	-0.3	0.4
3. Absence	-0.3	-0.7	-0.2	-0.5	-0.6	0.2	0.1
4. Labour force measured by number of hours	-0.5	-1.3	-1.3	0.7	0.2	0.6	0.9

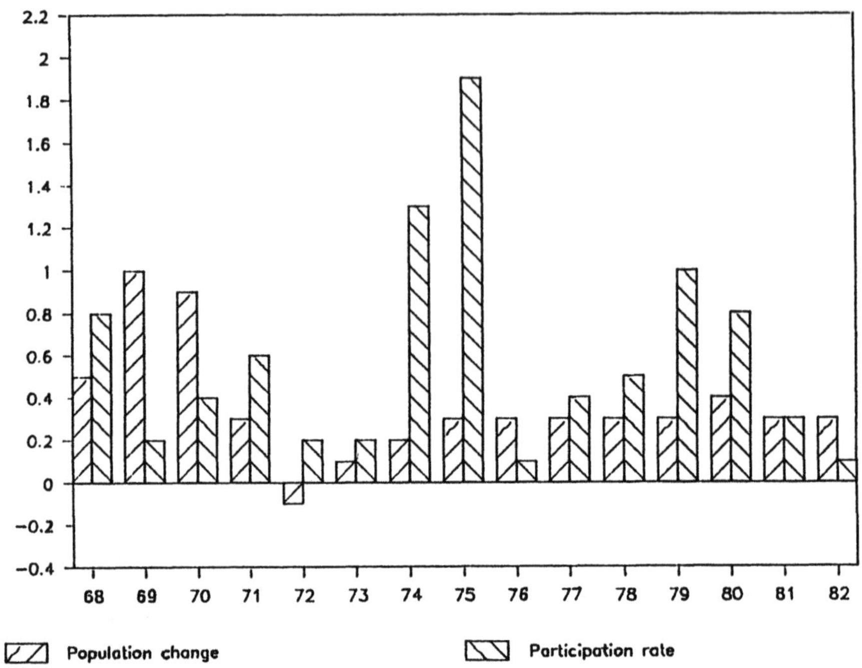

Fig. 9.4. Labour supply changes
Source: Statistics Sweden, 1968–1982 Labour Force Surveys, Basic Tables, Yearly Averages

supply. But in order to be consistent with other variables like migration, unemployment and vacancies, we have chosen to measure labour supply in terms of number of people.

During the same period, the labour force participation rate classified by sex and by region had the following characteristics: The male participation rates decreased proportionally less than the increase in female participation rates, and the regional distribution of the female labour force participation rates became more even than the male participation rates.

There are two types of unemployed workers, those covered by unemployment insurance and those who are uninsured. The two groups do not have equal bargaining power against the employers concerning the acceptance of the existing vacancies (reservation wage level and location). Workers who are eligible for unemployment compensation, have lower search costs and therefore their unemployment duration can be expected to be longer, and thus they are less mobile. The unemployment compensation in 1982 was about 80 percent of the average wage of an industry worker, and the maximum compensation period was 300 working days (450 days for older workers). During the period of interest, the proportion of workers covered by insurance has increased from 50 to 75 percent. The number of insured unemployed increased mainly because of the increase in the proportion of older unemployed persons. Heterogeneous behaviour in the mobility of unemployed workers is to be expected, depending on the relative share of these groups and their relation to vacancies. The scaled regional unemployment of Sweden by region is shown in Figure 9.5.

Labour Demand

Industrial employment increased by 38,000 during the 1970–1975 period but decreased by 112,000 in the 1975–1982 period. This decrease was most apparent in textile, shipbuilding and iron-ore mining industries. In the period 1980–1982 metal industry, non-metallic mineral products manufacture, wood products and rubber and plastic products manufacture have reduced their employment. In this context, we can observe that government supported areas (South Middle, North Middle and Upper North) have had constant or increased employment in industry.

Industrial employment declined not only because of the decline in industrial sales on the world market, but also because of the increasing use of capital intensive technologies. A more detailed description of employment growth by region and by branch can be seen in Table 9.3.

The number of employed in the agriculture/forest sector declined during the period 1970–1982 by 25 percent on the average; except in the Upper North and Lower North regions where the numbers show an increase by 35 percent. The smallest decline was observed in the Stockholm region – a 15 percent decrease. The growth of the private service sector was above the national average of 25 percent in

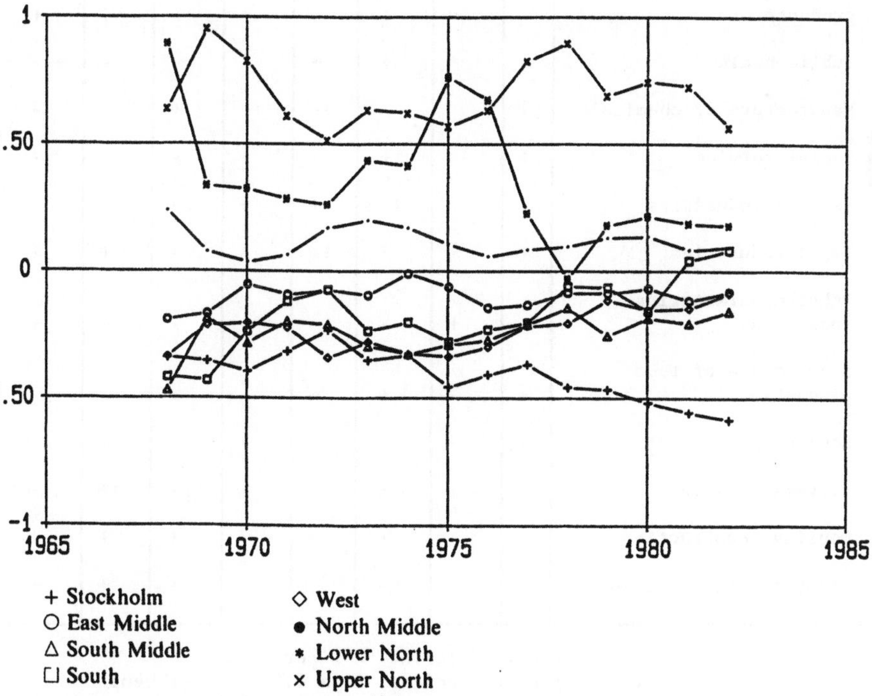

Fig. 9.5. Scaled regional unemployment of Sweden
Source: Statistics Sweden, 1968–1982 Labour Force Surveys, Basic Tables, Yearly Averages

Table 9.3. Trends in employment distribution by region and by sector in the period 1972–1979
Source: Derived from F. Snickars' calculations (1981) using Labour Force Survey data of Statistics Sweden.

Sector	West	North Middle	South Middle	South	Lower North	Upper North	East Middle	Stockholm
Public administration	+	+	+	+	-	+	+	-
Whole sale	+	+	+	-	+	+	-	-
Transport	+	+	+	+	-	-	+	-
Iron and steel industries	-	-	+	-	+	+++	+	-
Manufacture of fabricated metal products	-	+	+	+	+	-	-	+
Retail trade	+	+	-	+	+	+	-	-
Forestry	+	+	+++	-	+	+	-	++
Agriculture	+	+	-	+	-	-	+	-
Manufacture of wood products	-	+	+	-	+	-	-	+
Public health	+	-	+	+	+	-	-	---
Manufacture of chemicals	-	-	-	--	+	+	+	-
Social service	+	-	+	-	+	-	-	-
Textile industries	-	+	++	+	+	-	-	-
Paper industries	+	-	-	++	-	-	+	-
Other manufacturing industries	-	+	-	-	+	-	+	+
Manufacture of food products	+	-	+	+	-	-	-	--
Education	+	-	-	-	-	-	++	-
Private service	+	-	-	-	-	-	++	-
Housing construction	-	-	-	+	-	+	+	+
Other public services	-	-	-	-	-	-	+	+++

Notes: +++ (---) = growth (decline) > 1 per cent
++ (--) = growth (decline) from 0.5 to 1 per cent
+ (-) = growth (decline) from 0 to 0.5 per cent

the Upper North. The Lower North and West regions showed an increase of approximately 35 percent in the same period, while the South region had a slow increase of only 12 percent. In the public sector a regional employment redistribution from more dense to less dense regions is deliberately pursued. The public sector employment in the Stockholm region has been reduced from 22.3 to 20.9 percent of the total public sector employment in Sweden in the above period while the primary centers increased their share with 1.1 percent. The growth of scaled public sector employment (see Figure 9.6) was quite high, 1.6 percent yearly in the 1970s, while it decreased by 0.8 percent in 1982.

In Figure 9.6 the scaled public sector employment distribution of Sweden is shown. The county regions of Upper North, Lower North, South Middle, North Middle, and South are relatively constant and at a low level during the whole period. East Middle is also constant but at a relatively high level of public sector employment, while Stockholm county region has lost relative employment in the second half of the period. The county region of West is the only region which has increased its regional share.

In the model application we measure labour demand in terms of vacancies because the demand can be expressed as the sum of the number of people employed and the number of vacancies. On the other hand, the labour supply is the sum of the number of people employed and the numer of people unemployed, which means

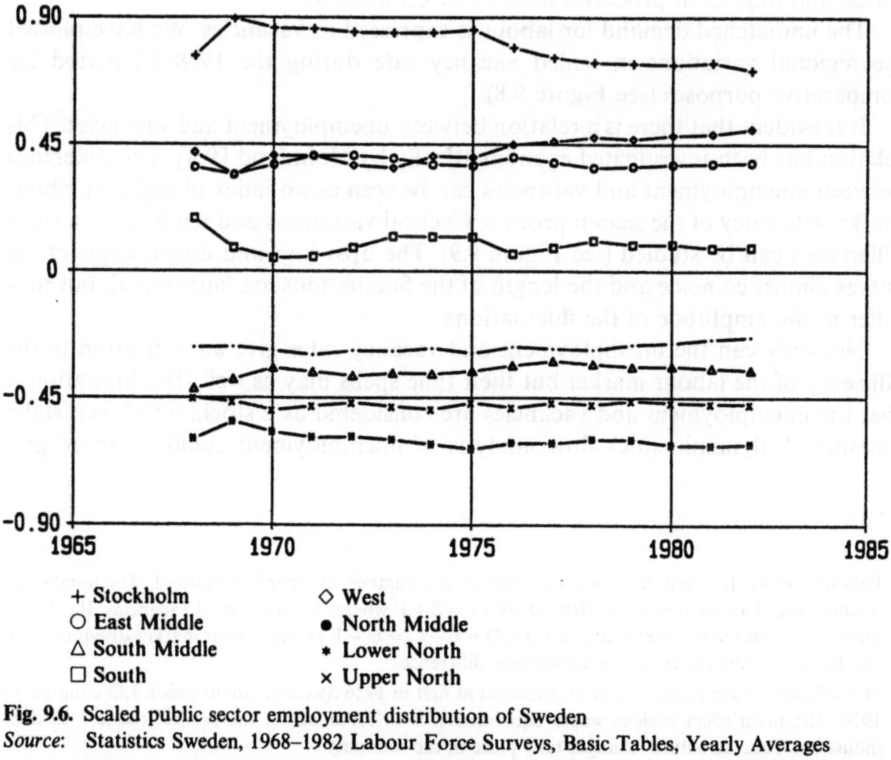

Fig. 9.6. Scaled public sector employment distribution of Sweden
Source: Statistics Sweden, 1968–1982 Labour Force Surveys, Basic Tables, Yearly Averages

that the labour supply can be expressed by the number of people unemployed for the study of labour market disequilibria.[2]

Labour Market Disequilibria

In traditional economic analysis (see Richardson [9.2]), inter-county migration reflects the differences in regional labour markets, in terms of wages, vacancies and unemployment. In Sweden, the regional wage differential for the same work has been reduced by the so called solidarity wage policy: The same wages should be paid for the same work regardless of sector, region etc. The solidarity wage policy would lead to a long-run equilibrium wage structure in the labour market.[3] Of course, there are still differences in fringe-benefits, but such benefits could not be accounted for, due to the lack of statistical data. Consequently, interregional migration is probably more associated with work changes in labour markets, unemployment and vacancy levels rather than regional income differentials (see Figure. 9.7)

Economic theories emphasize two labour market states: employed and unemployed. A third state, not in the labour force has mostly been neglected. The transition rates between these three states are highly dependent on the age-sex structure of the population. Women do not stay as long as men in unemployment, they rather leave the labour force (see Persson-Tanimura [9.3]).

Possible relationships between the transition probabilities of the labour market states, and migration processes need to be established.

The unmatched demand for labour is expressed as vacancies. We have plotted the regional variations in scaled vacancy rate during the 1968–82 period for comparative purposes (see Figure 9.8).

It is evident that there is a relation between unemployment and vacancies. This relation has been investigated econometrically by Holmlund [9.4]. The difference between unemployment and vacancies can be seen as an index of regional labour market efficiency of the search process. Cyclical variations and tendencies in these differences can be studied (see Figure 9.9). The upswings and downswings of the curves almost coincide and the length of the fluctuations are fairly equal, but they differ in the amplitude of the fluctuations.

Not only can the unemployment and vacancy rates give an indication of the efficiency of the labour market but their time spells may as well. The limitation is that the unemployment and vacancies are considered as a stock, which is a static measure. A dynamic stock-flow analysis of unemployment could be more pre-

---

[2] Labour supply is given by $L = E + U$ where: E = number of people employed, U = number of unemployed. Labour demand is defined by $L_D = E + V$ where: V = number of vacancies. By elimination of E from the two identities, we get $LM = L - L_D = U - V$ i.e. the labour market disequilibrium can be expressed as unemployment-vacancy difference.

[3] The solidarity wage policy has been discussed at first in 1936 Swedish labour union LO congress in 1936. The main effect besides wages equalization is to force out of the market the less efficient industries. Thus, structural changes take place in the economy.

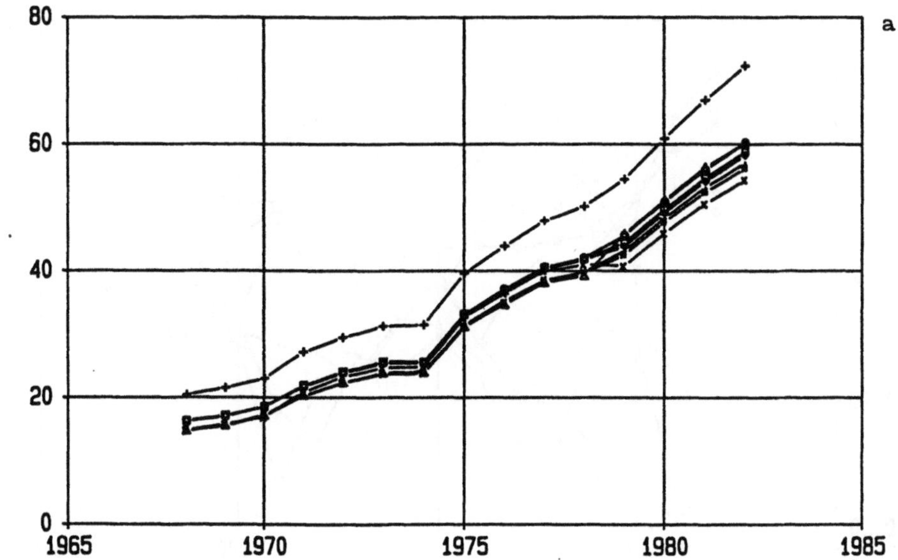

+ Stockholm
○ East Middle
△ South Middle
□ South
◇ West
● North Middle
* Lower North
× Upper North

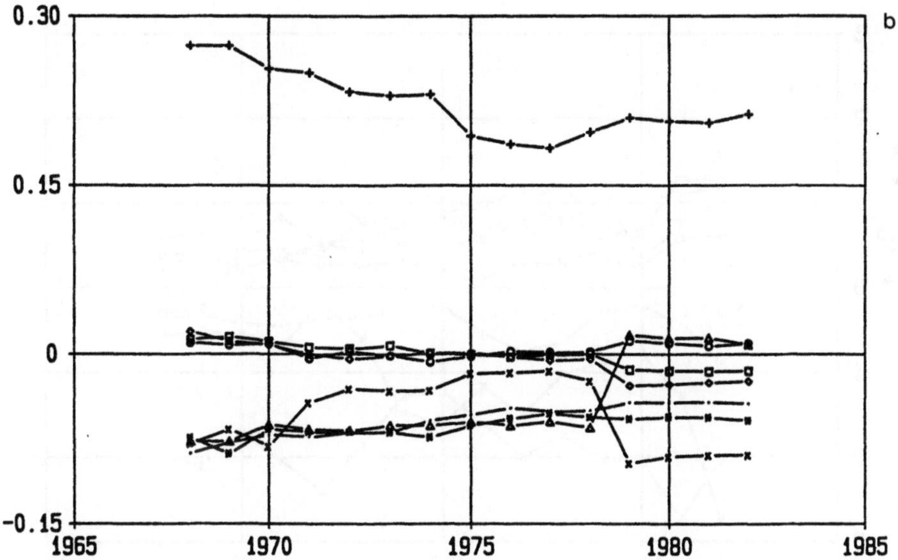

+ Stockholm
○ East Middle
△ South Middle
□ South
◇ West
● North Middle
* Lower North
× Upper North

Fig. 9.7a, b. Regional income per capita of Sweden (a) in 1000 Kr., (b) scaled
Source: Statistics Sweden, SOS, Inkomst och förmögenhet (income and wealth) 1968–1973, and Statistical Messages, N series 1974–1982

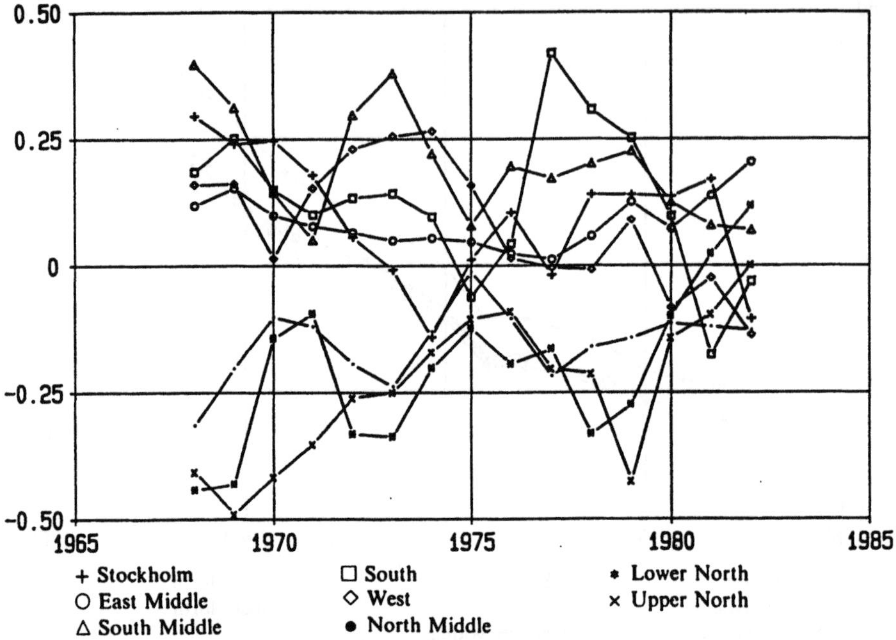

Fig. 9.8. Scaled regional vacancy rate of Sweden
Source: AMS, Arbetsmarknadsstatistik (Labour Market Statistics)

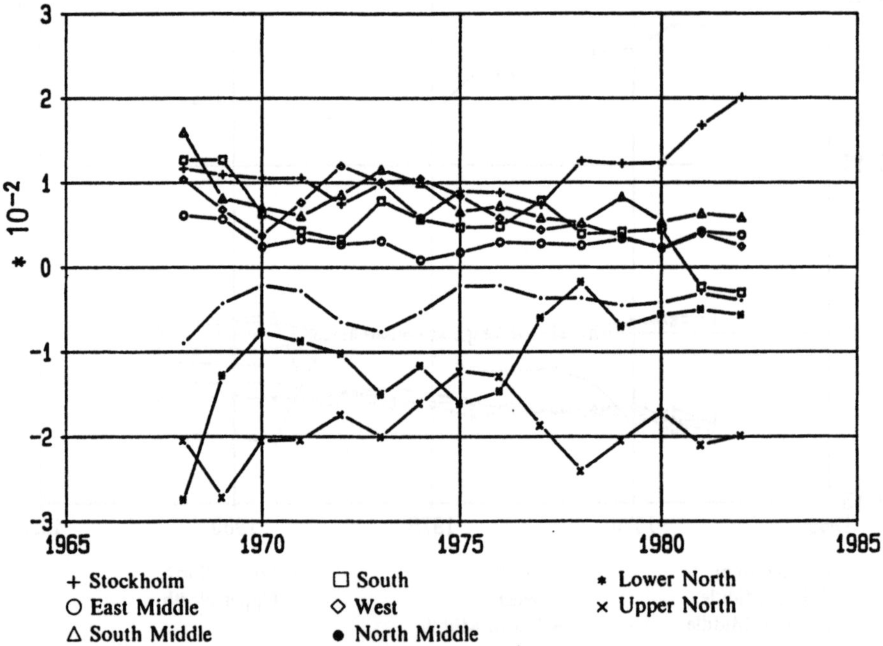

Fig. 9.9. Scaled difference between unemployment and vacancies by region in Sweden
Source: Statistics Sweden, 1968–1982 Labour Force Surveys, Basic Tables, Yearly Averages and AMS, Arbetsmarknadsstatistik (Labour Market Statistics)

ferable theoretically, because it explains both the distribution of unemployment and the unemployment spells.[4]

In less densely populated areas, and especially in Northern Sweden, the unemployment and vacancy spells are much longer than in the rest of the country.

Despite the distortions caused by labour market regulations and interventions unemployment and vacancies rates can be seen to be highly dependent on variations in the growth of GNP.

The most interesting question in the study of labour market disequilibria concerns the extent to which migration eliminates the regional unemployment and wage differentials, under the assumption of a perfect and a free information system. The consensus is that in the long-run equilibrium state, the expected wages (i.e. employment share times wages) will be equalized in all regions. But in reality, the long-run regional labour market disequilibrium, which creates migration flows, does not necessarily lead to a general equilibrium. There are quite a few arguments to advocate for such a statement.

The main micro-economic argument lies in the dichotomy of migrants as producers and consumers. The large out-migration from a high unemployment region does not only reduce the demand and supply for local goods and services and threatens the existence of the old enterprises but it also leads to negative market signals that might deter new enterprises to locate in the region.

The regional market consists of sub-markets of different professions, in which the mobility is limited by the time-money costs of on-the-job training or education. The number of these sub-markets, or the degree of heterogeneity of the labour market, is negatively correlated with the propensity to migrate.

The main macro-economic argument (in a narrow sense), is based on the existence of an unequal distribution of location-specific public goods e.g. the existence of cultural and educational institutions and amenities e.g. favourable climate which may cause regional differences in attraction. As has been pointed out by Å. E. Andersson [9.5], the public goods enter in the production function of private firms as environmental accessibility factors which are important for increasing productivity. High technology industries, for instance, seem to locate close to education centers. In the literature, geographic movement to express one's preferences for public goods is referred to as "voting with one's feet" and it was originally developed by Tiebout in 1956[5].

---

[4] The length of these spells may be measured as $U = F * D$, where F is the frequency (percent of labour force per week) and the D is the duration (in weeks). Since D has increased while F has been rather steady, this means that more people have suffered from unemployment for longer periods

[5] Tiebout in his article argued that Samuelson and Musgrave thesis was not right, i.e. the impossibility of right allocation of the expenditure on public goods. "Policies that promote residential mobility and increase the knowledge of the migrant-voter will improve the allocation of government expenditures on public goods."

Labour Market Policy

Sweden has an active labour market policy in the form of programs for labour market training, relief work, sheltered workshop[6], relocation subsidies etc. Approximately 3 percent of Swedish GNP was the cost of labour market policy by the end of 1970s. In Statistics Sweden these workers are classified as employed. In reality, this is a kind of institutionalized unemployment. If the figures of institutionalized unemployment are added to the open unemployment the unemployment rate would be at least doubled. In Figure 9.10, the counter-cyclical character and the growing importance of unemployment policy measures can be observed. The aforementioned labour market measures influenced mostly the labour supply, and they are therefore classified as labour supply oriented policies.

In the 1960s Swedish labour market policy was labour supply oriented, in order to promote an optimal resource allocation. But in the 1970s it changed and became labour demand oriented, the main goal being to minimize lay-offs in the private sector by financial support to firms. The main measures were stock-piling support, industrial orders, wage and hiring subsidies, The aim of this policy was to protect

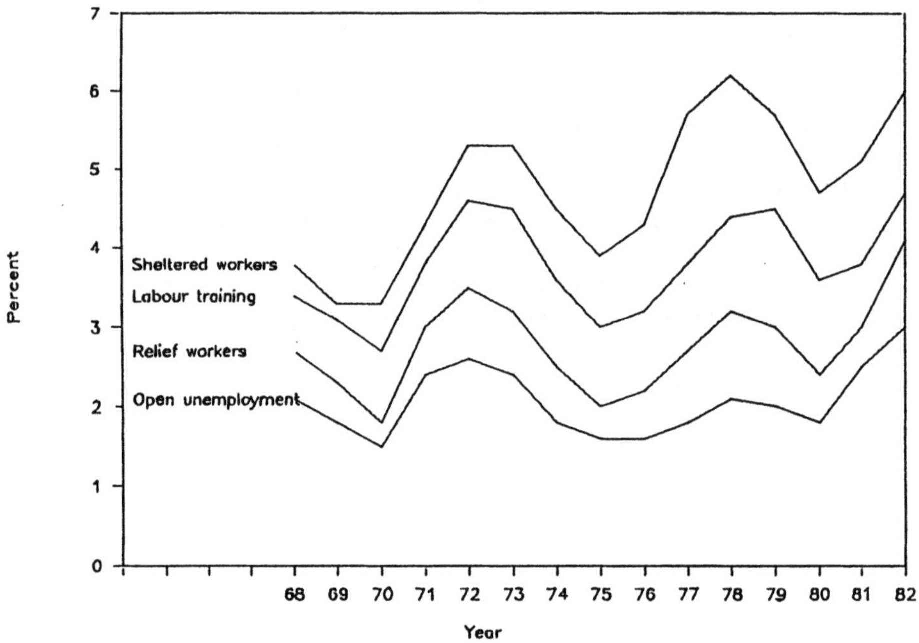

Fig. 9.10. Employment creation measures in Sweden
Source: Sheltered workers, labour market training, and relief workers are from AMS, Arbetsmarknadsstatistik (Labour Market Statistics). Open unemployment is from Statistics Sweden, Labour Force Surveys:, Basic Tables, Yearly Averages.

[6] Employment in sheltered workshops can be arranged for persons with occupational handicap who cannot obtain a job in the open market.

the existing labour force but it did little to facilitate new entrants e.g. youth and married women with long labour absence.

To sum up, the labour supply oriented policy increased labour mobility, but enlarged the regional attractivity differentiations, while the opposite is true for demand oriented policies. Because of the simultaneous existence of these labour policies, the effects on labour mobility and regional attractivity differentiations are not clear.

Generally, a regional employment policy might be able to catalyse the equilibrization process by encouraging firms to move to high unemployment regions at first, and by encouraging people to migrate later.

Labour Mobility and Global Mobility of Sweden

Labour mobility can be measured in three different ways; viz.; job mobility, labour turnover and migration. Job mobility, i.e. the number of employees changing jobs, divided by the total number of employees is a measure of employments changes. Labour turnover is consisting of quits, layoffs and new recruitments. Migration is an index for spatial changes of labour force.

There are two well-known theoretical approaches in the study of labour mobility. The first one is the search theory, according to which a worker can increase his earnings only by changing job, even at the cost of unemployment. According to this theory unemployment gives a person the possibility to search full-time in the labour market for a new job. Distinction has been made between stepping-stone jobs and permanent jobs, and how they affect the mobility under the assumption that human capital is ignored and the search and moving costs are given. A worker is mobile only if the discounted expected income of a new work is larger than the sum of his search costs and the discounted income of his present work.

The second approach is the human capital theory according to which a worker can increase the earnings through human capital accumulation. Education may be considered as either a general or a firm specific investment, depending on whether or not the worker's productivity increase is general or firm specific. The basic hypothesis is that job changes and thus mobility are increasing functions of a worker's general level of human capital, but decreasing functions of his level of firm-specific experience.[7]

However, one may consider labour mobility as a function of both the search process and the level of human investment. It is also worth mentioning the so-called contractarian approach, i.e. employers and employees agree on a contract, which guarantees higher wages at the cost of uncertain employment. Such contracts are motivated by reduced information and transaction costs. This type of contractual arrangement may become more common in the near future and the contractual approach therefore, become a more representative framework for analysing labour mobility.

[7] For a deeper analysis of labour mobility in Sweden see B. Holmlund [10].

As we have seen earlier, global mobility is the mean value of all interregional mobilities. It is a measure of labour mobility in space. In the period 1968–1982, the global mobility was going down with an average rate of 1.7 percent per year, as can be seen in Figure 9.11.

The reasons for a declining mobility of population may be attributed to changes in distribution between casual and ordinary workers, in the age distribution of the labour force, in vacancy trends, in the production technology, in social norms and institutions, and in commuting and migration propensities.

It is a well-known fact that casual workers are more mobile than ordinary workers. During the 1968–1982 period the number of casual workers has decreased (from about 1 to 0.8 million), and so had their mobility index. The number of ordinary workers increased and on the other hand their mobility index has remained constant.

The mobility index is a declining function of age. In 1974 the mobility index for ages 20–24 was 6.5 times higher than the index for the 55–65 age group in 1974. The number of employed in the oldest working age-group (55–65) has declined because of changes in the retirement age. The decrease in the youngest working age group (16–19 years) could be attributed to better educational opportunities and for institutional reasons (e.g. youths under 18 years old can not be employed in two-shift jobs).

The demand for labour as expressed by the number of vacancies, and job mobility are mutually dependent. An increase of the vacancy rate, will lead to an increase of work changes and vice versa. There is a stable relationship between job vacancies and work changes, and a multiplier[8] of 3.5 has been observed. In other words, for every 100 vacancies, 350 job changes were generated up to the end of the 1970s (see SOU [9.6]). Labour demand has decreased in urban but increased in rural areas as a consequence of labour market support programs like early pensions, public works etc.

There may be a relation between global mobility and the rate of growth of GNP. Generally, a positive relation has been observed in pooled cross-section and time series data for different countries.

It has been shown by Sääski [9.7] that the elasticity of global mobility with respect to GNP is around 0.5. However, the direction of causality in this observation has also been questioned. The reduced mobility may rather be the consequence than the cause of the slower growth of GNP in the 1970's, resulting in a less efficient allocation of the labour force.

The technological advancement of the industry has eliminated simple and monotonous jobs, and on-the-job-training, internal recruitments and internal education have become even more important as costs connected with new recruitments increased. As a consequence, the employers screen the labour force closer than previously, searching for long-time-employments and avoiding the earlier prevalence of unskilled workers. The marginal workers (teenagers, students, house-

[8] Multiplier is a coefficient showing how great an increase in mobility results from each increase in vacancies.

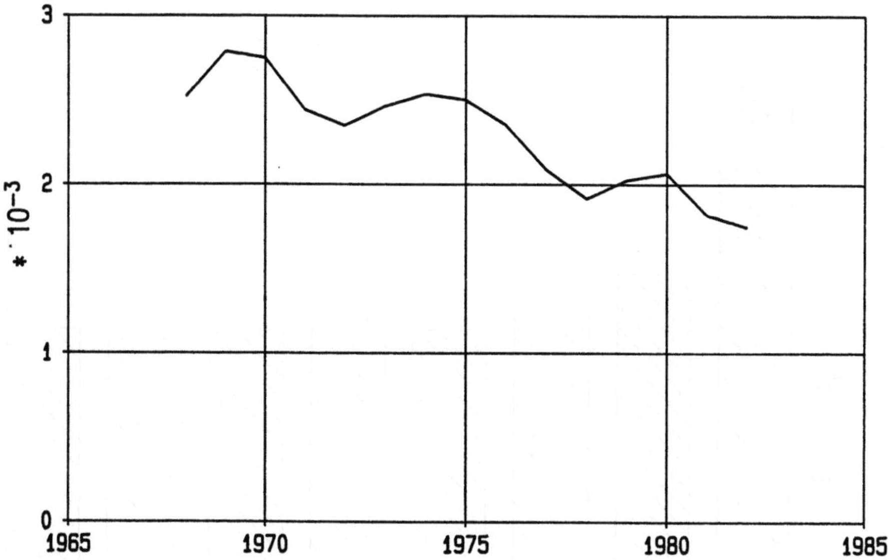

Fig. 9.11. The global mobility of Sweden

wives seeking gainful employment) have experienced increasing difficulties when entering the labour market.

Among unemployed workers a gradual change in the social norms can be observed. The unemployed (especially the young) nowadays prefer temporary relief work in home communities rather than a permanent job in far away industries[9]. The employment security laws (the Åhman laws of 1974) drastically decreased the new recruitments[10].

The transition from one-bread winner families, to two-bread winner families will reduce family migration probabilities, calculated as the product of the husband's migration probability and the wife's migration probability. Finally, the growing ownership of single family houses inreases mobility transaction costs.

Especially a person moving from a village to the city faces the following two problems:

a) housing demand in his home region is very low
b) housing price at his destination region is much higher. The interregional price differentiaion for housing can be seen in Figure 9.12.

Commuting will also be taken into consideration as an equilibrium creating mechanism similar to migration. But in the Labour Force Investigation (AKU), however, the labour force is classified according to place of residence rather than

[9] This statement is from an attitude investigation made by Studieförbundet Näringsliv och Samhälle (SNS) and published in SNS Konjukturrads rapport 1979–1980. "Mot nya förlorade ar?"

[10] The employment security legislation reduced the employers' right to dismiss, and increased the compensation period up to six months.

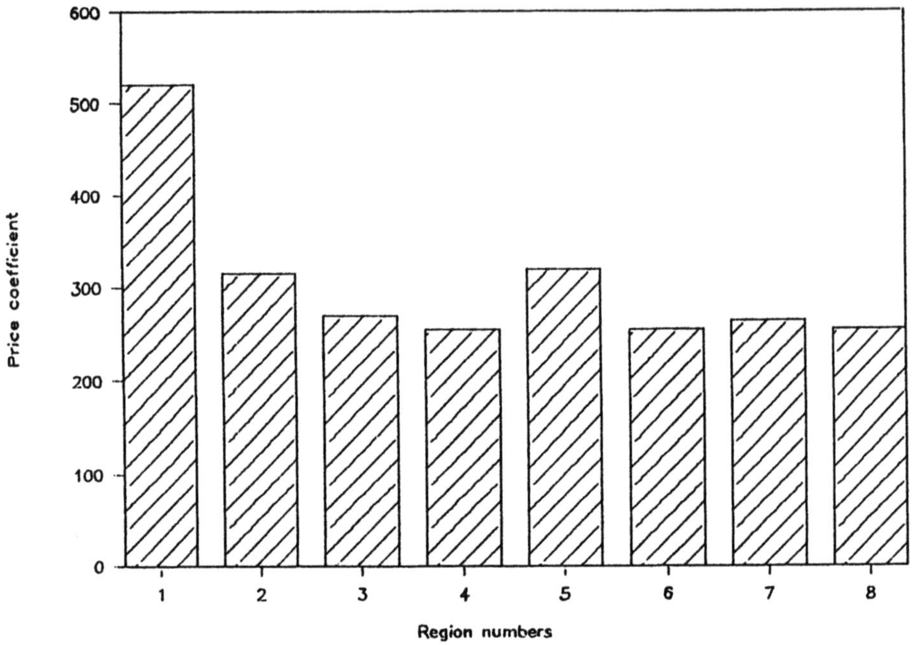

Fig. 9.12. Purchase price coefficient of sold detached and dwelling buildings in 1979
Source: Statistical Sweden; Statistical messages, Series P.

place of work. To quantify the interaction of migration with commuting is impossible with direct data, because commuting data are only collected in connection with the censuses (at 5 year periods) while migration data are collected in connection with labour force surveys (every month). It may be fruitful to use an indirect method for transformation of the commuting data to yearly data, and then to compare those data with the data on yearly migration. Taking commuting into account will increase the mobility index, because the increase in commuting overcompensates the decrease in migration (see Figure 9.13). This means that relations between migration and various explanatory variables found in other countries may not be possible to confirm in the case of Sweden.

Long distance commuting exists in Sweden, for instance between Uppsala-Stockholm and Skaraborg-Gothenburg. This is very important for regional analysis because the development of Uppsala and Skaraborg areas depends primarily on the labour markets of neighbouring metropolitan areas.

In the 1970s the total number of intra-municipality migrants and commuters has increased from 800,000 to 1,000,000.

The propensity to migrate between counties is independent of fluctuations in the economy and has declined during this period. The migration propensity by age and sex is shown in Figure 9.14.

Rather than utilizing economic factors as distinguishing variables in migration analysis, studies by Graves and Linneman [9.8] have revealed that a more appropriate variable may be the stage of the life cycles for each household as a

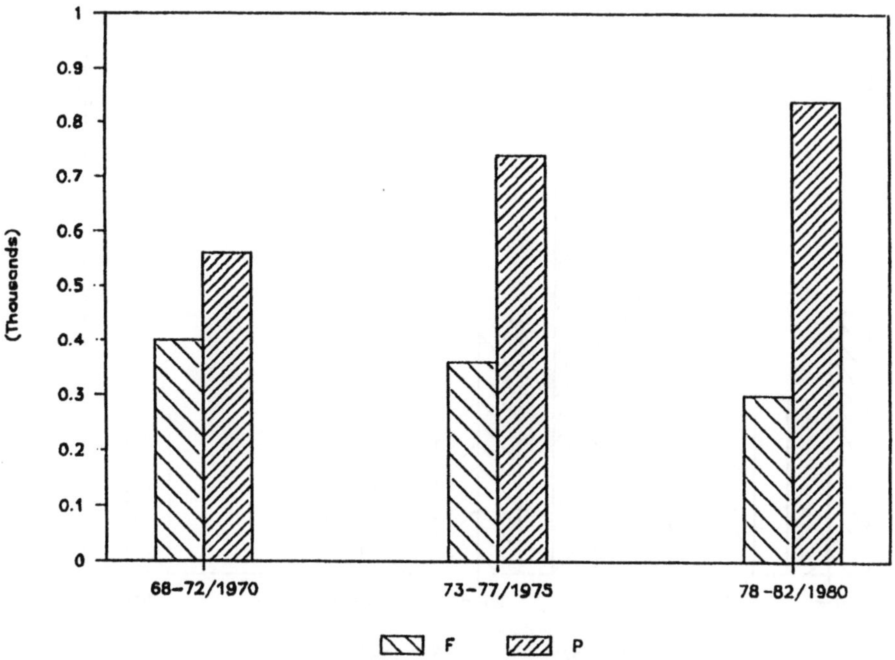

Fig. 9.13. Commuting and migrating. F = number of migrants, P = number of commuters, (F) = number of working migrants
Source: SOU 1984: 74.

descriptor of the migratory process. (By stages in household life cycles is meant for instance the times when basic education is finished, a job has been found, one is married or divorced, has children at home, or away from home, is retired, etc.). Age, fertility and marital situation are together usually taken as "proxies" for "stages" in the household life cycle.

The "life cycle" approach is portrayed in Figure 9.14. The age group 18–24 is the most mobile group for both sexes and the propensity is gradually reduced for older age groups. Women show a higher tendency to migrate than men of the same age-group. But the most important observation is that the declining propensity to migrate during 1972–1982 period is independent of disaggregation level by age and sex. This invariance of aggregation level for population mobility gives us the basis to consider the population homogeneity assumption as more realistic as it seems to be at first sight.

9.3.2 The Housing Market of Sweden

Two types of housing demand can be distinguished: a) Revealed demand is the number of people who have a dwelling plus the number of people who are in queue for a dwelling. b) Potential demand is the revealed demand plus the additional hidden demand attributable to those who have not joined or have withdrawn from

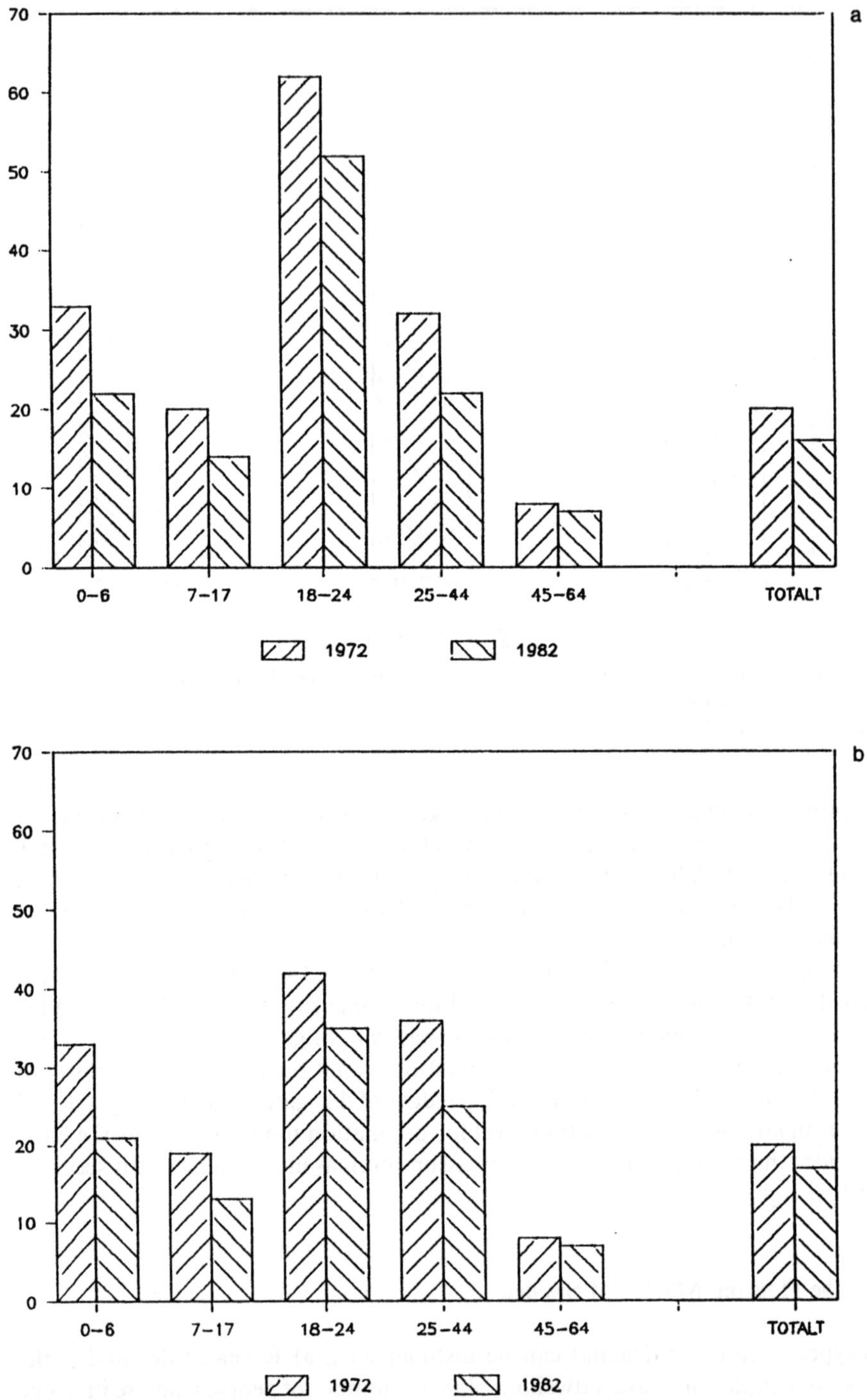

Fig. 9.14. Migration propensity by age and sex. a) women, b) men
Source: AMS, Meddelanden fran Utredningsenheten, 1984:10

the housing queue. Housing vacancy rates and housing quality are probably the most important housing-induced determinants of migration.

By observing the next Figures 9.15(a), (b), the following remarks can be made regarding housing construction:

- the downward trend for all regions in absolute numbers
- the scaled regional dispersion (Figure 9.15(b)) varies considerably
- the relatively large reduction of the Stockholm region (approximately 65 percent).

From a micro-economic point of view people primarily migrate in order to find a job and the availability of dwellings acts as a constraint on mobility. On the other hand, housing construction is of real importance for intra-regional migration on its own merits, since changes in household composition and income take place. But because of the difficulties involved in the collection of the relevant statistical data, we have used housing construction as a "proxy" for both vacancies and higher quality housing.

9.4 Regional Utilities, Variance of the Regional Utilities, Regional Preferences and Migratory Stress

9.4.1 Regional Utilities

The traces of the statistical estimations of the regional utility profiles are exhibited in Figure 9.16. The following comments seems warranted in the context of regional utilities.

The utilities of Stockholm region increased until the year 1972, then remained stationary until 1980, and then finally decreased again.

The two regions West and East Middle had a common growth path until they separated in the year 1974. After this they had stationary fluctuations around an average level. The region South showed the highest increase in utility among all regions until the year 1976 but after that year there were fluctuations around an almost constant level. The two northern regions, North Middle and Upper North, had a development in regional utility which was U-shaped with a minimum in 1975. The upward trend in utility after this year can be explained by the government location support to these regions after 1975. The two remaining regions, South Middle and Lower North, had regional utilities at almost the same level at the beginning of the period, but after 1971 the development diverged and we can observe fluctuations.

The variance of the regional utility and the regional preferences can also be computed from the above estimation, see Figure 9.17(a), (b). It had an upward trend for the utilities (Figure 9.17(a)), but a downward trend for the regional preferences (Figure 9.17(b)). However, the reference period is too small to allow deeper analysis of the relation between this variance and possible business cycles during the period.

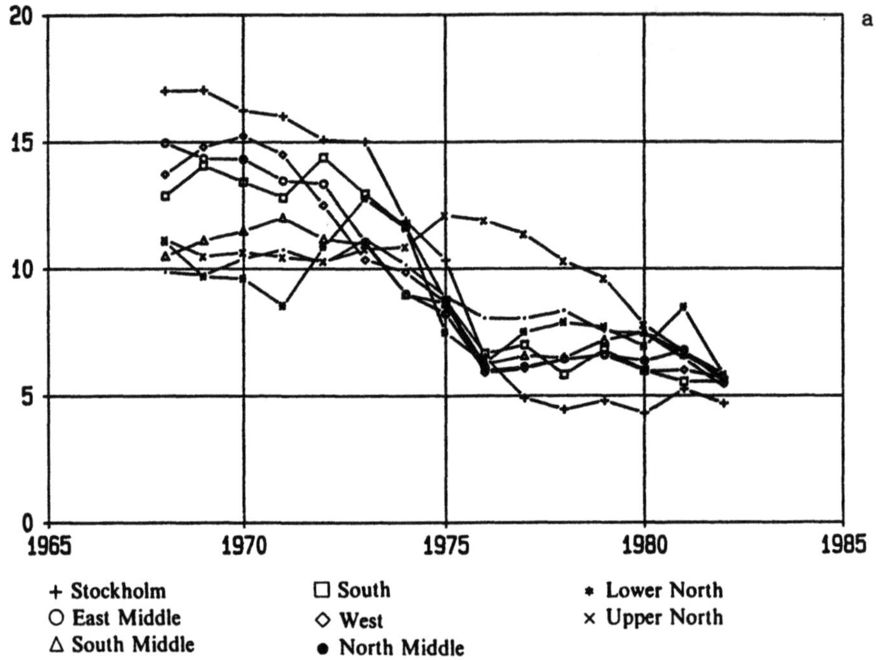

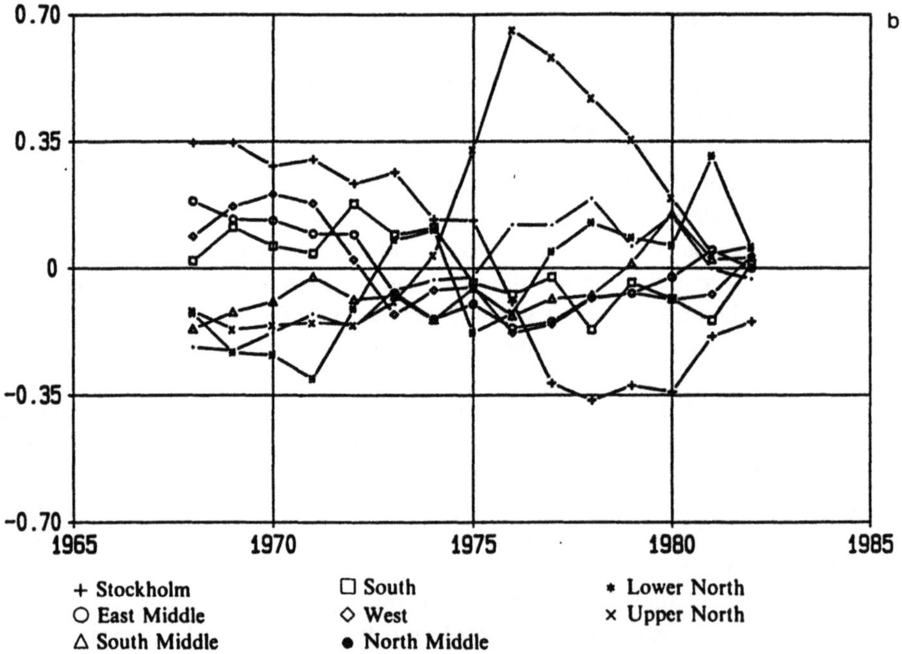

Fig. 9.15. Housing construction by region. a) per 1000 inhabitants, (b) scaled
Source: Statistics Sweden, SOS, Bostadsbyggande 1968–1974, and Statistical Messages, Bo series, 1975–1982.

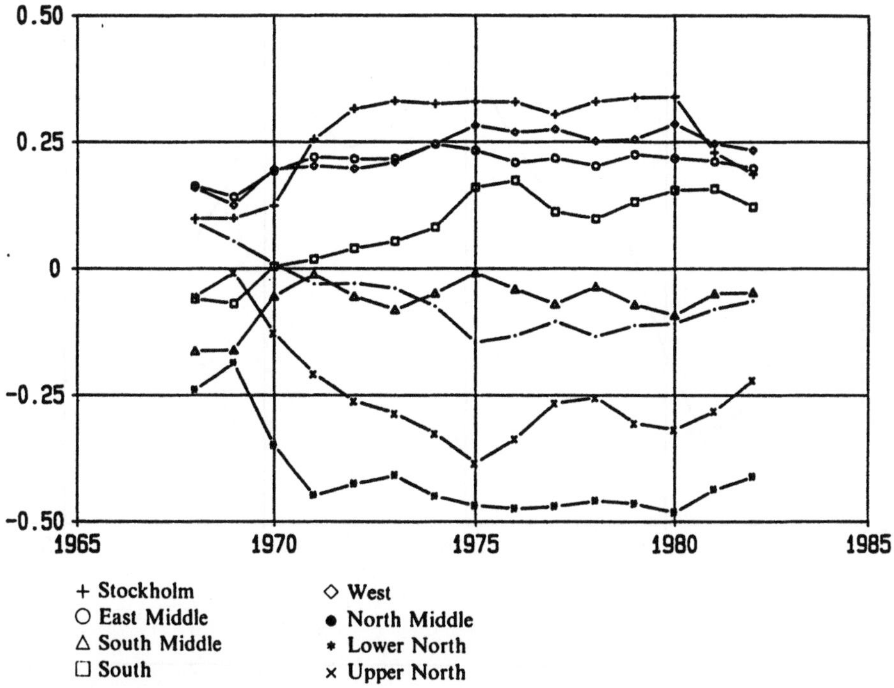

Fig. 9.16. Regional utilities of Sweden

In summary, the utility order reflects the population size order with the only exception the West which during most of the period lay higher in rank then warranted by population size.

9.4.2 Regional Preferences

We have seen in Chapter 1 that the population size independent part of the utilities is denoted as regional preferences.

The results of the estimation of the regional preferences are shown in Figure 9.18.

Regional preferences share a development which is almost parallel to the development of regional utilities. The main difference lies in the relation between the regions with respect to absolute levels.

The regions of high utility like Stockholm, West and East Middle had the lowest level of preferences, while regions with medium sized utility like regions North Middle, South Middle, and South obtain preferences that are among the highest. It is worthwhile to mention that changes in regional preferences $\delta_i(t)$ are mainly due to changes of the socio-economic conditions of the countries, since the population dependent part of the utilities $(\kappa n_i + \sigma n_i^2)$ is only slowly varying with time.

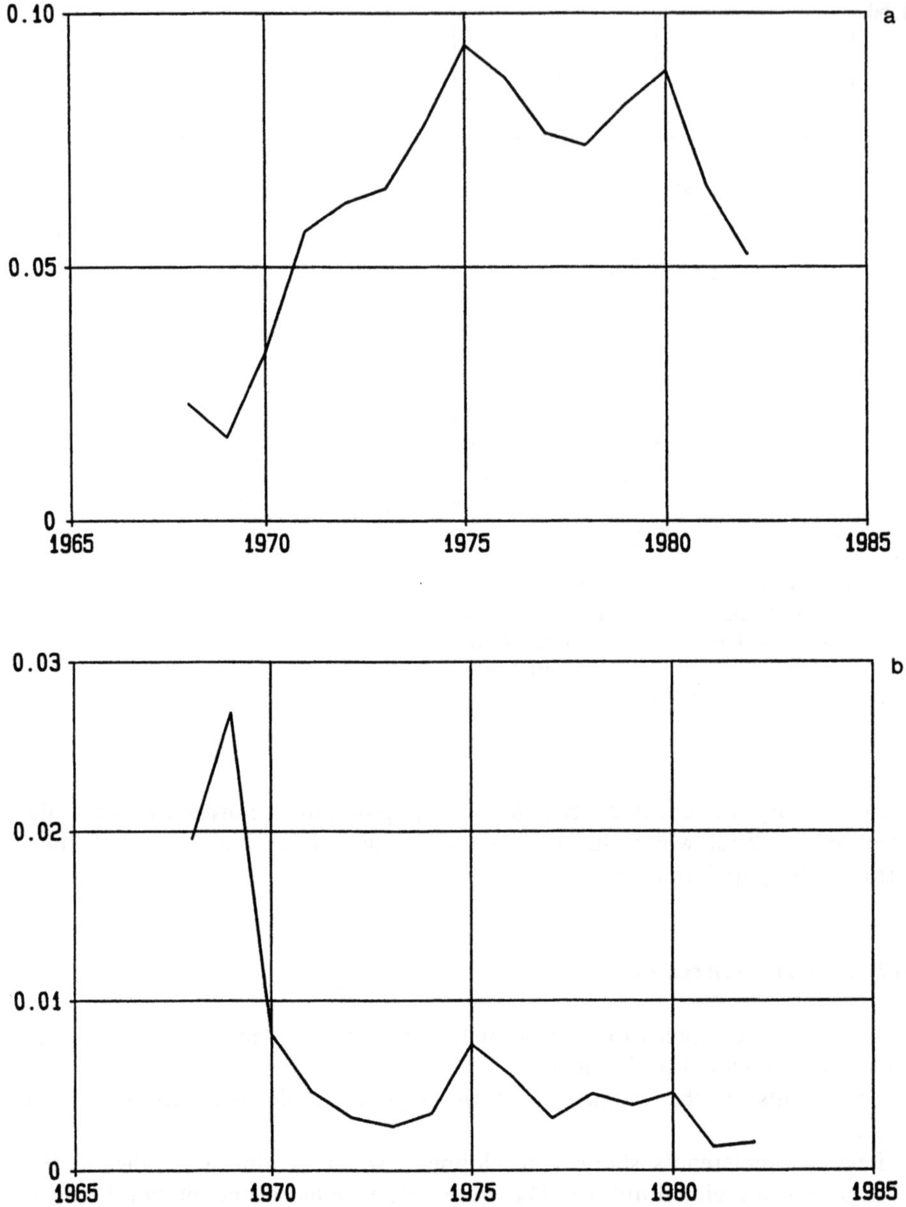

Fig. 9.17. (a) Variance of regional utilities of Sweden, (b) Variance of regional preferences of Sweden

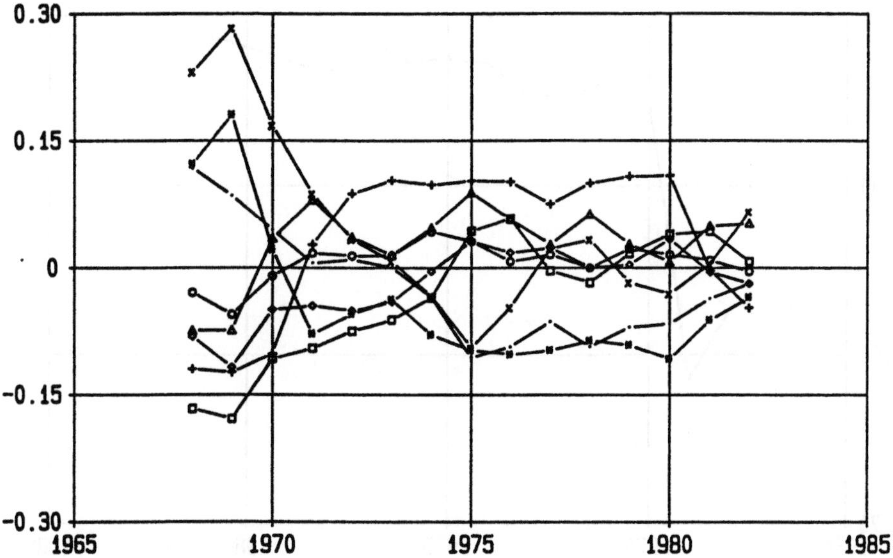

Fig. 9.18. Regional preferences of Sweden

9.4.3 Migratory Stress

The migratory stress, is a formulation of the deviation between the actual and the equilibrium population distribution (see Figure 9.19).

The migratory stress in Sweden tends to zero (in other words, the correlation between $\boldsymbol{n}(t)$ and $\boldsymbol{n}^{st}$ tends to one), which means that the system is approaching equilibrium. In contrary to other countries under investigation Sweden is very closed to its equilibrium state. This is a main conclusion of the case study of the Swedish migration dynamics.

9.5 Choice of Socio-Economic Variables

The master equation and the mean-value equations defined in part one of this book include two important parameters of migration, namely the regional dynamic utilities as a measure of regional attractiveness and the mobility factors. In this section we will introduce a set of socio-economic variables used in the application of the model to the study case of Sweden. The purpose of the next Section 9.6 is to investigate to what extent these parameters may be functions of the selected demographic and economic variables.

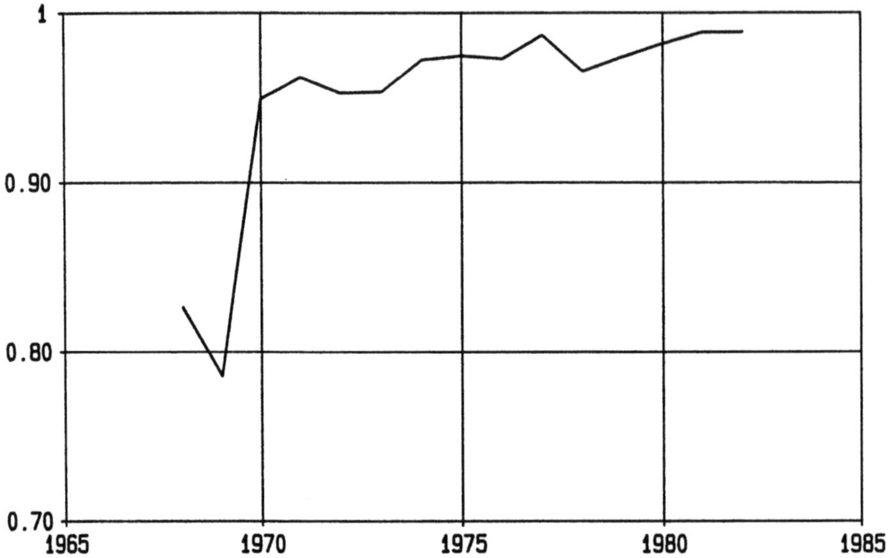

Fig. 9.19. Migratory stress of Sweden

To begin with, the basic hypothesis is that the population is homogeneous, i.e. the same migration propensity for all subpopulations.

The variables that have been chosen below to portray the development in population distribution, are considered to be the main determinants of regional attractiveness and mobility.

All data are spatially disaggregated and available on a yearly base, for the period 1968–1982. The variables are appropriate scaled as described in Chapter 3.

Size-effect Variables

$n_i(t)$ = population numbers *
$n_i^2(t)$ = square of population number *

Labour Market

$E(t)$ = employment (in 100s)
$U(t)$ = unemployment (in 100s)
$L(t) = E(t) + U(t)$ = total labour force (in 100s)
$V(t)$ = vacancies (in 100s)
$LM(t) = V(t) - U(t)$ = labour market variable (in 100s) *
$EPC(t) = E(t) * 100/N(t)$ = employment per capita (in %)
$UR(t)$ = rate of unemployment (in %) *
$LPC(t)$ = labour participation rate (in %)
$VR(t)$ = rate of vacancies (in %) *
$LMR(t) = LM(t) * 100/L(t)$ = labour market index

Public Sector

$PSE(t)$ = public sector employment (in 100s) *

Housing Market

$C(t)$ = new house constructions
$CPC(t)$ = new house constructions per 1000 inhabitants *

Living Standard

$I(t)$ = total income (in 1000 Swedish Kronor)
$IPC(t)$ = income per capita (in 1000 Swedish Kronor)

The variables with a (*) turn out (see Section 9.6) to be the most important key-factors in the evaluation procedure leading to the representation of mobility and utilities in terms of a few sequentially selected key-factors.

As we have seen earlier, in Sweden there are two types of unemployment: the official one and the unofficial one. The latter is an estimate of the total unemployment, being the official level plus institutionalized unemployment (relief work + labour market training + sheltered employment). Early retirement incentive schemes may have the effect that institutionalized component would gradually rise relative to the other forms of unemployment. In the application we have only used the official unemployment figures because:

a) it is available and more statistically detailed
b) it quite well portrays the cyclical variations which are sensitive to aggregate demand (short run unemployment)
c) it is reasonable to use because of our ignorance of the degree of substitutability between ordinary workers and relief workers.

The quality of socio-economic data is not the same for all variables during the whole period. For instance, vacancy registration became gradually compulsory by law after 1976.

9.6 Representation of Global Mobility and Regional Utilities in Terms of Key-Factors

9.6.1 Representation of the Global Mobility

The global mobility is an indication of interregional mobility. As mentioned in the theoretical discussion earlier, the global mobility $v_0(t)$ can be seen as the mean value of interregional mobility matrix $v_{ij}(t)$. We assume that the global mobility may depend on the above described set of socio-economic variables as well as a linear time trend which might appear lagged. The discussion above leads to the definition of the following function for global mobility $v_0(t)$.

$$v_0(t) = a_0 + a_1\ CPC\ (t - T_1) + a_2\ VR\ (t - T_2) + a_3\ UR(t - T_3)$$
$$+ a_4\ I(t - T_4) + a_5\ PSE\ (t - T_5) + a_6\ t\ . \qquad (9.1)$$

Of course, not all variables taken into account in (9.1) are appropriate to describe the time path of the global mobility. In order to find out the key-factors with their appropriate time-lags, the procedure described in detail in Chapter 15 is applied. The analysis is exemplified for the case of Sweden. Therefore we will only present the final results of the estimation.

From the Tables 9.4 and 9.5 the following conclusions can be drawn. The housing commencements, *CPC*, are positively related to the global mobility because they generate housing vacancies and generally raise the quality of the existing housing stock.

The rate of work vacancies, *VR*, measured as the ratio between vacancies and labour force, influences positively the global mobility. It is to be regarded as a "pull" factor. On the other hand, the rate of unemployment is positively related to mobility. This "push" factor is represented by the ratio of unemployment, *UR*, to the labour force. The income *I*, is negatively related to the global mobility, and so is, also, the public sector employment. That entity is represented by the abbreviation *PSE* in the above formula.

The estimated parameters of the mobility function are listed in Table 9.4 and 9.5 below.

a) Using Two Variables
b) Using Three Variables

If we compare Table 9.4 with Table 9.5, the following remarks can be pointed out: i) all *t*-statistics for coefficients are significant except *UR* which is marginally significant (significant at 0.05 level but insignificant at 0.025 level), ii) the *F* statistics are much larger than the tabulated values of *F*. The representation b) gives only slightly better results, than the representation a), because of low *t*-value 1.9 of the last variable *UR*, and the extremely marginal improvement of adjusted R^2. In other words, both the separate contributions of all variables to the explanation of the variation of global mobility and their joint contributions are rather strong, whereas the existence of multicollinearity is obvious.

Is this degree of multicollinearity harmful for the estimation of parameters? A criterion to test this is the following (see Kmenta [9.9]). If the *F* statistics are significant while all *t*-statistics are not significant, then the estimates are imprecise (very large variances and covariances) because round off errors are likely to occur due to multicollinearity. But according to this criterion, multicollinearity is not harmful for our estimation. A possible explanation for the time-lags found in some of the coefficients could be the increase in commuting flows over time. People may move away from the neighbourhood of the work place because of the housing scarcity. On the other hand, the time-lag found in the rate of unemployment can not be explained as an information lag, because Sweden has a very effective labour market information system. One reason for this time-lag could be a corresponding lag in the decision to move. It could also be the effect of unemployment com-

Table 9.4. Mobility function parameters

variable	time lag	coefficient	t-value
CPC	2	7.48×10^{-5}	20.6
VR	0	4.47×10^{-4}	10.7
constant	-	1.04×10^{-3}	-
correlation $R^2 = 0.989$ adjusted $\bar{R}^2 = 0.987$ Fishers $F = 451$ Durbin-Watson $= 2.2$			

Table 9.5. Mobility function parameters

variable	time lag	coefficient	t-value
CPC	2	7.33×10^{-5}	23.1
VR	0	4.18×10^{-5}	11.4
UR	2	5.78×10^{-5}	1.9
constant	-	9.63×10^{-5}	-
correlation $R^2 = 0.993$ adjusted $\bar{R}^2 = 0.990$ Fishers $F = 407$ Durbin-Watson $= 2.4$			

pensations. The time-lag in housing construction could be explained by administrative lags on the part of the housing authorities.

Figure 9.20 shows the development of the global mobility in Sweden between 1968 and 1982. In the same figure the estimated time path via Table 9.5 is depicted. The fit for the estimation period is very close.

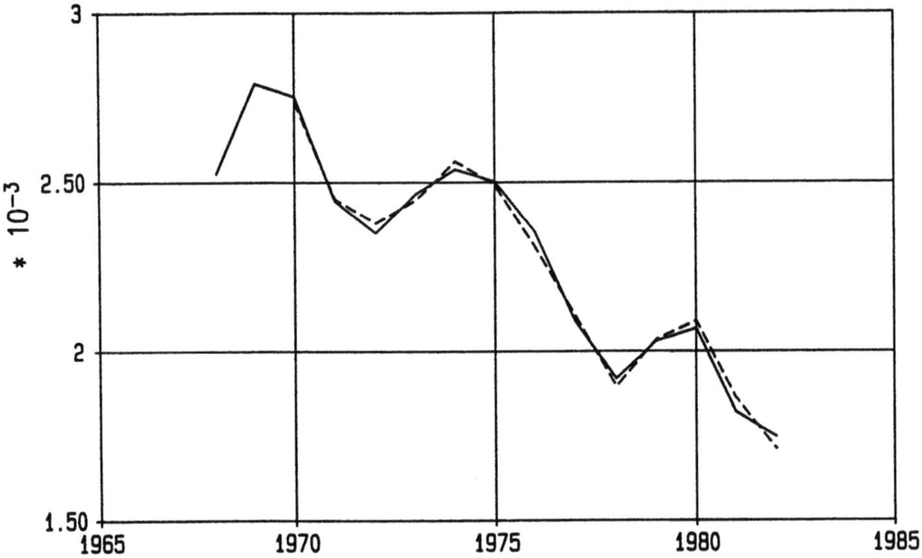

Fig. 9.20. Actual (solid line) and estimated (dashed line) change of the global mobility of Sweden

9.6.2 Representation of the Regional Utility

The measure of regional utility (u_i) may be assumed to be sensitive to a subset of the introduced socio-economic variables in Section 9.5. The motivation for this specification has been given earlier.

The estimated parameters of the utility function are given in Table 9.6.

The table shows that there is a strong correlation between utility and population size. However, the population exponent term was also formed to be significant.

The population is often assumed to be positively related to the regional utility as for instance in the gravity theory, where population size is often taken as a "proxy" for general economic opportunity.

The second power of population, is here as an indication of saturation effects, for instance traffic congestion, high transport costs, as well as costs of living.

The difference between unemployment and vacancies, i.e. $LM = (U - V)$ is a push factor and shows the efficiency of the search process in the labour market. It is negatively related to regional utility. The vacancy rate – which is a pull factor – is positively related to regional utility.

In Sweden, the public sector has been the fastest growing sector during the 1970s in terms of new employment opportunities. Here it is positively related to the regional utility

It is also logical that the stock variables *PSE* and *VR* would exhibit a lagged influence on the regional utility distribution.

The fact that regional income per capita was not found significant confirms the hypothesis that migration is more sensitive to regional labour market differences

Table 9.6. Estimated utility function parameters

variable	time lag	coefficient	t-value
Δn_i	0	1.071	105.3
Δn_i^2	0	-0.355	-10.6
LM	0	-0.014	-7.8
PSE	1	0.166	5.5
VR	2	0.093	3.2
correlation R^2 = 0.971 adjusted $\bar{R}^2$ = 0.969 Fishers F = 653.3			

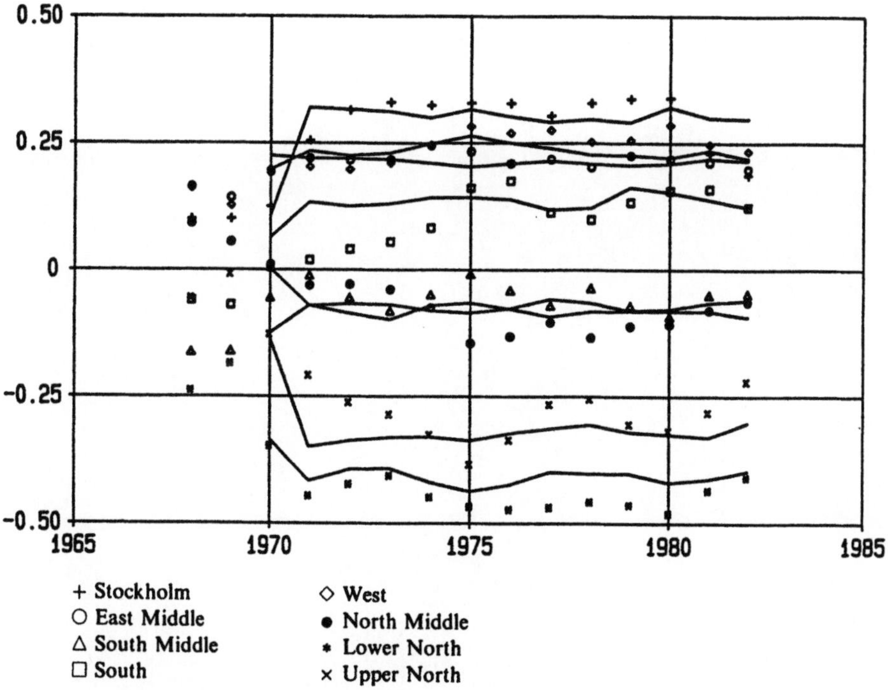

Fig. 9.21. Actual (symbols) and estimated (solid lines) change of the regional utilities of Sweden

than to differences in the regional income. The most probable explanation is that the variation in income between the county-regions is very small. Only the Stockholm region deviates to any large extent. The regional incomes would become even more equal if they are deflated by the cost of living index. Unfortunately this hypothesis could not be tested for lack of data.

The actual and estimated trajectories of regional utilities are depicted in Figure 9.21.

It is interesting to see, that the spatial ranking of the utilities as well as their temporal evolution is relatively satisfactory described. Especially the divergence of the trajectories $u_i(t)$ at the beginning of the period under investigation is correctly traced by the chosen set of socio-economic variables.

III. Comparative Studies

Synopsis of Part III

Part III is devoted to a comparative study of migration in two different respects: Chapter 10 intends to give a representative – though not exhaustive – survey over models treating several aspects of demography, migration and economy, whereas Chapter 11 gives a comparative analysis of interregional migration in the six countries Canada, FRG, France, Israel, Italy and Sweden, which have been studied individually in Part II.

Let us begin with Chapter 10, throughout which two cross-related lines of argumentation are pursued.

Firstly, a variety of *design principles* are exemplified by the models exhibited in this chapter. We mention some examples for such differences of approach. Some of the dynamic models work with *discrete* time steps, others use differential equations with *continuous* time steps. For models focussing in migration, however, this difference should not be decisive.

Some of the approaches – as well as the model used throughout the other chapters of the book – consider the population as a *homogeneous* one. Other models focus on its composition of *inhomogeneous* subgroups as for instance the regional age group model of Section 10.2.

In some of the models the demographic and migratory population evolution is described by *linear* equations, whereas the more complex models of Sections 10.3 and 10.4 include *nonlinear* evolution, in particular of the logistic type.

Finally one must distinguish between *deterministic* and *stochastic* models, where the latter, treated in Sections 10.5 and 10.6 include random fluctuations of the evolution process. The most general stochastic description of a dynamic process like the population evolution seems to be the master equation approach treated in Section 10.6. It is also the basis of the model used throughout this book and is taken up once more in Part IV, Chapters 12 and 13.

The second line of argumentation simultaneously exhibited in Chapter 10 shows, how migration must be seen as part of increasingly complex demographic and economic contexts.

Starting with the linear demographic/migratory model for the evolution of regional age groups in Section 10.2, it is shown in Section 10.3, how nonlinear demographic models can be constructed so, that the effects of economic productivity are integrated. These models are further generalized in Section 10.4 by including interregional migration, so that the interrelated effects of migration, birth/death processes and economy can be studied.

On the other hand the analysis of Chapter 11 starts from a complementary point of view: It takes advantage of the fact that the results of Part II have been obtained by using one and the same migratory model and hence are fully comparable.

Since regional utilities and the global mobility govern the interregional migration in each country according to this model, several natural questions arise; for instance: Can the effect of economic developments of either international scale, like the oil shock, or national scale, like business cycles, be identified in the time dependence of the national utilities and mobilities. What are the appropriate measures to describe, under comparative view, the regional differentiation of a country? Which regional socio-economic factors should be chosen in order to make their correlation to the migratory process comparable between the countries?

Whereas such questions are answered in Section 11.1, Section 11.2 is devoted to the explicit comparative interpretation of the quantitative results. Again the time dependence of the national global mobilities, regional utilities and preferences is the central object of comparison. In this way characteristic analogies but also differences between the countries can be demonstrated in quantitative detail. Furthermore it can be shown in Sections 11.2.2 and 11.2.3 that in all countries only a few key-factors belonging to the same classes of socio-economic variables correlate to the migratory process, whereas other socio-economic variables prove to be irrelevant with respect to migration.

10 Comparative Analysis of Population Evolution Models

Åke E. Andersson

10.1 Introduction and Survey of Modelling Approaches

Demographic analysis is concerned with the evolution of human (and other) populations over time and space. Demography can be subdivided into two branches, descriptive and analytical demography.

Descriptive demography is normally concentrated on classification and statistical characterization of populations according to age, sex, location, income, education, occupation, etc.

Analytical demography is concerned with theories and models of the evolution of populations in time, space, and in the different structural classes of descriptive demography.

Demometrics provides a link between descriptive and analytical demography. It is concerned with the construction and computer application of quantitative models of demographic evolution and the development of statistical methods, useful for estimation of demographic model parameters.

In recent years, demographers, sociologists, and economists have returned to the basic demo-socio-economic problems formulated by Malthus in the late 18th century. Malthus then claimed that productivity improvements would never solve the over-population problem in any permanent way. In the first edition (1798) of "Essay on the Principle of Population" he claimed that the population always tends to grow at a faster rate than the rate of growth of food production. The equilibrium population growth would always tend to be close to a point of permanent disasters due to wars, hunger catastrophes and pests. These disasters would with technological progress cause short periods of improvement in the standard of living, but increased fertility would soon bring the population back to a precarious equilibrium.

The only permanent improvement would, according to Malthus, be possible if the positive feed-back between productivity and fertility could be broken. In later publications Malthus therefore recommended late marriages and celibate.

At the end of the 19th century his analysis formed the basis of the Neo-Malthusian movement, which recommended family planning as the only viable solution to the over-population problem.

The importance of Malthus' contributions consisted in stressing the *interactions* between economic, behavioural and demographic variables. These studies of interactions have developed into two directions:

1. *Non-spatial* econometric and other statistical studies of the relation between the number of births per woman and the income, education, and other social and individual background variables.

Most of these studies have shown that a rising real income increases the number of children per woman, while the level of education has an *opposite* effect [10.1].

Other studies have shown that the consumption of tobacco and other "non-health" products is negatively related to the level of education [10.2].

Education should accordingly both decrease fertility and mortality and consequently have effects both on the growth and structure of populations.

2. *Spatial* econometric studies of migration flows and their economic, social and other environmental determinants. The number of such studies has been increasing during the last three decades. In general these studies have focused on "push and pull factors" or on the influence of transportation, communication and other costs of mobility.

Very often these econometric migration studies have been carried out in "splendid isolation" from demographic model building. This implies that estimation procedures do not permit proper identification of parameters.

Proper demo-economic analysis requires estimation of mobility parameters and modelling of demographic change in a coordinated research effort. The comparative study presented in this book is an example of such a coordinated research effort. This study concentrates on interactions in space between economic, social, mobility costs etc. and population change. Classical demographic analysis has a similar research program, but concentrates on the interactions between age-classes in a non-spatial context. Rogers, Willehens and Ginsberg have provided examples of integration efforts in age-class migration modelling within a similar framework.

Practically all of the demographic or economy-demography models proposed are Markov-models. This means that the time derivative $d\boldsymbol{n}/dt = \dot{\boldsymbol{n}}$ of the population vector $\boldsymbol{n}(t)$ only depends on the momentary parameters, but not on the explicit history. An exception is the Ginsberg model, which is "Semi-Markov". Full Non-Markov models would have to be formulated by integro-differential equations of the type

$$\dot{\boldsymbol{n}} = \int_{-\infty}^{t} K(t-s)\, \boldsymbol{n}(s)\, ds\,. \tag{10.1}$$

10.2 Linear Evolution of Age Groups in Regions – Classical Demographic Analysis

Classical demographic analysis has rarely been oriented to interactions between regions. Rather it is concentrated on the dynamics of *age-class interactions* in one region, closed off from the surrounding world. In order to simplify matters it is often assumed that the temporal inter-dependencies can be captured by a linear one-period recursive system

$$\boldsymbol{n}(t+1) = M\boldsymbol{n}(t); \quad \boldsymbol{n} = \{n_1, \ldots, n_k, \ldots, n_L\} \tag{10.2}$$

where n_k = size of population of age class k and $M = \{m_{hk}\}$. Because of the highly structured pattern of age-class interactions $\{m_{hk}\}$ the matrix M is necessarily sparse and non-negative. It is often called a Leslie-matrix. Slightly generalizing we have

$$\boldsymbol{n}(t+k) = M(t+k-1) \ldots M(t)\, \boldsymbol{n}(t) \tag{10.3a}$$

where M is a Leslie-matrix. If $M(t)=M$, then

$$\boldsymbol{n}(t+k)=M^k\,\boldsymbol{n}(t)\,. \tag{10.3b}$$

A steady state solution[1] requires that

$$\lambda\boldsymbol{n}=M\boldsymbol{n} \tag{10.4}$$

for which a meaningful solution ($\boldsymbol{n}\neq 0$) exists if

$$\det\,[\lambda I-M]=0\,. \tag{10.5}$$

It can be shown that the maximal λ^* is positive if $M\geq 0$ and that the $\boldsymbol{n}$ associated with λ^* has positive components: $\boldsymbol{n}=[\mathrm{n}_1\geq 0,\ n_2\geq 0,\ldots,\ n_L\geq 0]$ (see Frobenius-Perrons Theorem in [10.3]).

The eigenvector equation (10.4) for $\boldsymbol{n}^*$ implies with $\lambda^*=\exp(\beta)>0$, that

$$\boldsymbol{n}^*(t+k)=M^k\boldsymbol{n}^*(t)=e^{\beta k}\,\boldsymbol{n}^*(t)\,. \tag{10.6}$$

That means, the total population grows for $\exp(\beta)>1$ and decreases for $\exp(\beta)<1$, although the ratio between the different age groups remains constant.

If the classical system is generalized to a finite set of regions, (a multiregional system, see [10.4]), the populations of the L different regions are connected by interaction between age-groups as well as by interregional parameters. The regional population vector $\boldsymbol{n}$ is now subdivided into a number of sub-vectors giving the regional distribution of each age-group. Assuming an age classification in 5-year-groups a linear specification of the demographic model requires a generalized Leslie-matrix of the form

$$G=\begin{bmatrix} 0 & 0 & B(a-5) & \cdots\cdots & 0 & 0\\ S(0) & 0 & 0 & \cdots\cdots & 0 & 0\\ 0 & S(5) & 0 & \cdots\cdots & 0 & 0\\ \vdots & \vdots & \vdots & & \vdots & \vdots\\ \vdots & \vdots & \vdots & & \vdots & \vdots\\ 0 & 0 & 0 & & S(z-5) & 0 \end{bmatrix} \tag{10.7}$$

The G-matrix thus contains a larger number of sub-matrices representing the regional migration and survival rates of each age group. If we for simplicity assume two interconnected regions, then we have a representative $S(a)$:

$$S(a)=\begin{bmatrix} S_{11}(a) & S_{21}(a)\\ S_{12}(a) & S_{22}(a)\end{bmatrix} \tag{10.8}$$

[1] It should be mentioned, that this is a more general concept of "steady state" than used in pure migration theory, where the total population over all regions remains constant, and where "steady state" or equivalently "stationary state" means that all n_i are independent of time.

in which $S_{ij}(a)$ is the share of the $(a, a+4)$-aged of region i, at time t, who will be $(a+5, a+9)$ years of age and residing in region j, at time $t+5$. An estimate of $S(a)$ is given by:

$$S(a)=[I+Q(a+5)]\,Q(a)\,[I+Q(a)]^{-1} \tag{10.9}$$

where $Q(a)$ is a matrix of given survival probabilities. For the age class $(z-5)$, we have

$$S(z-5)=2/5\,M^{-1}(z)\,Q(z-5)\,[I+Q(z-5)]^{-1} \tag{10.10}$$

where M is a matrix of given age-specific death-rates. The first row of the G-matrix contains the sub-matrices $B(a)$, which in the two-region-case are:

$$B(a)=\begin{bmatrix} B_{11}(a) & B_{21}(a) \\ B_{12}(a) & B_{22}(a) \end{bmatrix} \tag{10.11}$$

where $B_{ij}(a)$ is the number of children, born in the period and who are living in region j at the end of the period, per $(a, a+4)$-aged of region i at the beginning of the period. $B(a)$ is estimated by:

$$B(a)=5/4\,[Q(0)+I]\,[F(a+5)\,S(a)] \tag{10.12}$$

where $F(a)$ is a matrix of fecundity rates.

All the elements of the generalized Leslie-matrix are constant and thus it is possible to calculate the steady state distribution and growth of the regionally differentiated age-groups by forming the eigen-equation:

$$\lambda \boldsymbol{n}=G\boldsymbol{n} \tag{10.13}$$

where we are again sure to find a steady state solution

$$(\boldsymbol{n}^*, \lambda^*) \quad \text{with} \quad \lambda^*>0\,,$$

and $\boldsymbol{n}^*$ having positive components only. The model discussed so far is a *linear* one. That means, the right hand side of the evolution equation (10.3) is linear in $\boldsymbol{n}(t)$. Therefore, in the linear case, the Leslie-matrix M *must not depend* on $\boldsymbol{n}$. In distinction, *nonlinear models* are defined by the fact, that the Leslie-matrix M, or more generally the transition rates, do depend on the population configuration $\boldsymbol{n}$.

The classical Lotka-Keyfitz population model [10.13] focuses on interactions between age-groups and the multi-regional Rogers' population model is essentially a generalized classical model that *regionalizes* the age-interaction effects. It, however, takes into account the slow-down of migration that occurs after long periods of out-migration when the mean age of the population passes beyond the very typical age-class peak of migration probability as illustrated in Table 10.1 for Sweden (1971–1975).

The classical one-regional and multi-regional model is in a very limited sense a probabilistic model. A proper probabilistic approach requires a probability distribution $P(\boldsymbol{n}, t)$ over the population $\boldsymbol{n}$, whose equation of motion is the *master equation* (see Section 10.6). In this case $P(\boldsymbol{n}, t)$ does *not only* reflect the most probable (mean) $\boldsymbol{n}=\bar{\boldsymbol{n}}(t)$, but also the probability of deviations from the mean-values.

Table 10.1. Dependence of the out-migration frequency on age

Age class	Relative outmigration frequency, 1971-1975
0 - 4 years	11 per cent
5 - 9 "	10 "
10 - 14 "	5 "
15 - 19 "	8 "
20 - 24 "	19 "
25 - 29 "	19 "
30 - 34 "	11 "
35 - 39 "	6 "
40 - 44 "	3 "
45 - 49 "	2 "
50 - 54 "	2 "
55 - 59 "	1 "
60 - 64 "	1 "
65 - 69 "	1 "
70 - 74 "	1 "
75 - 79 "	0 "
80 - ... "	0 "
Sum	100 per cent

10.3 Economy and Demography

10.3.1 Logistic Evolution

In the study of populations the assumption of linearity becomes rather unreasonable in any extended time perspective. The exponential growth pattern inherent in most demographic models cannot be more than an approximation valid for rather limited periods of time. In biological applications (bacteria populations, etc.) a long time perspective often is less than a month. For such population systems the linear approximation approach with detailed age class interaction representations often is unreasonable. In biological applications these considerations have led to *non-*

linear differential (or difference) equations of the rather simple but often realistic logistic form:

$$\dot{N}(t) = aN(t)\,[\sigma - N(t)] \tag{10.14}$$

where $N(t)$ = population at time t and σ = "saturation population" or "carrying capacity".

In the case of bacteria and other simple populations the carrying capacity is mostly determined by the availability of food and space. For a human population the size of σ is much harder to calculate and to motivate. The factor a indicates the almost exponential rate of growth close to zero population. The shape of the growth function is given in Figure 10.1.

At equilibrium holds:

$$a\sigma N_{st} = aN_{st}^2, \quad \text{or} \quad N_{st} = \sigma\,. \tag{10.15}$$

It is obvious that σ is a stable equilibrium point. Whenever $N \neq \sigma$, N approaches the σ-level asymptotically, either from above or from below.

10.3.2 Population and Production

In human populations, the individual is both a *user* and a *producer* of the means of subsistence. This implies that the logistic population model (10.14) is not a proper representation of reality for human populations. We can represent the human population effect upon the means of subsistence by a production function:

$$q = F(N, l) \tag{10.16}$$

where F maps positive combinations of N, l into positive q.

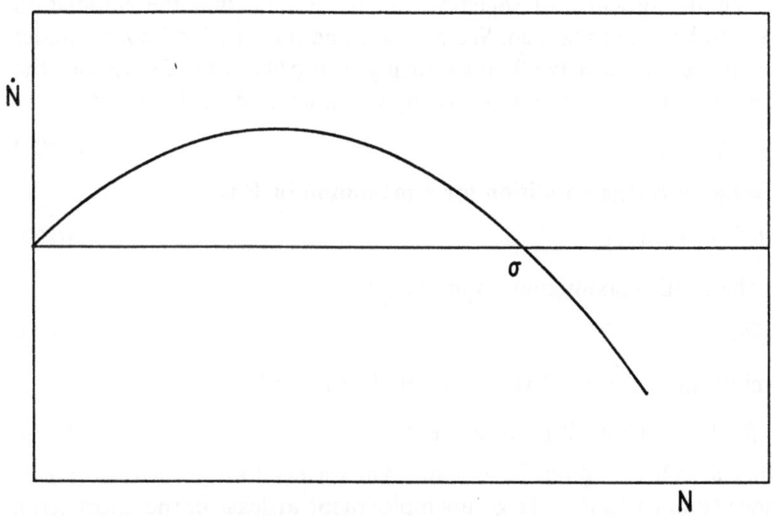

Fig. 10.1. The shape of the logistic growth function

q = production of a homogeneous commodity (e.g. corn)

N = population

l = land

Assuming the amount of land given, we have $q=F(N, \bar{l})$. If we assume that production increases but at a decreasing rate with increasing population (labour) we could approximate the production function by:

$$q=\alpha \bar{l} N^{\beta} \quad \text{where} \quad 0<\beta<1 . \tag{10.17}$$

A simple population dynamics would now be given by

$$\dot{N}=a(\alpha \bar{l} N^{\beta}-kN) \tag{10.18}$$

where the demand D is given by

$$D=kN \tag{10.19}$$

The equilibrium population is equal to

$$N^{*}=[\alpha \bar{l}/k]^{1/(1-\beta)} . \tag{10.20}$$

Thus the equilibrium population is given larger:

- the larger the amount of land (l)
- the higher the productivity (α)
- the lower the propensity to consume (k)

Like in the logistic model the equilibrium of this model is asymptotically stable. This model of interactions between the economic production system and the population captures parts of the classical analysis of Malthus according to which any improvement of production technology will *only temporarily* give an improved standard of living.

This model can also be used to explain the effect of profit (or rent) maximization on unemployment in agricultural societies. Let us assume that the population equilibrium (10.20) has been reached. We now assume that the land comes under the control of an agricultural profit maximizing authority, who determines the equilibrium population N^{**}, as determined by maximization of the profit V.

$$V=\alpha \bar{l} N^{\beta}-wN \tag{10.21}$$

with w = real wage rate. The condition for a maximum of V is

$$dV/dN=\beta\alpha \bar{l} N^{\beta-1}-w=0 \tag{10.22}$$

implying that the profit maximizing population is

$$N^{**}=[\beta\alpha \bar{l}/w]^{1/(1-\beta)} \tag{10.23}$$

Further assuming the real rate of wages w equal to k, we have

$$N^{**}/N^{*}=\beta^{1/(1-\beta)} \ll 1 \text{ for all permissible } \beta . \tag{10.24}$$

This means that a shift of regime from a survival society to a profit maximizing agricultural society must lead to large unemployment at least in the short term, before the profits could have been used for capital investment.

In the switching process it is possible that the original productivity α^* could be increased to $\alpha^{**} > \alpha^*$ as a consequence of a reorganisation into a capitalistic form. In that case the original high population N^* might survive, but now nourished by the smaller rural population $N^{**} < N^*$. This interpretation would imply three equations:

$$N^* = [\alpha^* \bar{l}/k]^{1/(1-\beta)}, \tag{10.25}$$

N^* is the stationary survival population calculated with original productivity α^* (see (10.20)).

$$N^{**} = [\beta\alpha^{**}\, \bar{l}/w]^{1/(1-\beta)} \tag{10.26}$$

N^{**} is the profit maximizing rural population with new productivity α^{**}.

$$\alpha^{**}\bar{l}N^{**\beta} = k\, N^* = \alpha^*\bar{l}N^{*\beta}\,. \tag{10.27}$$

The first equation in (10.27) describes the equilibrium between the supply produced by N^{**} and the demand from N^*, and the second equation makes use of (10.25). The equations (10.25) and (10.26) yield (with $k = w$, and $0 < \beta < 1$)

$$N^{**}/N^* = [\beta\alpha^{**}/\alpha^*]^{1/(1-\beta)} \tag{10.28}$$

and (10.27) gives

$$N^{**}/N^* = [\alpha^*/\alpha^{**}]^{1/\beta} \tag{10.29}$$

From equations (10.28) and (10.29) one easily obtains:

$$\alpha^*/\alpha^{**} = \beta^\beta < 1, \quad \text{and} \quad N^{**}/N^* = \beta < 1\,. \tag{10.30}$$

That means: After the productivity shift from α^* to $\alpha^{**} > \alpha^*$ a new equilibrium is established in which the profit maximizing rural population $N^{**} = \beta N^* < N^*$ now produces the supply of food for the total population N^*.

10.3.3 Models Combining Economic and Demographic Change

Based on an article by Hotelling [10.5], Puu [10.6] has proposed an endogenous determination of the saturation level of the logistic population model according to the principles of production theory.

Puu further assumes the production function to be a polynomical approximation of a more general production function, exhibiting increasing returns at low levels of population and decreasing returns at high levels. Such a production function can be of the form

$$q = \alpha_0(\alpha_1 N^2 - N^3) \tag{10.31}$$

with increasing returns for $N < 2\alpha_1/3$ and decreasing returns beyond that level. α_0 is a measure of technological efficiency, land and other factors exogenous to the model. We can further assume a given availability of a natural supply of means of subsistence c_0. Thus, total means M of produced and natural means of subsistence are

$$M = c_0 + \alpha_0(\alpha_1 N^2 - N^3) . \tag{10.32}$$

Further assume k to be the subsistence consumption (real wage) rate. Then we can assume that the saturation level (see (10.15)) is

$$\sigma = [c_0 + \alpha_0(\alpha_1 N^2 - N^3)]/k \tag{10.33}$$

so that k also reflects demographic behaviour. This further implies that the population growth model of Section 10.3.1 can be transformed into

$$\dot{N} = N[c_0 + \alpha_0(\alpha_1 N^2 - N^3) - k\, N] . \tag{10.34}$$

This differential equation can have up to four equilibrium (stationary) solutions. One of these is $N=0$, and this solution is always unstable. The other, different solutions are distinguished by the value of the discriminant of the equation (after scaling $c_0 = 1$)

$$N^3 - \alpha_1 N^2 + (k/\alpha_0)\, N - 1/\alpha_0 = 0 \tag{10.35}$$

which is given as

$$D = 0.037\ (\alpha_1^3/\alpha_0) - 1/108\ (\alpha_1 k/\alpha_0)^2 - 3/18\ (\alpha_1 k/\alpha_0) + 1/27\ (k/\alpha_0)^3 + 1/2\alpha_0^2 . \tag{10.36}$$

If $D<0$ then we have three distinct real roots.
If $D>0$ then there is only one real root.
If $D=0$ the solutions are degenerate.

Puu further shows that this model is equivalent to a "swallowtail catastrophe" model, implying that the equilibrium solutions will bifurcate as the parameters α and k change slowly, due to changes in technology of consumer demographic behaviour.

10.4 Economy, Demography and Migration

10.4.1 Logistic Population Growth and Migratory Diffusion

Hotelling in his original model argued that migration could be introduced by an assumption that populations tend to diffuse from more densely to less densely populated regions, either as a consequence of random movements or as a consequence of diminishing returns to labour, which favours a thinly distributed population. Such a density difference is measured by the Laplacian

$$\nabla^2 N = \partial^2 N/\partial x^2 + \partial^2 N/\partial y^2 . \tag{10.37}$$

Generalizing the standard growth equation of Section 10.3 by adding a diffusion term $k\nabla^2 N$ we obtain the combined migration–population growth equation in continuous two-dimensional space.

$$\partial N/\partial t = k\, \nabla^2 N + a\, N(\sigma - N) . \tag{10.38}$$

From this starting point Puu proceeds to combine the Hotelling model with the endogenous determination of σ by the use of the polynomial production function (10.32). In this more complicated model it can be shown that non-homogenous population patterns in space easily develop.

10.4.2 Migration and Public Resources

In the economy-demography models discussed above, consumer goods are private and produced in a production system that can be captured by a production function. It is shown that inclusion of production and population interaction leads to phenomena in the population model, that can be captured by *non-linear* partial differential equations with highly complicated solution properties, including phase transitions and other bifurcations. In a series of papers the public ([10.7], [10.8], [10.9] and [10.15]) or collective phenomena have been shown to give rise to non-linear phenomena as well.

The starting point is a finite set of regions ($i=1, 2, \ldots, L$). In each one of the regions there is an availability of nodal public capacity Q_i (parks, museums, housing etc.). Contacts with people (relatives, etc.) are also assumed to be of value and are reflected by

$$a_i=\sum_j f_{ij} n_j \tag{10.39}$$

where f_{ij} is a spatial discount factor for anyone in region i, contacting the population n_j of region j.

The utility u_i of a representative individual located in region i can be represented by

$$u_i=H\left[(Q_i/n_i);\ \sum_j f_{ij} n_j\right]. \tag{10.40}$$

A simple example for the choice of u_i is

$$u_i=(Q_i/n_i)^{\alpha_1}\left(\sum_j f_{ij} n_j\right)^{\alpha_2}. \tag{10.41}$$

A population equilibrium can now be defined by

$$u_i=u_j=\lambda \tag{10.42}$$

i.e. a situation of no migration gain.

With $\alpha_1=\alpha_2=\alpha>0$, the equation (10.42) reduces to the eigen-value equation

$$\boldsymbol{Mn}=\mu \boldsymbol{n} \tag{10.43}$$

with $M=[M_{ij}]=[Q_i f_{ij}]$ and $\mu=\lambda^{1/\alpha}$.

This implies that any change of functions between a pair of regions (i, j), (that would increase a f_{ij}) must lead to an increased level of equilibrium utility *and* to a redistribution of population by migration. Furthermore any increase of public capacities in a region will have a similar general effect of increasing utility and redistribution of population. Leonardi and Casti [10.10] have shown that model (10.41) is equivalent to the logit-like model:

$$x_i = \frac{Q_i\left(\sum_j f_{ij} x_j\right)^\alpha}{\sum_k Q_k\left(\sum_l f_{kl} x_l\right)^\alpha} \tag{10.44}$$

with $\alpha = (\alpha_1/\alpha_2)$ and x_i defined as relative population shares $x_i = n_i \Big/ \sum_j n_j$. Indeed, from (10.41) and (10.42) one obtains

$$(Q_i/x_i)^{\alpha_1}\left(\sum_j f_{ij} x_j\right)^{\alpha_2} = \tilde{\lambda} = \lambda\, N^{(\alpha_1 - \alpha_2)} \tag{10.45a}$$

with $$N = \sum_{j=1}^{L} n_j$$

or equivalently

$$\tilde{\lambda}^{1/\alpha_1}\, x_i = Q_i\left(\sum_j f_{ij} x_j\right)^{\alpha_2/\alpha_1}. \tag{10.45b}$$

Because of $\sum_{k=1}^{L} x_k = 1$ it follows from (10.45b), that

$$\tilde{\lambda}^{1/\alpha_1} = \sum_k \left[Q_k \sum_j f_{ij} x_j\right]^{\alpha_2/\alpha_1}. \tag{10.46}$$

Reinserting of (10.46) into (10.45b) yields (10.44). Equation (10.44) which is of the logit form defines a family of maps of a convex set D to itself, i.e.

$F_2 : D \to D$, where $D : \boldsymbol{x} \in R^L$, $x_i \geq 0$

and $\sum_{i=1}^{L} x_i = 1$.

It can be concluded that this type of model may exhibit many different equilibria, depending upon the size of α and the initial distribution.

Leonardi and Casti show that micro maximization of random utility of migrants (see Section 10.5) leads to equation (10.43) and that the equilibrium (10.44) is equivalent to a rational expectations migration equilibrium.

A *dynamic generalization* of the static model is of even greater interest in this context.

The population dynamics equation for the x_j to be associated with the static version of the model is:

$$\dot{x}_j = \sum_i p_{ji} x_i - \sum_i p_{ij} x_j \tag{10.47}$$

with $x_j = n_j/N$, $N = \sum_i n_i$

This equation agrees with the equation for the *meanvalues* of the relative population numbers. In the present model the transition rate p_{ji} from i to j is:

$$p_{ji}=\beta Q_j\left(\sum_k f_{jk}x_k\right)^{\alpha}. \tag{10.48}$$

According to (10.48), p_{ji} does *not* depend on the origin region i and could therefore be denoted as $p_{ji}\rightarrow p_{j\cdot}$ (whereas in the general case p_{ji} *does depend* on destination – *and* origin region as well!).

Because of the special form of $p_{ji}=p_{j\cdot}$ equation (10.47) reduces to

$$\dot{x}_j=p_{j\cdot}\sum_i x_i-x_j\sum_i p_{i\cdot} \tag{10.49}$$

The stationary solution $\hat{x}_j$ of (10.49) is given by:

$$\hat{x}_j=\frac{p_{j\cdot}}{\sum_i p_{i\cdot}}=\frac{Q_j\left(\sum_k f_{jk}x_k\right)^{\alpha}}{\sum_i Q_i\left(\sum_k f_{ik}x_k\right)^{\alpha}} \tag{10.50}$$

and agrees with the static version (10.44) of the model.

A more general way of formulating the dynamics is by the following argument: Let there be L alternatives to be chosen by a population and let $x_k(t)$ be the relative population adopting k at time t. Let us assume that the rate p_{ji} of moving from i to j is proportional to the accessibility of j to the movers from all other alternatives. Thus a plausible form of probability of switching is

$$p_{ji}=\varepsilon h_{ji}Q_j\left(\sum_k f_{jk}x_k\right)^{\alpha} \tag{10.51}$$

where ε is a speed adjustment parameter, and $h_{ji}=h_{ij}$ a symmetric function of the distance from i to j. Inserting the rates (10.51) into the population dynamic equation (10.47), one now obtains:

$$\dot{x}_j=\varepsilon\left[\sum_i h_{ij}Q_j\left(\sum_k f_{jk}x_k\right)^{\alpha}x_i-x_j\sum_i h_{ij}Q_i\left(\sum_k f_{ik}x_k\right)^{\alpha}\right]. \tag{10.52}$$

It is seen by direct insertion of (10.50) that $\hat{x}_j$ is *still* the stationary solution of (10.52), although the transition rates have been generalized from (10.48) to (10.51). This generalization of p_{ji} changes the dynamics *towards* the stationary state, but not the stationary state *itself*. Thus equations (10.49) or (10.52) are true dynamic generalizations of the static version of the model.

Let us now discuss the solutions of equation (10.47) with the transition rates (10.48) for the case of two regions! If the regions are primordially equivalent, one can put

$$f_{11}=f_{22}=1;\qquad 0\leq f=f_{21}=f_{12}<1;\qquad Q_1=Q_2=Q\,. \tag{10.53}$$

The equations (10.47) and (10.48) then take the form

$$\dot{x}_1 = p_{1.} x_2 - p_{2.} x_1$$
$$\dot{x}_2 = p_{2.} x_1 - p_{1.} x_2 \tag{10.54}$$

with

$$p_{1.} = \beta Q (x_1 + f x_2)^{\alpha}; \qquad p_{2.} = \beta Q (f x_1 + x_2)^{\alpha} . \tag{10.55}$$

Expressing x_1 and x_2 by the difference variable $x = x_1 - x_2$, where $-1 \leq x \leq +1$,

$$x_1 = (1+x)/2; \qquad x_2 = (1-x)/2 \tag{10.56}$$

one obtains from (10.54) the equation of motion for x:

$$dx/dt = -(p_{1.} + p_{2.})x + (p_{1.} - p_{2.}) \tag{10.57a}$$

or, inserting (10.55),

$$dx/d\tau = (\bar{\alpha} + x)^{\alpha} (1-x) - (\bar{\alpha} - x)^{\alpha} (1+x) \tag{10.57b}$$

with

$$\tau = \bar{\beta} t; \ \bar{\beta} = \beta Q[(1-f)/2]^{\alpha}; \qquad \bar{\alpha} = (1+f)/(1-f) \geq 1 . \tag{10.58}$$

Plotting $\dot{x}$ versus x in the phase-diagram of Figure 10.2, one observes an interesting phase-transition behaviour of the dynamics: For small $\alpha < \bar{\alpha}$ – corresponding to weak influence of the populations x_1, x_2 on the transition rates – the point $x = 0$ with equal populations in both regions is the stable equilibrium. On the other hand for large $\alpha > \bar{\alpha}$ – corresponding to strong dependence of the transition rates on the populations x_1 and x_2 – the points $x = x_+$ or $x = x_- = -x_+$ are stable equilibria. They correspond to the agglomeration of the population in one of the two regions!

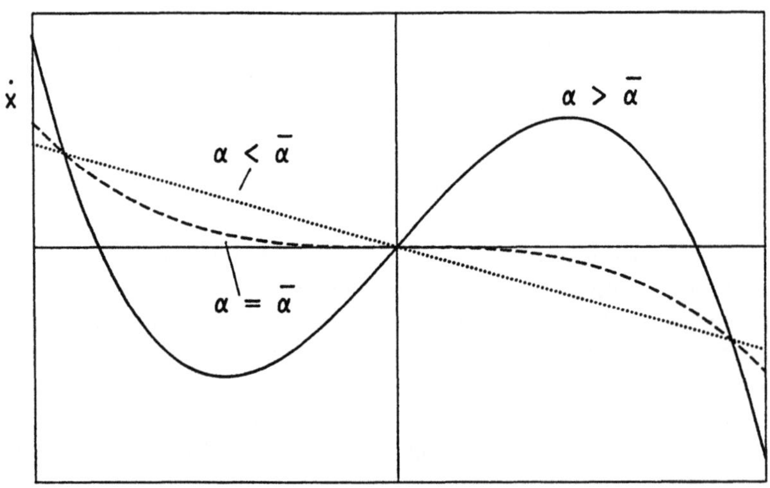

Fig. 10.2. Phase diagram of the variable $\dot{x}$ versus x, for different values of the parameter α: $\alpha > \bar{\alpha}$ (straight line), $\alpha = \bar{\alpha}$ (intermitted line), $\alpha < \bar{\alpha}$ (dotted line)

This behaviour of the solutions demonstrates in a simple manner a general phenomenon observed in nonlinear dynamics: The breaking of symmetry. This means, that the solutions need not have the symmetry inherent in the model. In our case, each of the stationary solutions $x=x_+$ and $x=x_-$ distinguishes one region before the other, although the model equations and model parameters are symmetrical in both regions. The symmetry is however re-established on a higher level, namely if the total set of solutions of the nonlinear system is considered.

10.5 Stochastic Choice Theory and Migration

Many analysts have argued that migration cannot be modelled as a purely deterministic phenomena. Rather it ought to be viewed as a rational search process with stochastics either entering the utility functions (or preferences) or the informations about choice alternatives (regional population characteristics). In the static choice between two regions, i and j, of a representative potential migrant, his choice might be determined by strict non-stochastic utility, V, and by chance factors, ε.

Assuming "independence of irrelevant alternatives", "choice rationality", and "transitivity of preferences" we can view the whole locational choice process as a series of pairwise choices according to the stochastic choice rule

$$P(j|i)=\mathrm{Prob}[(V(Z_j,f_{ji})+\varepsilon(Z_j,f_{ji}))>(V(Z_i)+\varepsilon(Z_i))] \tag{10.59a}$$

Rearranging we have:

$$P(j|i)=\mathrm{Prob}[(V(Z_j,f_{ji})-V(Z_i))>(\varepsilon(Z_i)-\varepsilon(Z_i,f_{ji}))]\,. \tag{10.59b}$$

Under the assumption that $\varepsilon(Z_i,f_{ji})$ are independently and identically distributed according to a Gnedenko or Weibull (negative exponential) distribution, it has been shown (by McFadden [10.11]), that the choice probabilities can be expressed in the multinomial logit form

$$P(j|i)=\frac{\exp[V(Z_j,f_{ji})]}{\sum_k \exp[V(Z_k,f_{ki})]} \tag{10.60}$$

The basic problem is then to determine a reasonable set of regions and characteristics, $Z=[Z_1,\ldots,Z_m]$, so that the well-known "Green bus-Blue bus" problem can be avoided[2].

[2] We assume that a person is indifferent between choosing a perhaps red bus or a train, thus P(Red Bus) $=P$(Train)$=0.5$. If we now introduce a blue bus, the person should be indifferent between this blue bus and the red bus. Thus P(Blue Bus)$=P$(Red Bus)$=P$(Train)$=1/3$. The colour-distinction is obviously not a characteristic of importance for the subdivision into choice alternatives.

10.6 The Master Equation Approach to Population Analysis

10.6.1 Comparison of Two Approaches

The master equation approach is capable of an explicit consideration of stochastic phenomena from the very beginning. The master equation is always an equation of motion for a probability distribution. The special form of the master equation however depends on the space of states chosen.

Let us begin with a first form of the master equation. Here one starts with the space of states c, s, x of *one individual.* The variables c, s, x mean class, status and location of the individual, respectively. Correspondingly, the probability

$$P(c, s, x; t)\, dc\, ds\, dx \tag{10.61}$$

is introduced, to find one individual at time t in state c, s, x within the interval $dc\, ds\, dx$. If R is the domain of states, the probability normalization leads to

$$\int_R P(c, s, x; t)\, dc\, ds\, dx = 1 \qquad \text{for any time } t\,. \tag{10.62}$$

Now let

$$dc'\, ds'\, dx'\ p_t(c'\, s'\, x' | csx) \tag{10.63}$$

be the probability transition rate (probability transfer per unit of time) from state $c s x$ into the interval $dc'\, ds'\, dx'$ of states $c'\, s'\, x'$. Making use of the probability transition rates (10.63) the master equation for the evolution of $P(c, s, x; t)$ then can be written in the form

$$\begin{aligned} \partial P(c, s, x; t)/\partial t = &-\iiint_R dc'\, ds'\, dx'\ p_t(c'\, s'\, x' | csx)\ P(c, s, x; t) \\ &+\iiint_R dc'\, ds'\, dx'\ p_t(csx | c'\, s'\, x')\ P(c', s', x'; t)\,. \end{aligned} \tag{10.64}$$

The first term on the r.h.s. describes the probability flow from state csx into all other states $c'\, s'\, x'$, and the second term the probability flow from all states $c'\, s'\, x'$ into state csx. Considering an ensemble of N individuals, each behaving according to probability (10.61), the *mean number of individuals* in state csx in interval $dcdsdx$ is given as

$$n(c, s, x; t)\, dc\, ds\, dx = N\, P(c, s, x; t)\, dc\, ds\, dx\,. \tag{10.65}$$

Multiplying (10.64) with N, one immediately obtains the equation of motion for the *mean population densities* (10.65), namely

$$\begin{aligned} \partial n(c, s, x; t)/\partial t = &-\iiint_R dc'\, ds'\, dx'\ p_t(c'\, s'\, x' | csx)\ n(c, s, x; t) \\ &+\iiint_R dc'\, ds'\, dx'\ p_t(csx | c'\, s'\, x')\ n(c', s', x'; t)\,. \end{aligned} \tag{10.66}$$

If N is time dependent with

$$\partial N/\partial t\ P(c,s,x;\ t)=B(c,s,x;\ t)-D(c,s,x;\ t) \tag{10.67}$$

the terms $B(c,s,x;\ t)-D(c,s,x;\ t)$ must be added on the right hand side of (10.66). The equation (10.66) is considerably simplified, if instead of a continuous state space $\{c,s,x\}$ discrete states i are introduced:

$$\begin{aligned} &n(c,s,x;\ t)\ dc\,ds\,dx \rightarrow n_i(t) \\ &p_t(c'\,s'\,x'|csx)\ dc'\,ds'\,dx' \rightarrow p_{ji}(t)\ . \end{aligned} \tag{10.68}$$

Then instead of (10.66) it follows that:

$$\begin{aligned} \dot{n}_i(t)= &-\sum_j p_{ji}(t)\ n_i(t)+\sum_j p_{ij}(t)\ n_j(t) \\ &+B_i(t)-D_i(t)\ . \end{aligned} \tag{10.69}$$

This is, of course, the usual rate equation for regional populations $n_i(t)$.

In the approach just formulated, however, there remains the problem, that the transition rates of the N *independent individuals* may depend on origin and destination state, csx and $c's'x'$, respectively, *but not* on the origin and destination population $n(c,s,x;\ t)$ and $n(c',s',x';,t)$, or more generally on the whole population configuration!

Therefore we now consider a second master equation approach capable of implying such cases, too, (and this is the approach followed throughout this book).

In contrast to the first approach, the state space now consists of the population configuration $\boldsymbol{n}=\{n_1, n_2, \ldots, n_L\}$, where n_i is the number of individuals in state i. Correspondingly, the configurational probability

$$P(\{n_1, n_2, \ldots, n_L\};\ t) \tag{10.70}$$

to find, at time t, the population configuration $\{n_1, n_2, \ldots, n_L\}\equiv\boldsymbol{n}$, is now introduced! Defining further the *configurational transition rates*

$$\begin{aligned} &w_{ji}(\boldsymbol{n}) \text{ from } \boldsymbol{n} \text{ to} \\ &\boldsymbol{n}^{(ji)}=\{n_1, n_2, \ldots, (n_j+1), \ldots, (n_i-1), \ldots, n_L\} \end{aligned} \tag{10.71}$$

one can prove (see Part I of the book), that

$$w_{ji}(\boldsymbol{n})=p_{ji}(\boldsymbol{n})\ n_i \tag{10.72}$$

where $p_{ji}(\boldsymbol{n})$ is the individual probability transition rate from i to j, which now may (but need not necessarily) depend on the configuration $\boldsymbol{n}$ itself!

The master equation for the configurational probability $P(\boldsymbol{n};\ t)$ now reads (see [10.14])

$$\begin{aligned} dP(\boldsymbol{n};\ t)/dt= &-\sum_{i,j} w_{ji}(\boldsymbol{n})\ P(\boldsymbol{n};\ t) \\ &+\sum_{i,j} w_{ji}(\boldsymbol{n}^{(ij)})\ P(\boldsymbol{n}^{(ij)};\ t) \end{aligned} \tag{10.73}$$

derived in the same manner as above (see also Part I).

The *meanvalue* of n_i at time t must now be defined as

$$n_i(t) = \sum_{\boldsymbol{n}} n_i \, P(\boldsymbol{n}; t) \tag{10.74}$$

where the sum goes over all configurations $\boldsymbol{n}$. For $n_i(t)$ there has been derived in Part I the following equation of motion:

$$dn_i(t)/dt = -\sum_j \langle w_{ji}(\boldsymbol{n}) \rangle_t + \sum_j \langle w_{ij}(\boldsymbol{n}) \rangle_t \tag{10.75}$$

where $\langle \quad \rangle_t$ are again meanvalues at time t. Inserting (10.72), equation (10.75) may approximately be written as:

$$dn_i(t)/dt = -\sum_j p_{ji}(\boldsymbol{n}(t)) \, n_i(t) + \sum_j p_{ij}(\boldsymbol{n}(t)) \, n_j(t) \, . \tag{10.76}$$

This is again the rate equation (10.69), if also birth-death-terms are added, but with the decisive generalization, that p_{ji} may now depend on the configuration $\boldsymbol{n}(t)$, with the consequence, that (10.76) is in general a *nonlinear* system of differential equations.

It should be mentioned, that the configurational probability $P(\boldsymbol{n}; t)$ of course also contains the probabilities of deviations of $\boldsymbol{n}$ from the meanvalues $\boldsymbol{n}(t)$.

In the model used throughout this book, the individual probability transition rate is put equal to

$$p_{ji} = v_{ji} \exp(u_j - u_i) \tag{10.77}$$

where u_j may or may not depend on n_j. If (10.77) is inserted in (10.76), it can be easily seen, that

$$n_k^* = \frac{N \exp(2\, u_k)}{\sum_i \exp(2\, u_i)} \tag{10.78}$$

is the stationary solution of (10.76).

Reviewing the results, it can be seen, that the basic advantages of the master equation approach are that

1. it ensures consistency of calculations
2. it is flexible and general
3. it does not require local interactions only, which is implicit in the partial differential equation models.

The critical problem lies in the formulation of the behavioural assumption, which must be reflected by the transition probability functions.

10.6.2 The Model of this Book and its Relation to Economics

The economic causes of population redistribution are essentially captured u_i as functions of a set of determinants. We have seen above that most of the economic models of population distribution and migration are based on three principles:

1. The effect of the population as a pool of labour upon production and income and the effect of income on demographic behaviour.
2. The effect of migration distances between pairs of regions (as a migration cost item) upon migration propensities and thus upon migration flow patterns.
3. The consequences of regional natural resources or public good capacity endowments for the attractiveness of a region and with its effect upon population distribution and its change by migration.

These three factors have been observed in the model, in particular by the specification and estimation procedures employed in the country studies.

The mobility v_{ij} has in general been decomposed into two factors, the first reflecting the interregional flow effect of migration costs and the second being an intertemporal global mobility factor, being regressed upon yearly indicators of national economic activity, such as employment or unemployment, total income or labour income, housing production, or service activity.

Similarly, the utility of a region is decomposed into two parts. The first is in general related to the size of the population of the region and the second is a regional preference indicator. The latter is constructed to be the difference between the regional utility index u_i and the part of it that is explained by the population size. The regional preferences δ_i are thus residuals in nature and constrained to

$$\sum_i \delta_i = 0 . \tag{10.79}$$

The regional preferences are then regressed against different economic variables. In five of the six case studies the most important economic variables determining regional preferences is an indicator of the conditions in the labour market. The second most important factor is the situation on the housing or service markets.

The most surprising result of the studies is the consistency of population effects upon regional utilities. For the sample the average value of α_1 (in the equation $u_i = \delta_i + \alpha_1 n_i + \alpha_2 n_i^2$) is 1.22 (standard deviation 0.22) and of α_2 is -0.24 (standard deviation 0.07). Thus the model has to some extent captured the interaction between population increases (decreases) and economic factors influencing migratory decisions. To some extent this is too implicit to bring this effort to its final state.

It is well known from economic theory that unemployment, real wage rates and migration are interacting in a complicated way, highly dependent upon institutional conditions [10.12].

If for instance, the wage rates are determined by central negotiations to stay at some uniform level $w_j = w_j = \bar{w}$, then *all* the imbalances of the labour market will be registered as vacancy and unemployment figures. This implies that the whole pull/push mechanism must be modelled in terms of vacancy-unemployment pools.

If on the other hand a pure market equilibration rule is approximated by a country, then figures of vacancies and unemployment would be of no interest for the modelling effort, as these figures would be indicators of *short term* frictions only. In this case the relevant procedure would be to focus on wage and cost-of-living differences in the modelling of migration decisions.

This means that there is either a non-linear interaction between wage rates in regions or unemployment/vacancy rates and populations of the regions. The next steps in the refinement of the model presented therefore should be to take these higher order interactions into account. But there are even greater possibilities of development of this model of migration and population evolution in space. In the present analysis public and private capital formation is outside the scope of the model. In a separate study Haag and the author of this chapter have proceeded to analyse the evolution of capital stocks in space in a non-linear master equation model. In the longer term one could envisage a fusion of these non-linear evolutionary population and capital models into a proper theory of non-linear economic evolution.

11 Comparative Analysis of Interregional Migration

Günter Haag, Martin Munz, Rolf Reiner and Wolfgang Weidlich

11.1 Purposes and Problems of the Comparison of Interregional Migration in Different Countries

The interregional migration processes within the six countries (FRG, Canada, France, Israel, Italy, Sweden), which have been described in detail in Part II, Chapters 4 to 9, are now studied under comparative aspects in this chapter.

It is the natural purpose of such a comparison to find out firstly analogies and/or differences of the migratory process as such between these countries and secondly correspondence or disagreement between the interpretations of the migratory process in terms of "explanatory" socio-economic variables. This kind of investigation is of course facilitated by having used one and the same migratory model and hence immediately comparable methods of evaluation in all six cases.

In Section 11.2 some general problems arising in a comparative study of interregional migration are discussed in a qualitative manner. Thereupon, Section 11.2 is devoted to the analysis of quantitative results of Chapters 4 to 9, obtained with the model used in this book, from a comparative point of view.

11.1.1 Interregional Migration and Total Population Evolution

Before beginning the comparative analysis it is worthwhile to remember that interregional migration is only one – more or less substantial – share of the total migration, and that migration is only one part of the total evolution of the population.

At first let us consider the relative importance of interregional versus intraregional migration by a concrete example. The German federal states are subdivided into counties, and the registered total migration volume consists of three parts:

- the intra-county migration
- the migration between counties of one federal state, and
- the migration between counties of different federal states.

Only the latter is interregional migration by definition, whereas the two other shares contribute to intraregional migration only.

For the year 1983 the partial migration volumes are:

intra-county migration	727796
migration between counties of one state	1330643
migration between counties of different states	674186
total migration volume	2732625

Hence the relative share of interregional migration is $674186:2732625 \approx 0.247$. The order of magnitude of this relative share is about the same in all countries under consideration.

It is also interesting to note, that the trends of intraregional and interregional migration need not to coincide. In the FRG for instance one observes over the years a slow increase of intraregional migration, but a slow decrease of interregional migration. Thus, in the course of time short distance adaptations of the place of residence seem to be preferred against long distance moves.

Secondly we have to recall, that the total change of population per year in region i decomposes as follows:

$$\frac{\Delta n_i(t)}{\Delta t} = (W_i^+ - W_i^-) + (\tilde{B}_i - \tilde{D}_i) + (M_i^+ - M_i^-) \tag{11.1}$$

with

$$W_i^+ = \sum_{j \neq i} w_{ij} = \text{immigrants per year into } i \text{ from regions } j \neq i$$

$$W_i^- = \sum_{j \neq i} w_{ji} = \text{outmigrants per year from } i \text{ into regions } j \neq i$$

$$M_i^+ = \quad \text{immigrants per year into } i \text{ from foreign countries}$$

$$M_i^- = \quad \text{emigrants per year from } i \text{ into foreign countries}$$

$$\tilde{B}_i = \tilde{\beta}_i n_i = \text{births per year in region } i$$

$$\tilde{D}_i = \tilde{\mu}_i n_i = \text{deaths per year in region } i$$

The last two terms on the r.h.s. of (11.1) can be combined into

$$(\tilde{B}_i - \tilde{D}_i) + (M_i^+ - M_i^-) = (B_i - D_i) = \rho_i n_i \tag{11.2}$$

with

$$B_i = \tilde{B}_i + M_i^+ = \beta_i n_i; \qquad D_i = \tilde{D}_i + M_i^- = \mu_i n_i$$

$$\rho_i = (\beta_i - \mu_i) \tag{11.3}$$

where ρ_i is the (effective) rate of natural increase in region i (see also Chapter 2, eq. (2.29)).

It is only the first term on the r.h.s. of (11.1) which refers to interregional migration and which is explicitly analyzed in this book. There may however exist an indirect influence of natural increase on interregional migration. Immigrants or newborn children arriving inhomogeneously over the regions may finally settle in other regions and thus distort the "normal" interregional migration. In countries with waves of immigrants into regions of arrival, for instance, one has to expect a time-delayed wave of interregional migration of these immigrants into their final regions of destination.

Only in the case, when the effective rate of natural increase is independent of the region i, that is for $\rho_i(t) = \rho(t)$, the pure migratory process can be separated off by going over to the relative regional populations $x_i(t) = n_i(t)/N(t)$, as shown in Chapter 2 (see eq. (2.39)).

J. Ledent has shown in Chapter 5, that this assumption of regional independence of ρ_i is not fulfilled in the case of the regions of Canada; in other words, Canada has a regionally inhomogeneous rate of natural increase so that migration and natural increase cannot be treated separately and independently.

What are the consequences of the nonseparability of both processes? On the one side it is clear, that for applying the equations of motion to purposes of forecasting the full set of equations (2.29) instead of the pure migratory equations (2.39) must now be solved with migration matrix $w_{ji}(t)$ and regional rates of natural increase $\rho_i(t)$.

On the other hand the regression analysis for the representation of the $w_{ji}(t)$ in terms of mobilities and regional utilities remains valid also in this more general case. But the regional preferences and the mobility now also reflect the effect of regionally inhomogeneous immigration, emigration or birth/death rates. An extreme example is Berlin, before the wall was built in 1961. The immigrants from the GDR to Berlin-West showed up in the interregional outmigration to the states of the FRG, with the effect of a strongly negative preference of Berlin at this time.

11.1.2 Problems in the Choice of Regions

In order to make the interregional migration within different countries comparable, some general criteria for the subdivision of the country into regions had to be (and have been) obeyed. These criteria in particular refer to the *size* and the *homogeneity* of regions.

On the one hand there exist arguments for choosing *not too small* regions. Since the model used here is based on meanvalues of migration flows and equations for the meanvalues of regional populations, the relative deviations from the meanvalues by random fluctuations should be small. This is only guaranteed if the regional populations and migration flows are not too small numbers. A further practical reason is the availability of spatially disaggregated data.

On the other hand the relative importance of migration versus birth/death processes is growing the smaller the regions are! This is due to the fact, that a higher percentage of the total migration volume becomes interregional migration, if smaller regions are chosen. Thus for instance, the average number of in-migrations per birth is about 5 for a county, but 1.7 for a federal state [11.1].

Therefore, in order to keep interregional migration as a relevant factor of population evolution, the regions should also not be chosen too large. In particular they should be chosen *as* intraregionally *homogenous as possible* in economic, ethnographic, climatic and other respects. Indeed, their description in terms of regional utilities becomes more meaningful and interpretable if structural characteristics ($\Rightarrow$key-factors) can be attributed to one region as a whole, and not to disparate subunits. In this case the individual will consider the region more or less as a unit with respect to his migration decisions.

There exist other special problems in the choice of regions. For instance, the boundaries of regions should not divide cities, since interregional migration then would be biased by very short distance migrations.

In all six countries such general requirements for the choice of regions have been taken into account in an adequate and also comparable manner, although the historical and geographical conditions are rather different.

For the *Federal Republic of Germany* the system of regions consists of the 10 federal states and the region Berlin (West). It is getting more and more "natural", although it was considered as provisional at the foundation of the Federal Republic of Germany, 1949. Only some of these states (for instance Bavaria) had a long history of their own; others, for instance Baden-Württemberg were established as new units after the second world war. Due to the high population density and the historical urbanisation process the towns and cities are distributed in a relatively homogeneous manner over the whole territory. Since the regional governments have the authority of planning a certain homogeneity of economic development has been reached within each federal state.

In *Canada*, the historical growth of the population as well as its geographical redistribution were strongly promoted by successive waves of European immigrants. In the first two decades of this century European immigrants again arrived in great numbers. The creation of New France in 1608 – now the province of Quebec –, the settlements of the British immigrants after 1763 in the Maritimes and especially in Lower Canada (Ontario) and the development of the Prairies carried ahead by the railway link from the Atlantic to the Pacific, were stations in the development of the regional system. By the early 1920th, the geographical distribution of the Canadian population by province (and territory) had reached a state which has been maintained more or less up to now. The historical evolution of settlement patterns had lead to a rather inhomogeneous distribution of the population being mainly concentrated along the American border. Thus it can be said that "Canada is a linear country".

France has a relatively low average population density as compared to the neighbouring European countries; but the population is unevenly distributed over the country, with 18% concentrated in the single metropolitan area Paris, and on the other hand with sparcely populated rural or mountainous areas. The contrasts in settlement density however are widely smoothed out by the subdivision of France into 21 administrative regions recently designed (1955) for planning purposes. Essentially they correspond to functional regions, most of them surrounding their largest city. Being relatively homogeneous in size and with respect to internal structure, there exist strong differentiations in economic structure and socio-cultural characteristics between these regions. Thus the 21 administrative "régions de programme" establish an appropriate subdivision for the study of interregional migration.

The exceptional history of *Israel*, founded 1948, is of course also reflected in migration and in the formation of regions. Huge waves of immigration lead to the metropolisation of the Tel-Aviv, Haifa and Jerusalem areas, and it can be said, that the cohortes of immigrants shaped the main structure of the population distribution, which is high in the coastal plane and low in the Beer-Sheva region. Since 1970 the migratory process turned from a catastrophic to a more natural one. The official administrative division of the country into 6 districts and 14 quasi-functional subdistricts takes into account the results of the preceding turbulant migration as well as natural geographical structures, so that the 14 regions system seems to be most appropriate for the study of present migratory processes.

The population of *Italy* is rather inhomogeneously distributed over the territory, with a population density varying from 35 to 350 inhabitants per square kilometre. This inhomogeneity of population is caused by the geomorphical structure of Italy being a mountainous country, and on the other hand by the urbanisation process.

The regional system of Italy, being a united Republic since 1861, still reflects this inhomogeneity, although in recent years a strong homogenisation process has taken place due to the interregional migration.

The system of twenty administrative regions of Italy, which can be aggregated into five clusters known as "grande ripartizioni" has been chosen because of the following reasons: it is commonly used in demographic and socio-economic studies of Italy and thus guarantees the comparability of our results with other studies; furthermore, the availability of economic data is just limited to the twenty regions system. These advantages overcompensate some weaknesses of the chosen zoning system, namely the still existing considerable intraregional inhomogeneity of the population and the rather different population size of the regions.

The old and very comprehensive population statistics of *Sweden* dates back to the 18th century. The system is based on 2570 parishes, aggregated into 280 municipalities or 24 counties. For the purposes of the present analysis it is however sufficient to consider 8 county regions, since the latter are internally homogeneous in economic, geographic and ethnographic respects.

11.1.3 General Conclusions from the Comparison of Mobilities and Utilities

If the evolution of a country is partially reflected in the migratory behaviour of its population, it should be possible to draw conclusions from the comparison of their mobilities and utilities, if these are available for the same period of time. Such a common set of data was available from about 1960 to 1982, but unfortunately the immediate post-war period could not be included because of lack of data in some countries.

According to their definition, one can expect, that the global mobility reflects global evolutions of all regions of the whole country, whereas regional utilities or their size-independent parts, the regional preferences, are indicators of the evolution of regional differentiation.

Thus naturally the increase or decrease of the global mobility reveals, that the whole country is in a period of setting out or of calming down. Furthermore, economic developments of international scale – like the oil shock and other international economic crises and shifts – should be identifiable by their synchronous effect on the global mobilities. Business cycles could also have an impact on the global mobility, so that vice versa migration can be taken as an indicator for market conditions. It cannot be expected, however, that booms and slumps are synchronous in all countries.

But also the regional differentiation, hence utilities and preferences, could be influenced by international interactions: The Dollar exchange rate for instance has another influence on export-orientated regions than on regions with domestic market orientation. Economic booms may increase the prosperity of successful regions, and deteriorate the situation in disadvantaged regions.

Two quantities (to be discussed in detail in Section 11.2) now seem to be appropriate measures for the regional differentiation of a country. The variance $\sigma_u(t)$ of utilities (see (11.10)) describes the spread of regional attractiveness including the effect of the regional population sizes, whereas the variance $\sigma_\delta(t)$ of preferences (see (11.11)) measures the spread of regional attractiveness excluding size effects.

It is obvious, that in general various factors can influence $\sigma_u(t)$ and/or $\sigma_\delta(t)$: Governmental national policy normally aims at the assimilation of regions, that means at diminishing $\sigma_\delta(t)$ for instance by development programs for economically less developed or climatically disadvantaged regions. The self organizing economic process may either lead to agglomeration effects in highly industrialized regions (leading to an increase of $\sigma_u(t)$), or in other situations to de-concentration effects, for instance if industries, in search of labour force reserves, wander into less developed regions and thus contribute to the assimilation of regional preferences. Thus the quantities $\sigma_u(t)$ and $\sigma_\delta(t)$ can give rise to a comparative consideration of the regional homogeneity of countries and of the eventual impacts of governmental development programs as well as of the regional differentiation of the economies.

11.1.4 Problems in the Choice of Comparable Socio-Economic Factors

The main step in analyzing the interrelation between migration and the socio-economic process in the frame of the model used throughout the book, consists in representing the global mobility and the regional utilities in terms of socio-economic variables. Hence it would be ideal for comparative purposes to have at one's disposal the same standardized socio-economic variables in all countries.

Unfortunately, however, the data sets available in different countries do not coincide in this strict sense. The way out of this difficulty consists in considering the individual socio-economic variable as a proxy of some more general class of influences, on which migration may depend. In this sense one can identify classes of variables, so that the variables available for every country belong to one of the following classes:

Size and Population Structure (SP)
Labour Market (LM)
Housing Market (HM)
Industry, Investment (II)
Public Sector, Politics (PP)
Living Standard (LS)
Climate (C)

The case of the Federal Republic of Germany (see Section 4.3) can be taken as an example, how all socio-economic variables fall into one of these classes. It is also easily possible to reach agreement between this and other classifications, used in the individual case studies.

The comparative analysis of the representation of mobilities and utilities in terms of socio-economic key-factors to be implemented in Section 11.2 then primarily will focus on the question which class of variables (and not which specific variable) proves to be most important in that representation (see Tables 11.1 and 11.3).

Finally a general comment should be made, whether or not all correlations found between utilities, mobilities and certain socio-economic factors must necessarily be in agreement with "direct intuition". To answer this difficult question it

Table 11.1. Key-factors and *t*-values of coefficients (in parantheses) of the global mobility of five countries

country	1. key factor	2. key factor	3. key factor	4. key factor
Federal Republic of Germany	(LS) real income (-49.3)	(LM) vacancies (12.7)	(II) investment structure index (8.4)	(LM) employment (-7.3)
Canada	(LS) income per capita (-34.8)	(LM) employment index (6.2)	(LM) unemployment rate (2.3)	-
France	(HM) housing construction (34.5)	-	-	-
Israel	-	-	-	-
Italy	(SP) population (-86.3)	(LM) agricultural employment (-9.2)	(LS) gross national product (7.0)	-
Sweden	(HM) housing constructions (20.6)	(LM) rate of vacancies (10.7)	-	-

Table 11.2. Agglomeration parameters κ and saturation parameters σ and their t-values (in parantheses) of six countries

country	κ	σ	stability of homogeneous distribution (s=stable, u=unstable)	Symbols
Federal Republic of Germany	1.300 (+151)	-0.258 (-27)	u	+
Canada	1.546 (+105)	-0.313 (-15)	u	o
France	0.831 (+90)	-0.124 (-15)	u	△
Israel	1.250 (+92)	-0.179 (-16)	u	□
Italy	1.285 (+73)	-0.245 (-14)	u	◊
Sweden	1.106 (+66)	-0.295 (-5)	u	·

must be kept in mind, that utilities and mobilities govern the probabilistic behaviour of individuals who consider a migratory act. Socio-economic factors on the other hand characterize the mean state of regions. Although both are correlated, utilities and mobilities belong to the microlevel of decisions, whereas socio-economic factors are macrovariables. This relation leaves space for different interpretations, and sometimes the "true" interpretation can be somewhat counter-intuitive and lead to intriguing results.

Let us illustrate this point by an example. Assume that the regional income per capita turns out to be a key-factor. In a direct interpretation this key-factor must correlate positively to the regional utility, since "higher regional income" means "higher utility or attractiveness of the region". This interpretation however naively presumes that the income of individuals participating in migration is directly linked to the average regional income. If, instead, the migrating persons (e.g. specialists) could typically expect higher salaries in the poorer destination region, the lower mean income of the latter region would be of no relevance for them, and the correlation of utilities with mean income could be reversed. More generally, the

Table 11.3. Key-factors and *t*-values of coefficients (in parantheses) of regional preferences for six countries

country	1. key factor	2. key factor	3. key factor	4. key factor
Federal Republic of Germany	(LS) (C) overnight stays per capita (17.7)	(II) export structure index (9.2)	(LM) unemployment rate (-5.7)	(LM) (PP) tertiary sector employment rate (5.0)
Canada	(LM) labour income per capita (12.3)	(LS) (PP) benefit payments (-5.6)	-	-
France*	(LM) unsatisfied job demand per capita (-26.2)	(HM) housing constructions per capita (18.9)	(C) temperature (15.5)	(LM) (PP) tertiary sector employment rate (11.1)
Israel	(SP) age structure (-3.2)	(LS) auto ownership (9.1)	(LS) persons per room (2.7)	-
Italy	(LM) employment index (12.4)	(HM) housing constructions (9.4)	-	-
Sweden	(LM) labour market variable (-7.8)	(PP) (LM) public sector employment (5.4)	(LM) rate of vacancies (3.2)	-

*as the fifth key-factor turned out: qualification (3.1)

correlation could be reversed to the counter-intuitive sign, if the regional utility for the average *migrant* does not coincide with the utility for the average *resident*.

In the comparative analysis of the following section we shall see, that almost all key-factors in all countries correlate to utilities and mobilities as expected by "direct" intuition. The few exceptions can be interpreted in a way indicated above.

11.2 Comparative Analysis of Quantitative Results

We begin the quantitative comparative analysis with a synopsis of the definitions, tables and figures used in this chapter.

11.2.1 Synopsis of Definitions, Tables and Figures

All analyses start with the individual probability transition rate from region i to j

$$p_{ji}(t)=v_{ji}(t)\exp[u_j(t)-u_i(t)]=v_0(t)\,f_{ji}\exp[u_j(t)-u_i(t)] \tag{11.4}$$

with mobilities $v_{ji}(t)=v_{ij}(t)$ and regional utilities $u_i(t)$. The global mobility is defined as the meanvalue of all mobilities over all regions

$$v_0(t)=\frac{1}{L(L-1)}\sum_{j,i}^{L}{}' v_{ji}(t) \tag{11.5}$$

or, equivalently, as a weighted average of the individual transition rates from all regions to all regions

$$v_0(t)=\frac{1}{L(L-1)}\sum_{j,i}^{L}{}' p_{ji}(t)\exp[u_i(t)-u_j(t)]\,. \tag{11.6}$$

The latter interpretation of $v_0(t)$ makes clear, that the value of the global mobility depends on the fineness of the regional subdivision with a scaling behaviour approximately proportional to $1/L$. In order to explain this, let us assume that each region i is divided into l equal parts i,α, with $\alpha=1,2,\ldots,l$, so that the total number of regions is now $L'=l\cdot L$ instead of L. How will $v_0(t)$ change to $v'_0(t)$ by this transition to a more fine-grained regional subdivision? One can assume, that the utility of each subregion i,α is again approximately $u_i(t)$. But the individual transition rate from any subregion i,α into one of the l subregions j,β of j will be reduced by the factor $1/l$, that means $p_{j\beta,i\alpha}\approx\frac{1}{l}p_{j,i}$. If also the newly arising transition rates between the subregions of one region, e.g. $p_{i\beta,i\alpha}$, are of the same order of magnitude as the $p_{j\beta,i\alpha}$, it can be concluded, that $v'_0(t)\approx\frac{1}{l}v_0(t)$.

In spite of its dependence on the fineness of the regional system, the global mobility thus turns out to be a directly interpretable and comparable quantity because it is an average individual probability transition rate from one to another region.

Since, however, the number and the distance of regions within the different countries does not coincide and since one is mainly interested in a comparison of the temporal evolution of global mobilities, the latter are represented in Figures 11.1a to 11.1f for the six countries in a form scaled to the temporal meanvalue 1.

The global mobility determined from migratory data by regression analysis in the frame of the model should not be confused with another measure of mobility, the total volume of interregional migration defined by

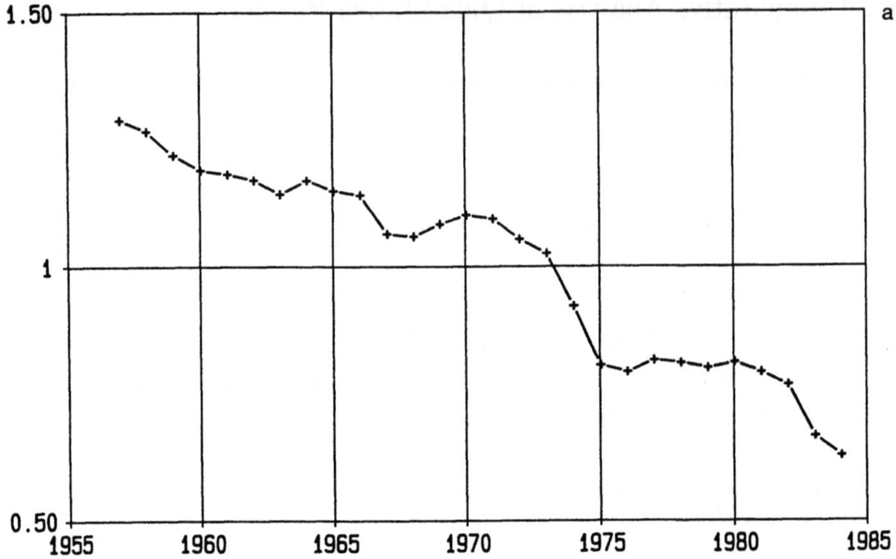

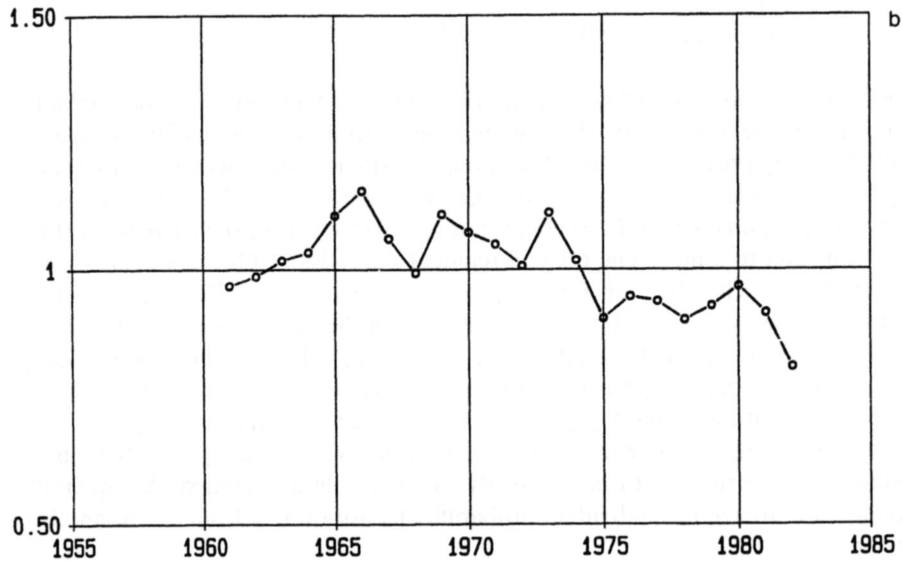

Fig. 11.1a–f. The global mobilities $\nu_0(t)$ of the six countries: a) Federal Republic of Germany (+), b) Canada (○), c) France (△), d) Israel (□), e) Italy (◇), f) Sweden (●) scaled to the temporal meanvalue 1

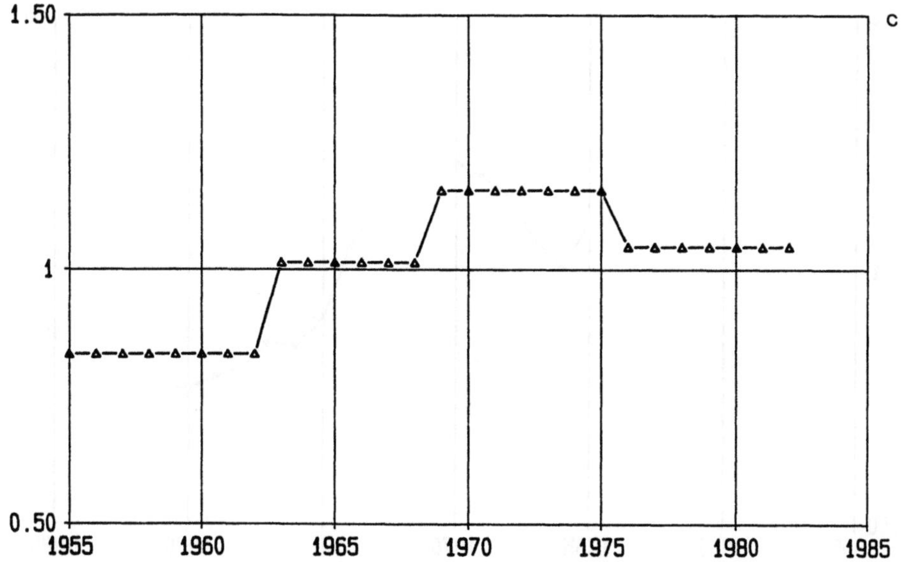
c
1.50
1
0.50
1955
1960
1965
1970
1975
1980
1985

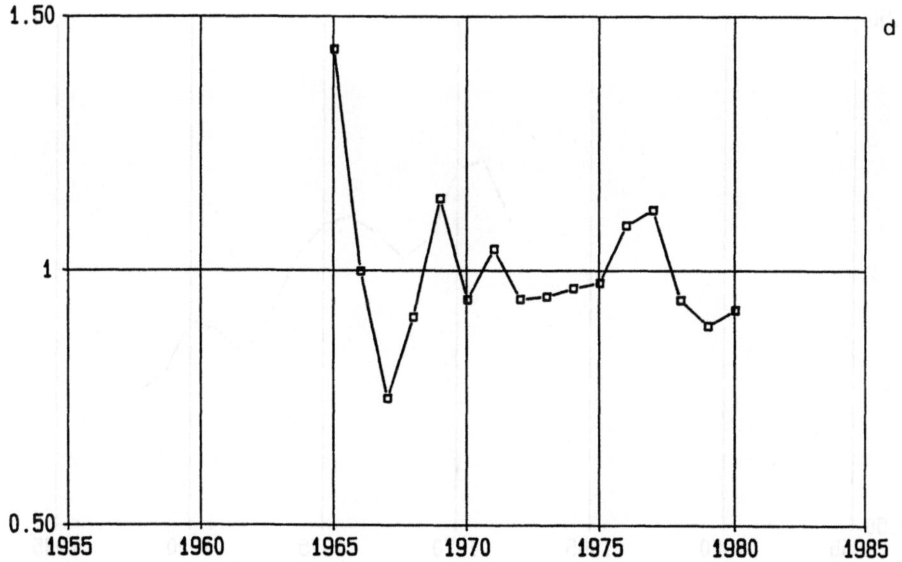
d
1.50
1
0.50
1955
1960
1965
1970
1975
1980
1985

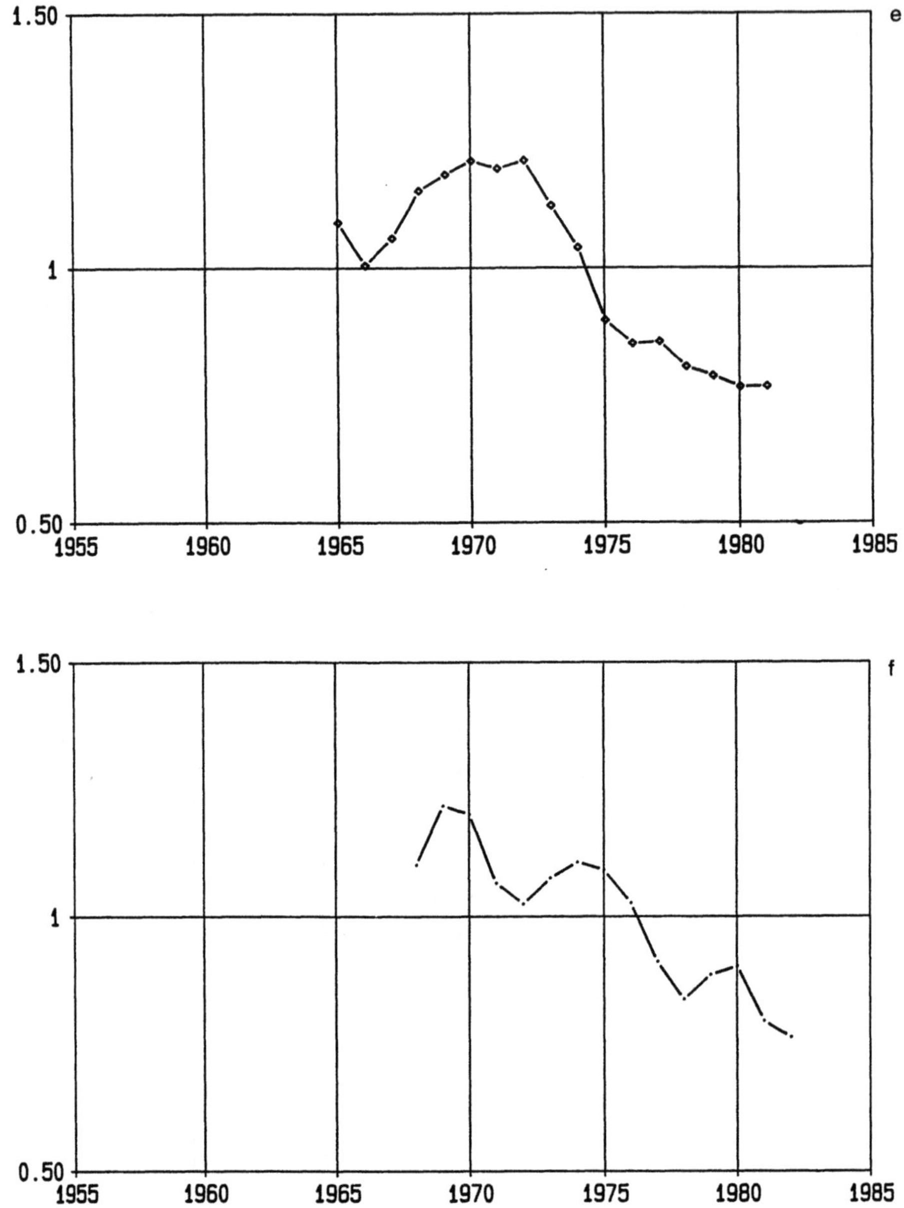

Fig. 11.1 e, f. Legend see p. 296

$$w_0(t) = \sum_{i,j}^{L}{}' w_{ji}(t) = v_0(t) \sum_{i,j}^{L}{}' f_{ji} n_i(t) \exp[u_j(t) - u_i(t)] \tag{11.7}$$

where

$$w_{ji}(t) = n_i(t) p_{ji}(t) = n_i(t) v_0(t) f_{ji} \exp[u_j(t) - u_i(t)] \tag{11.8}$$

is the (mean) number of migrants per unit of time from region i to region j. Since the sum on the r.h.s. of (11.7) is only weakly time-dependent, the evolution with time of $w_0(t)$ and $v_0(t)$ is proportional in practically all case studies. Deviations from this proportionality are due to the inclusion of the effects of natural growth in the case of $w_0(t)$ in contrast to $v_0(t)$. This can be seen for the German case by comparing Figures 4.2 and 4.4.

The global mobility has now been represented in terms of key-factors in five countries. The key-factors found to be relevant and the corresponding t-values of the coefficients of the multiple regression analysis are listed in Table 11.1. A regression analysis of the global mobility of Israel was not possible, because the socio-economic variables were not available on a yearly basis.

Let us now go over to the regional utilities. The stepwise regression of utilities on socio-economic variables has lead to the remarkable result, that in all six countries the regional population $n_i(t)$ and its square, $n_i^2(t)$, are the two most important key-factors. This means, that the regional utilities $u_i(t)$ can be decomposed as

$$u_i(t) = s_i(t) + \delta_i(t) \tag{11.9}$$

where $s_i(t)$ is the size-dependent part depending on $n_i(t)$ and $n_i^2(t)$ or on the scaled variables

$$\frac{[n_i(t) - \overline{n(t)}]}{\overline{n(t)}} = [x_i(t) - 1] \qquad \text{with} \quad x_i = \frac{n_i(t)}{\bar{n}(t)}$$

$$\frac{[n_i^2(t) - \overline{n^2(t)}]}{\overline{n(t)}^2} = \left[x_i^2(t) - \frac{\overline{n^2(t)}}{\overline{n(t)}^2}\right]. \tag{11.10}$$

The term $s_i(t)$ has the form

$$s_i(t) = \kappa[x_i(t) - 1] + \sigma\left[x_i^2(t) - \frac{\overline{n^2(t)}}{\overline{n(t)}^2}\right]. \tag{11.11}$$

The remaining terms $\delta_i(t)$ are the regional preferences describing the size-independent part of the utilities.

The agglomeration parameters κ and the saturation parameters σ for the six countries, together with their t-values, have been listed in Table 11.2, together with the symbols + (Federal Republic of Germany), ○ (Canada), △ (France), □ (Israel), ◇ (Italy), ● (Sweden), henceforth used in this chapter. Their values are slightly different from the corresponding coefficients in the tables of Part II, since the latter were slightly shifted by the use of further key-factors in the representation of $u_i(t)$. In Table 11.2 it is also noted, whether the homogeneous population distribution over the regions would be stable or unstable for the given values of κ and σ.

From (11.9) with (11.11) one can derive marginal utilities defined by

$$\frac{\partial u_i(t)}{\partial x_i} = \kappa + 2\sigma x_i(t) . \tag{11.12}$$

The marginal utility describes the strength of reaction of the utility of region i to changes of the scaled regional population $x_i(t)$. For a given country with fixed κ and σ the marginal utilities of its regions lie on a straight line with axis intercept $\kappa(>0)$ and slope $2\sigma(<0)$. The Figures 11.2a,b exhibit the situation; the points on the straight lines show the regions of the corresponding country.

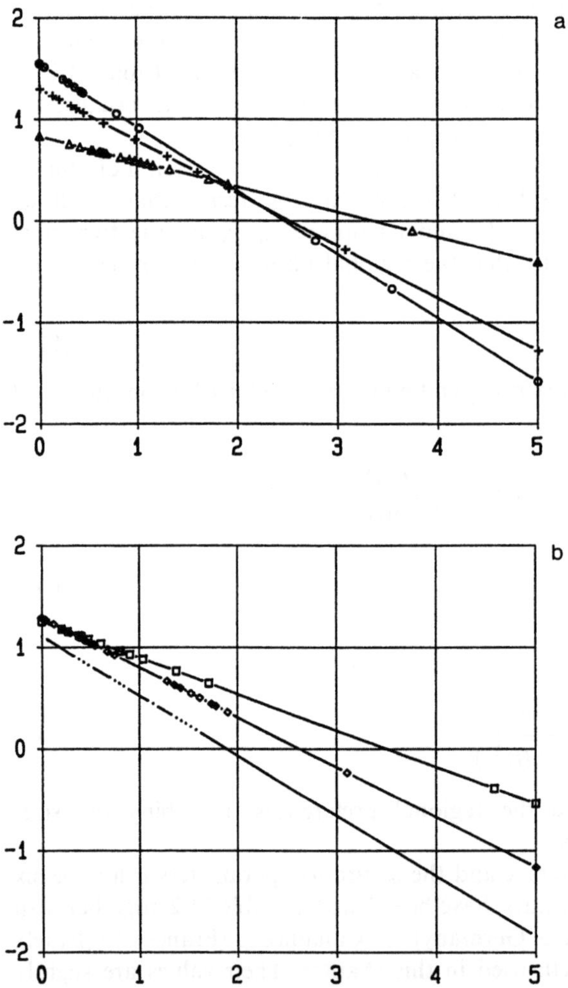

Fig. 11.2a,b. Marginal utilities for a) the countries FRG (+), Canada (○) and France (△), and b) the countries Israel (□), Italy (◇) and Sweden (●) as functions of $x_i = n_i/\bar{n}$. The points on the straight lines show the regions of the country

In order to characterize and to compare the regional differentiation of the countries one can introduce the variance of the utilities and of the preferences of each country, which are defined by

$$\sigma_u^2(t) = \frac{1}{L-1} \sum_{i=1}^{L} u_i^2(t) \tag{11.13}$$

and

$$\sigma_\delta^2(t) = \frac{1}{L-1} \sum_{i=1}^{L} \delta_i^2(t) \tag{11.14}$$

taking into account, that utilities and preferences fulfil the relations

$$\sum_{i=1}^{L} u_i(t) = 0; \sum_{i=1}^{L} \delta_i(t) = 0 \,. \tag{11.15}$$

Evidently, $\sigma_u^2(t)$ is a measure of regional differentiation including the effect of different sizes of regions, whereas $\sigma_\delta^2(t)$ measures regional differentiation without size effect, by only taking into account socio-economic differences.

Figure 11.3 compares the temporal evolution of the variances of the utilities of the six countries. On the other hand, in Figures 11.4a to 11.4f the variance of the utility is compared with the variance of preferences, separately for each of the six countries. The regional preferences have now been represented in terms of key-factors in the case studies for each of the six countries. The main results, namely the sequences of the most important key-factors and the t-values of the corresponding

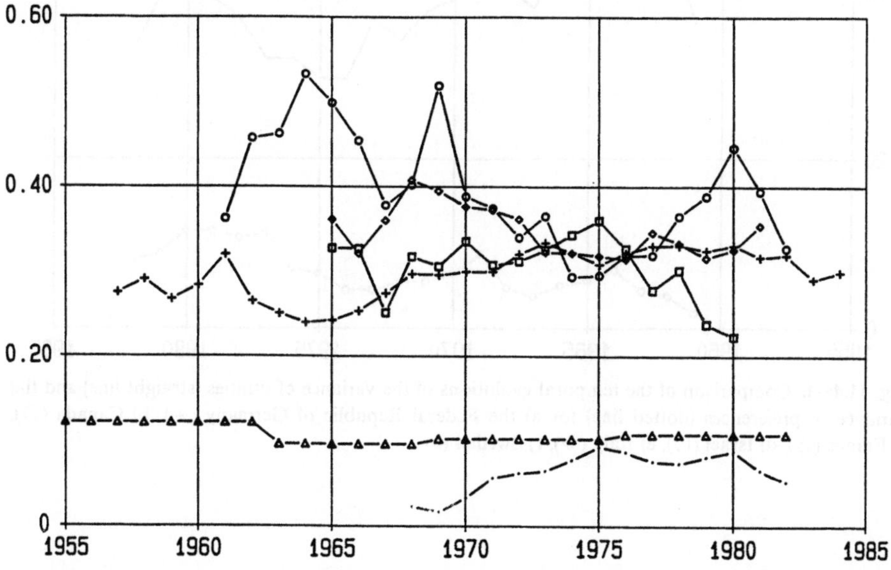

Fig. 11.3. Variances of utilities for the countries Federal Republic of Germany (+), Canada (○), France (△), Israel (□), Italy (◇) and Sweden (●)

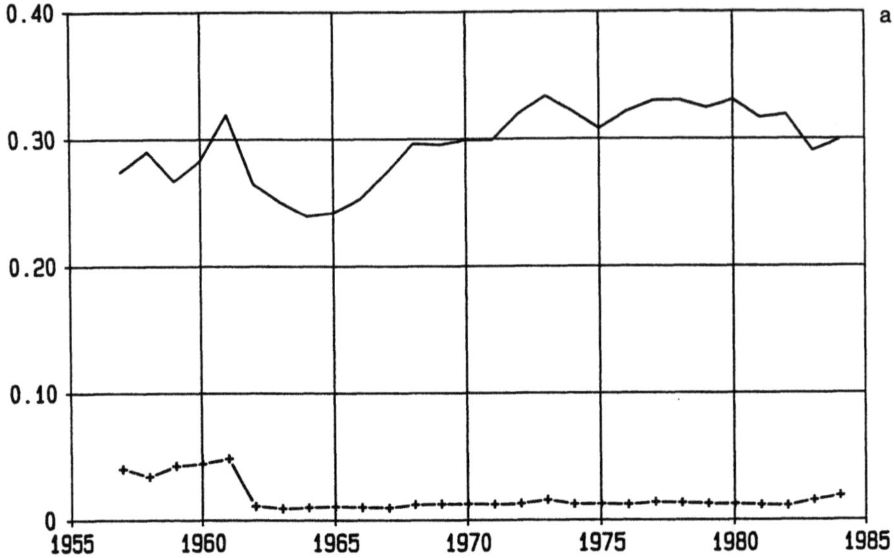

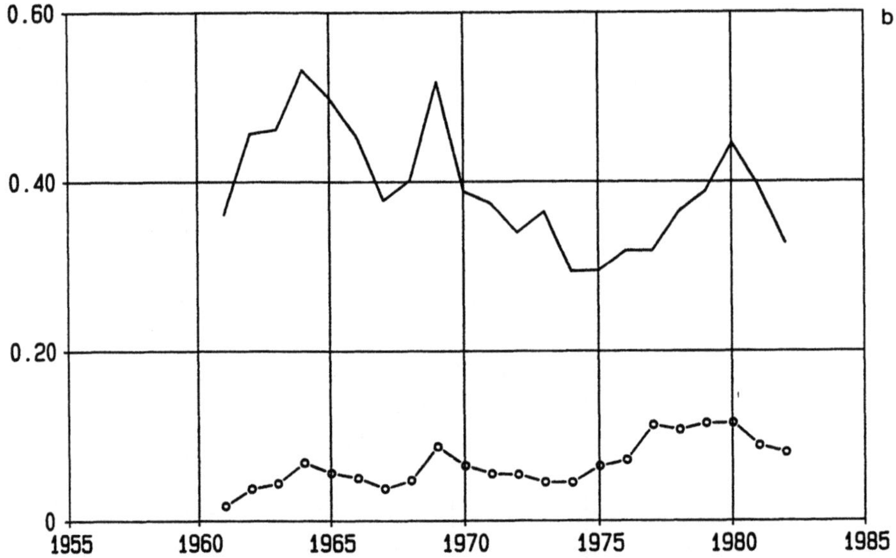

Fig. 11.4a–f. Comparison of the temporal evolutions of the variance of utilities (straight line) and the variance of preferences (dotted line) for a) the Federal Republic of Germany (+), b) Canada (○), c) France (△), d) Israel (□), e) Italy (◇), f) Sweden (●)

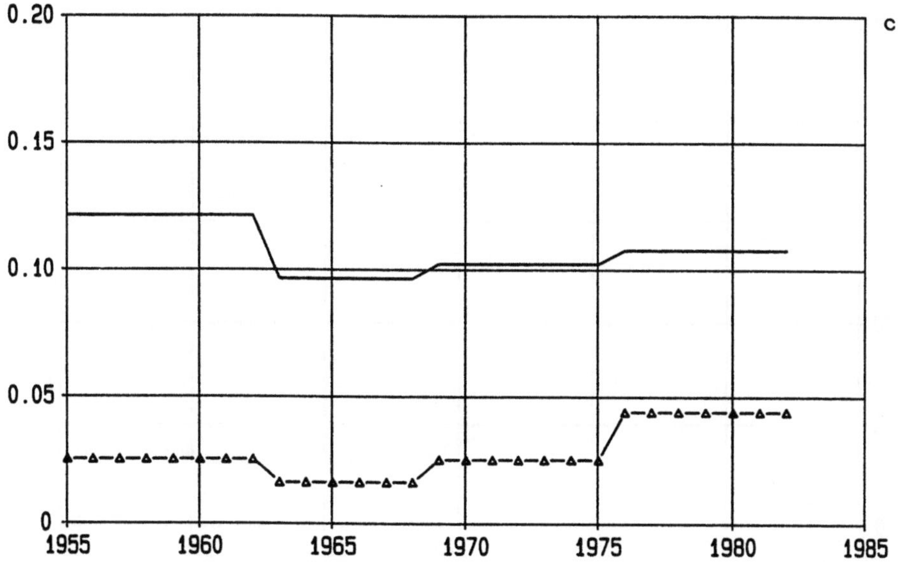

c
0.20
0.15
0.10
0.05
0
1955
1960
1965
1970
1975
1980
1985

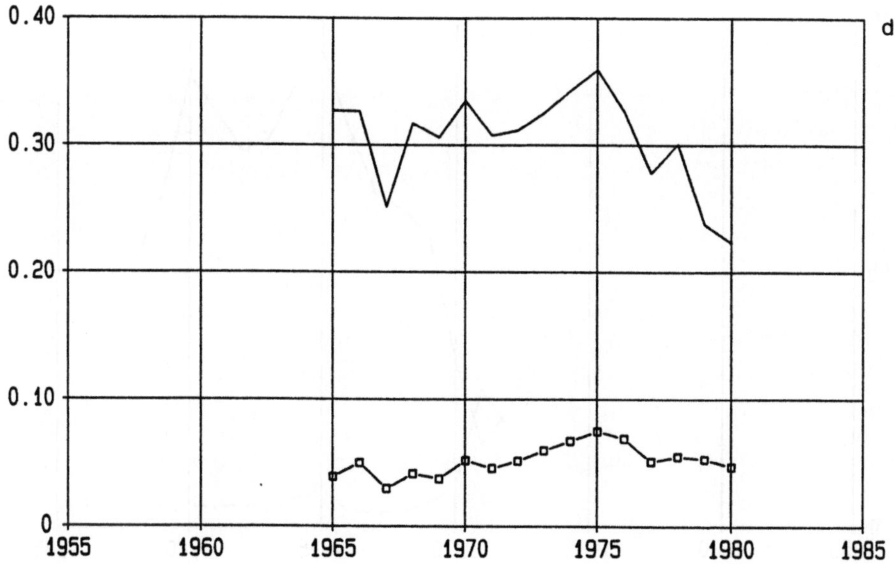

d
0.40
0.30
0.20
0.10
0
1955
1960
1965
1970
1975
1980
1985

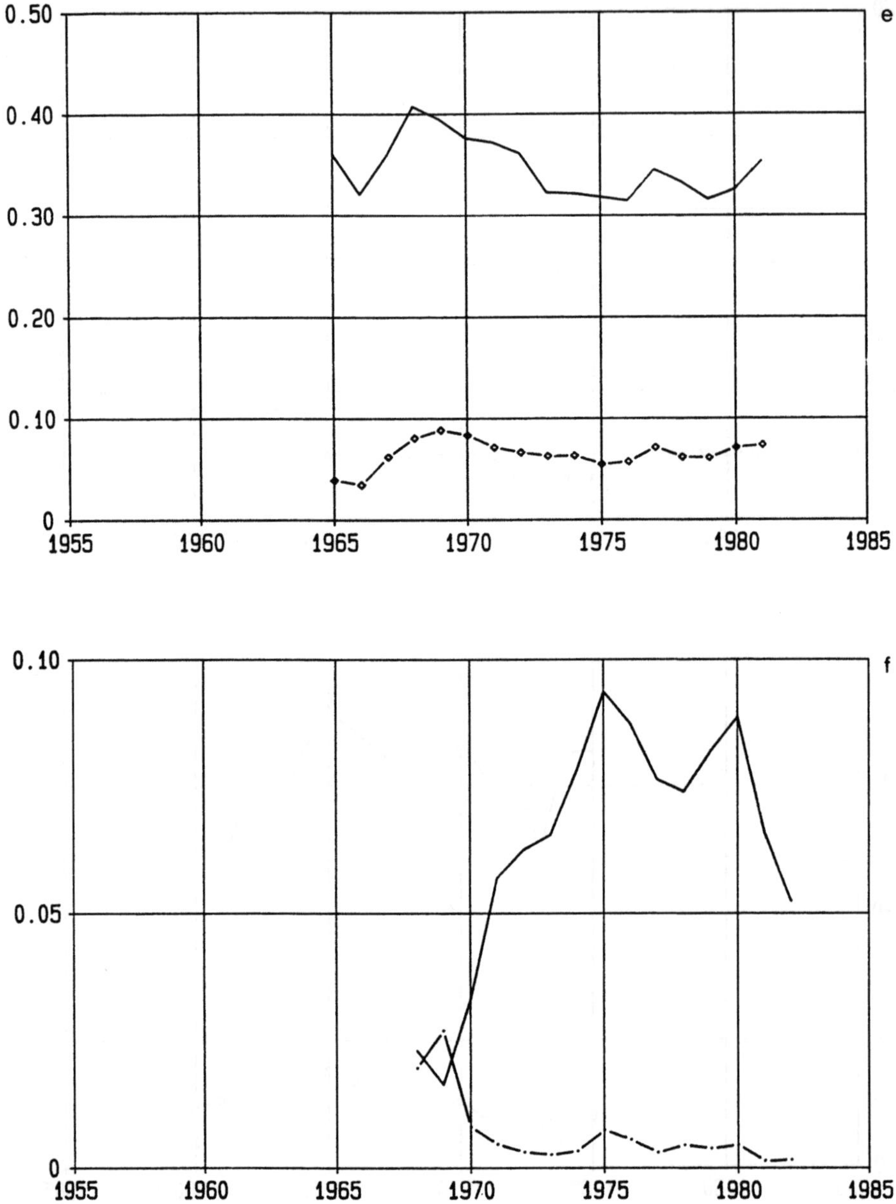

Fig. 11.4e, f. Legend see p. 302

coefficients are listed in Table 11.3. Furthermore it is indicated, to which class of socio-economic factors the key-factors belong.

Finally, the correlation coefficient $r(\boldsymbol{n}, \hat{\boldsymbol{n}})$ between the momentarily realized population distribution and the virtual equilibrium distribution is plotted in Figure 11.5 for all six countries. This correlation coefficient enters the definition of the "migratory stress"

$$s(\boldsymbol{n}, \hat{\boldsymbol{n}}) = \tfrac{1}{2}[1 - r(\boldsymbol{n}, \hat{\boldsymbol{n}})] \tag{11.16}$$

introduced in Chapter 2: The more $r(\boldsymbol{n}, \hat{\boldsymbol{n}})$ approaches 1, the smaller is the migratory stress, because then the population must already have approached the regional equilibrium distribution belonging to the momentary utilities.

11.2.2 Comparative Interpretations: The Global Mobility

We shall now try to interpret the material summarized in the synopsis under comparative aspects and in view of the general remarks of Section 11.1. In the "old" European countries Sweden, Italy and Germany a permanent long term decrease of the global mobility can be observed, which begins slowly after 1960 in Germany and after 1970 in Sweden and Italy. One might expect the same behaviour for France, in which however the mobility increases until 1975 and begins to decline only afterwards. Regional development programs may have lead to this trend differing from the other European countries.

The other reason for the prolongation of the increase of the mobility rate in France until 1975 is, that France comes later, as compared with countries like U.K.,

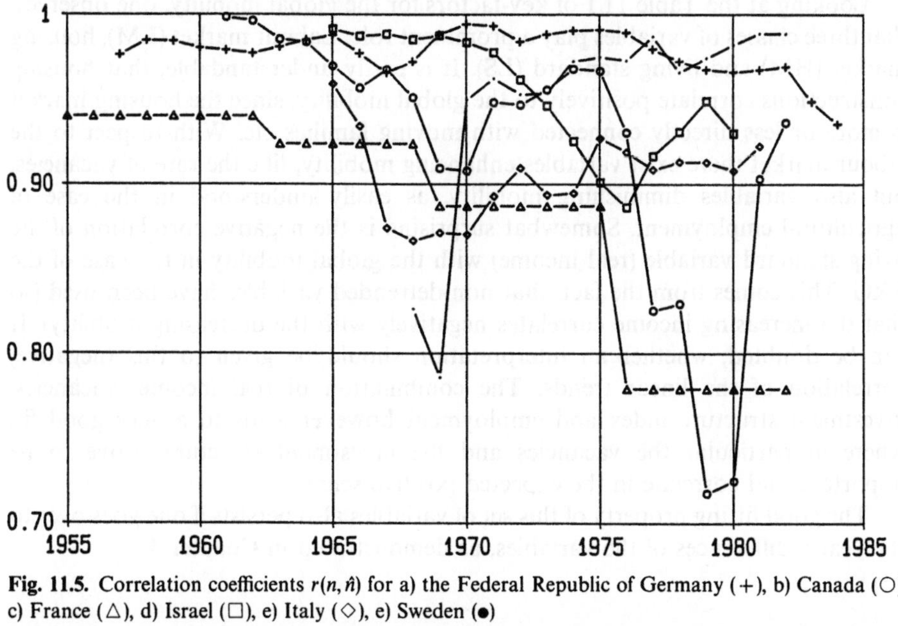

Fig. 11.5. Correlation coefficients $r(n, \hat{n})$ for a) the Federal Republic of Germany (+), b) Canada (○), c) France (△), d) Israel (□), e) Italy (◇), e) Sweden (●)

FRG or Sweden, in the urbanization process. Its urbanization rate reached only 57% in 1954 and 73% in 1975.

On the other hand, in Canada and Israel, whose histories both differ characteristically from those of European countries, only a slight decline or no decline at all of the mobility takes place. This may also be due to the fact, that the preceding immigration waves afterwards lead to a higher level of interregional mobility. Thus, Canada even showed a small but long prevailing trend of increased mobility, which was only stopped in the late sixties and reversed after the 1973 oil crisis. Israel has strong initial fluctuations of the mobility (mainly because of huge immigration waves) before it reaches a more stationary migration after 1970.

The finer details of the evolution of global mobilities superimposing the long-term trends are also of interest. (In France however such details cannot be analyzed because of the longer intervals between census data.)

Firstly the economic recession caused by the oil-prize shock seems to have everywhere influenced the global mobility in a negative sense. Its particularly steep decrease from 1972 to 1975 in Germany and Italy and its dip 1972 in Canada and Sweden may confirm this supposition.

Furthermore, short term business cycles have their effect on the mobility. Whenever economic booms can be identified in one, or simultaneously in several countries, as for instance 1969, the mobility exhibits a synchronous increase (see Figures 11.1). Since the correlations to labour market variables and to industrial investment are positive in all countries (see Table 11.1), it can be concluded more generally, that prosperity induces a higher level, and recession a lower level of mobility. The diminution of mobility in a period of recession of course may enhance the negative effects of the latter (if it should turn out, that the immobility refers to unemployed workers).

Looking at the Table 11.1 of key-factors for the global mobility, one observes, that three classes of variables play a prominent role: Labour market (LM), housing market (HM) and living standard (LS). It is easily understandable, that housing constructions correlate positively to the global mobility, since the housing market is more or less directly connected with moving families etc. With respect to the labour market there exist variables enhancing mobility, like the rate of vacancies, but also variables diminishing mobility, as easily understood in the case of agricultural employment. Somewhat surprising is the negative correlation of the living standard variable (real income) with the global mobility in the case of the FRG. This comes from the fact, that non-detrended variables have been used (so that the increasing income correlates negatively with the decreasing mobility). It can be doubted, whether an interpretation should be given to this (negative) correlation of the linear trends. The combination of real income, vacancies, investment structure index and employment however leads to a very good fit, where in particular the vacancies and the investment structure prove to be important and correlate in the expected positive sense.

The good fitting property of this set of variables also persists if one goes over to the yearly differences of the variables, as demonstrated in Chapter 4.

11.2.3 Regional Utilities and Preferences

As already mentioned the leading term in the representation of regional utilities is size-dependent and has the form (11.11). Since scaled population numbers have been used, the agglomeration parameter κ and the saturation parameter σ of different countries can immediately be compared (see Table 11.2). The regression analysis yields positive κ-values and negative σ-values for all six countries, which allow for the satisfactory and consistent interpretation, that the first term of $s_i(t)$ describes the agglomeration trend, and the second term the saturation trend for increasing values of $x_i(t)$.

Supplementary information about the size-dependence of utilities comes from the marginal utilities defined by (11.12) and depicted in Figures 11.2a,b. If the marginal utility is positive for a given region, the utility (attractiveness for migrants) of this region would still be growing with a further increase of population, whereas negative marginal utility means, that the region has already passed the saturation point, so that a further increase of its population would lead to a decline of its utility. The Figures 11.2a,b reveal, that the great majority of the regions of all countries have positive marginal utility, whereas there exist a few densely populated regions (namely Nordrhein-Westfalen (FRG), Ontario and Quebec (Canada), Ile de France (France), Tel Aviv (Israel) and Lombardia (Italy)), with slightly negative marginal utilities.

Now there follows a comparative view on the evolution of utilities and preferences of the six countries!

Since these quantities have been discussed in detail for the individual regions in the case studies of Part II, we now focus on the variances of utilities and preferences of the countries. They present global measures of regional differentiation, where, as already mentioned, the utilities include, and the preferences exclude the population size effect. Therefore, the comparative representation of the variances of utilities in Figure 11.3 partially only reflects the fact, that the spread of population sizes of the regions is different in different countries.

Thus the low and smooth value of this variance in France and the small but slowly increasing variance in Sweden seems to be due to a subdivision into regions of rather similar population size, in both cases done by administrators and planners in view of having a certain homogeneity. And on the other hand the higher variance values in the four other countries (FRG, Canada, Israel and Italy) comes from the natural inhomogeneity either between historically grown provinces or federal states, or between densely and sparsely populated regions.

But the temporal evolution of the utility variance seems to be of some interest: The variations with time are highest in the case of Canada. The peaks of this variance 1964, 1969 and 1980 coincide with peaks in the regional preferences of Alberta, British Columbia and Ontario, so that the fluctuations could be interpreted by a different pace of economic development between different regions, perhaps in connection with a prosperity depending on oil prospecting, at least in the case of Alberta.

In Italy one observes a decline of the utility variance from 1968 to 1975 remaining on this lower level afterwards. This trend to more homogeneity means, that migration, perhaps initiated by regional development programs, somewhat smoothed out the differences of regional utilities.

It is remarkable, that Israel's utility variance was increasing from 1967 to 1975, but declining afterwards until 1980. Looking at the regional utilities, one can see that the later decline is due to an increase of the utility of the non-metropolitan areas in the late seventieth, so that afterwards the country becomes regionally more homogeneous.

The Federal Republic of Germany has a practically stable regional differentiation with a very small long term increase, which is however not due to changes of preferences, but to population concentration effects.

Further information is derived from the comparison of the variances of utilities and of preferences for each of the six countries.

Two main conclusions can be drawn:

Firstly it can be seen, that for all six countries the variance of utilities is much higher than the variance of preferences. This confirms once more the importance of the population size term in the utilities and it allows for the conclusion, that the differences between the socio-economic situation of the regions within the countries are relatively moderate.

Secondly it turns out, that the variances of utilities and preferences exhibit a more or less parallel evolution with time, although utilities and preferences are not automatically and necessarily correlated. Nevertheless, it can be seen in most cases, in particular for Canada, France, Israel and Italy, that the strong relative fluctuations of the spread in preferences are accompanied by a somewhat milder, but parallel relative fluctuation in the spread of utilities, so that the population sizes must in general have reacted to the preference changes. Sweden seems to be an exception in this respect. It has an extremely low variance of preferences, which may be due to the fact, that incomes, social benefits and other regional socio-economic preference indicators are regionally equalized by government policy. Hence the (also small) utility spread is almost exclusively due to population size differentiation. In the case of the FRG one observes a particularly constant spread of preferences after 1961, and even the higher variance before 1961 is only due to the distorsion of preferences by the refugees coming from the GDR to Berlin West and migrating to the FRG before the wall in Berlin was built.

The Table 11.3 summarizing the key-factors of the regional preferences of all six countries is now useful in answering the question, which classes of socio-economic variables have recurrently – and therefore reliably – the highest weight as an indicator for the migratory process.

It is satisfactorily interpretable and not too surprising, that the few relevant key-factors selected by the ranking-procedure out of many variables for all six countries belong to the following classes, in sequence of their importance: Labour Market (LM), Living Standard (LS), and Public Sector (PP). Further variables of Industrial Structure (II) and of Housing Market (HM) are of complementary but not so prominent importance for describing the regional differentiation of migration in the six case studies under consideration.

11.2.4 Migratory Stress

Before comparing the correlation coefficients $r(\mathbf{n}, \hat{\mathbf{n}})$ between actual population distribution and the virtual equilibrium population, as depicted in Figure 11.5, one should recall the meaning of migratory stress defined by (11.16). For a given momentary set of regional utilities $\{u_1, u_2, \ldots, u_L\}$ there exist virtual equilibrium values

$$\hat{n}_i = c\exp(2u_i)$$

with (11.17)

$$c = N\left[\sum_{i=1}^{L}\exp(2u_i)\right]^{-1}$$

into which the actual population distribution would evolve, given that the utilities would persist for ever. The actually realized population distribution $\{n_i\}$ in general differs from the virtual equilibrium values $\{\hat{n}_i\}$ and the migratory stress is a measure for the "distance" between the actual and the equilibrium distribution.

The more the correlation coefficient $r(\mathbf{n}, \hat{\mathbf{n}})$ approaches 1, the smaller is the migratory stress (11.16) and the higher is the degree of coincidence of $\{n_i\}$ with $\{\hat{n}_i\}$. A growing migratory stress (or equivalently a decreasing correlation coefficient) means, that changes of preferences and a re-orientation of migratory trends have led to new regional utilities $\{u'_1, u'_2, \ldots, u'_L\}$ and a new virtual equilibrium $\{\hat{n}'_1, \hat{n}'_2, \ldots, \hat{n}'_L\}$, so that the actual distribution $\{n_i\}$ is now farther away from the new equilibrium $\{\hat{n}'_i\}$ than from the original equilibrium $\{\hat{n}_i\}$.

Another general remark refers to the comparison of models: The migratory stress can be taken as a measure for the expected agreement between the model used in this book and the gravity model: The reason is, that near equilibrium – but only in this case – the present model merges into a version of the gravity model. To see this, let us compare the form of the migration matrix of both models:

$$\text{(present model)} \quad w_{ji} = n_i p_{ji} = n_i \nu_0 f_{ji} \exp(u_j - u_i) \tag{11.18a}$$

$$\text{(gravity model)} \quad w_{ji}^G = g\frac{n_j^{\alpha_2} n_i^{\alpha_1}}{d_{ji}^{\alpha_3}} \tag{11.18b}$$

In the vicinity of the virtual equilibrium that is for $n \to \hat{n}$, it follows with (11.17), that

$$w_{ji} \Rightarrow \hat{n}_i \nu_0 f_{ji} \hat{n}_j^{1/2} \hat{n}_i^{-1/2} = \nu_0 f_{ji} \hat{n}_i^{1/2} \hat{n}_j^{1/2}. \tag{11.19}$$

This can be written in the form

$$w_{ji} = \nu_0 \frac{\hat{n}_j^{1/2+\beta} \hat{n}_i^{1/2+\beta}}{d_{ji}^{\gamma}} \tag{11.20}$$

if the "spatial deterrence factor" f_{ji} is identified with

$$f_{ji} = f_{ij} = \frac{\hat{n}_j^{\beta} \hat{n}_i^{\beta}}{d_{ji}^{\gamma}} \tag{11.21}$$

where $d_{ij}=d_{ji}$ is the geographical distance between regions i and j. That means, both models coincide for the case $\alpha_1=\alpha_2=\frac{1}{2}+\beta$, and $\alpha_3=\gamma$

It is satisfactory, that in a migratory system near equilibrium, namely the FRG, the quantitative analysis on the basis of the gravity model leads to the following values of the parameters listed in Table 11.4, so that $\alpha_1=\alpha_2$ is approximately fulfilled [11.1].

The concrete comparison of the correlation coefficients $r(\boldsymbol{n}, \hat{\boldsymbol{n}})$ of the six countries roughly leads to the result, that there exist three groups.

The first group, consisting of the FRG and Sweden, is very near to equilibrium in the observation period. This is understandable in the case of Sweden with a smooth, undisturbed and balanced development of regions. It is more remarkable, that in the case of the Federal Republic of Germany already after 1957 the migratory equilibrium has almost been reached, in spite of the over 10 million post war migrants from the former eastern parts of Germany. This can be interpreted as the early integration of these migrants into their new regional environment. One explanation of this phenomenon may be, that the Akademie für Raumforschung und Landesplanung (Hannover) made strong efforts from the very beginning of the FRG to elaborate plans for the distribution of the new citizens over the states.

The second group consists of France and Canada. Both countries have strong and growing deviations from virtual equilibrium.

France shows a steady decrease of the correlation coefficient to a value below 0.8 after 1975. This fact may be connected with a general reorientation of the migratory trends from centralization to a preference for "peripheral" regions. It is understandable that it needs time until the French system reaches the new equilibrium in accordance with the new trend. It is also interesting that a comparison of the gravity model with the model of this book leads to relatively strong discrepancies in the case of France (see Section 6.6). This is due to the fact, that both models do not coincide in nonequilibrium situations.

Canada also exhibits strong fluctuations of the correlation coefficient, with relative minima of 0.9 in 1970 and 0.7 in 1980. It can be observed, that the migratory stress is growing (the correlation coefficient is decreasing) parallel to peaks of the utility variance. The relatively sudden arising of economic booms in Alberta, British Columbia and Ontario has set the (not immediately adapting) migratory system into stronger deviations from its virtual equilibrium, hence into peaks of migratory stress.

Table 11.4. Estimated exponents of the gravity models

	1964	1969	1971	1975
α_1	0.950	0.855	0.861	0.857
α_2	0.967	0.945	0.920	0.905
α_3	0.949	0.952	0.942	1.036

The third group consists of Israel and Italy. Both countries exhibit moderate migratory stress having a relatively smooth evolution with time.

Italy produces a minimum of the correlation coefficient at 0.87 in the late sixtieth, and its slow permanent increase afterwards. Regional development programs set into effect in the sixtieth and initiating new migration trends may have caused this peak of the migratory stress, which slowly reduces on the way to equilibrium afterwards.

Israel also starts with a nearly equilibrium situation before 1970, until a maximal deviation from equilibrium is reached 1976 (correlation coefficient 0.9). Afterwards, a re-approach of the equilibrium begins. Similar to the Canadian case, this goes parallel to the evolution of the variance of utilities: Newly arising regional attractiveness have led to a momentary migratory nonequilibrium about 1976, which later by adaptive migration re-approaches equilibrium.

Acknowledgement

One of the authors (W.W.) wishes to thank Prof. Herwig Birg for intense discussions about this chapter. They were a great help in the preparation of the manuscript.

IV. Mathematical Methods

Synopsis of Part IV

This part is devoted to mathematical derivations in connection with the general model theory and to complementary details referring to the evaluation methods.

The master equation is fundamental for probabilistic dynamic processes with a range of applications reaching far beyond migration theory. In view of its importance, Chapter 12 gives an independent and selfcontained derivation of the master equation embedding it into general concepts of probability theory. In Section 12.3 the relation between individual and configurational transition probabilities is derived in detail for the migratory master equation, being the basic equation of the model of this book.

In Chapter 13 selected explicit solutions of the master equation are set up. If the transition probabilities fulfil the condition of detailed balance (see Section 13.1) – as in the case of the migratory master equation – the stationary probability distribution can be constructed explicitly, as exemplified in the case of the migratory system, where regional utilities still may depend on population numbers (see Section 13.2). For the case of time-dependent solutions at first in Section 13.3 the famous H-theorem and its relation to the concept of entropy is derived, stating that every time-dependent solution approaches the stationary solution for $t \to \infty$, whenever the transition probabilities do not explicitly depend on time. In Section 13.4 the special case of time-dependent transition rates linear in population numbers is discussed, for which an exact time-dependent solution of the master equation can be found, demonstrating the relation between the probability distribution and the meanvalues.

Chapter 14 gives an introduction to the statistical tests of significance made in all applications of Part II to secure the validity of results. After the definition of the coefficient of multiple correlation, in particular the F-test, the t-test and the Durbin-Watson-Test are briefly introduced and commented. Furthermore, the distribution of residuals are exhibited for different versions of the model in the case of the FRG to illustrate the degree of validity of the model.

Chapter 15 exemplifies the method of stepwise representation of migratory quantities in terms of socio-economic variables by the Ranking Regression Analysis of the global mobility. This chapter complements Chapter 3 of Part I, where the analogous analysis is done for the regional utilities. The explicit example of the global mobility of Sweden illustrates the kind of problems arising in the sequential choice of key-factors, and demonstrates how the systematic selection of two key-factors (out of seven socio-economic variables) leads to an excellent fit of the global mobility.

Finally, in Chapter 16 a FORTRAN program for the estimation of utilities and mobilities on the basis of the log-linear estimation including an explicit numerical example is listed with the aim of enabling the reader to test the program and perhaps to apply it to his own country.

12 Derivation of the Master Equation

Wolfgang Weidlich

This section provides a selfcontained derivation of the master equation approach in general, complementary to that given in Part I. Starting from a general comparison of the deterministic and probabilistic evolution of systems, some constitutive concepts of probability theory are introduced. From these the general master equation is derived. The latter can easily be specialized to assume the form (2.6) appropriate for the migratory system. Finally, the relation between individual and configurational transition probabilities is derived.

Let us first compare the deterministic with the probabilistic description of evolving systems.

Consider a system which can pass through different states in the course of time. For simplicity, we assume that the states of the system are discrete (noncontinuous), so that the index i characterizing each state is a discrete number or a set $i=\{i_1, i_2, \ldots, i_n\}$ of discrete numbers. Let us further assume that at an initial time t_0 the system is in state $i_0 = i(t_0)$. Two possible descriptions of the evolution with time are then feasible:

a) The information about the dynamics of the system can be *complete*. In this case the description is fully deterministic and leads to the unique determination of the states $i(t)$ at later times.

Any computer can be taken as an example for such a *deterministic system* with a finite – though very large – number of states i. Beginning with an initial state i_0 set by the program, the central processor, the memory units and the peripheral devices of the computer traverse a sequence of states $i_1, i_2, \ldots, i_N$ which are fully predetermined by the prescribed program. The unique result of the calculation represents the final state of the system.

b) The information about the dynamics of the system on the other hand can be *incomplete*. In this case the description of the time evolution can only be a *probabilistic* one. That means, an exact prediction of the state $i(t)$ reached by the system is not possible. Instead, the members of an ensemble of identical systems – each of them prepared in the same initial state $i(t_0)$ – will develop into *different* states $i(t)$ at time t. The best information available in this situation is the *probability* that a system reaches the state i at time t, given that it was prepared in state i_0 at time t_0. This special probability $P(i,t|i_0,t_0)$ is denoted as "conditional probability". The master equation will turn out to be the tool for determining this quantity.

Comparing the deterministic with the probabilistic evolution we must state that the latter is the more general form of dynamics, since the case of incomplete knowledge about the system comprises complete knowledge as a limit case whereas the converse is not true. The limit case of (almost) complete knowledge of the dynamics is revealed by the shape of the probability distribution $P(i, t)$ itself: In this case the master equation leads to an evolution of the distribution such that it develops one outstanding mode sharply peaked around the most likely state $i_{\text{Max}}(t)$. This means that the system assumes the state $i = i_{\text{Max}}(t)$ with overwhelming probability at time t, whereas all other states $i \neq i_{\text{Max}}(t)$ are highly improbable at the same time. Evidently this case describes a *quasi-deterministic evolution* of the system along the path $i(t) \approx i_{\text{Max}}(t)$.

12.1 Some General Concepts of Probability Theory

Before deriving the master equation it is useful to introduce some fundamental concepts of probability theory that especially apply to systems whose evolution is described in probabilistic terms, (see also [12.1, 2, 3, 4]).

As before, it is assumed that the system can be in one of mutually exclusive different states which are characterized by an index i consisting of one discrete number or a multiple of discrete numbers. In the course of time, transitions between different states can take place. Since in general one does not know with certainty in which state the system is, the probability distribution function

$$P(i, t) \tag{12.1}$$

is introduced.

By definition, $P(i, t)$ is the probability of finding the system in state i at time t. This probability has the following statistical interpretation: In an ensemble of a large number of initially equally prepared systems – so that each of them belongs to the same probability distribution – one would find systems in state i at time t with approximately the *relative frequency* $P(i, t)$. Going to the limit case of an *infinite* ensemble its systems are found in state i with *exactly* the relative frequency $P(i, t)$.

Since the system is with certainty in one of the states i at any time t, the probability function has to satisfy the condition

$$\sum_{\{i\}} P(i, t) = 1 \tag{12.2}$$

where the sum extends over all states i.

Furthermore, we now introduce the most important quantity for the time evolution of the system, the *conditional probability*

$$P(i_2 t_2 | i_1 t_1, \text{p.h.}) \ . \tag{12.3}$$

By definition it is the probability to find the system in the state i_2 at time t_2, given that it was with certainty in the state i_1 at time t_1. The conditional probability is fundamental for the dynamics of the system, since it describes how the probability spreads out in the time interval $(t_2 - t_1)$, given that it was concentrated on the state i_1 at time t_1. The letters p.h. in (12.3) indicate that the conditional probability may also depend on the *previous history* of the system, that is on states traversed *before* arriving at state i_1 at time t_1. In this general case the probability evolution process may become very complicated.

Fortunately, in many dynamic systems the so called *Markov assumption* holds, at least as a good approximation. The Markov assumption postulates that the evolution within time of the conditional probability $P(i_2, t_2 | i_1, t_1)$ *only depends* on the initial state i_1 at time t_1 and the state i_2 at t_2, but *not* on states of the system prior to t_1. In other words: After arriving at state i_1 at time t_1 the system has "*lost its memory*" so that states before t_1 do not matter in the process of the further evolution. Since in many cases the systems can be defined such that the Markov assumption holds we shall presume it below.

Let us now draw some conclusions about the properties of the conditional probability. As a consequence of the definition the following equations must be satisfied

$$\left.\begin{array}{lll} P(i_2 t_1 | i_1 t_1) = \delta_{i_2 i_1} & & \\ \text{where} \quad \delta_{i_2 i_1} = 1 & \text{for} & i_2 = i_1 \\ \text{and} \quad \delta_{i_2 i_1} = 0 & \text{for} & i_2 \neq i_1 \end{array}\right\} \tag{12.4}$$

and

$$\sum_{\{i_2\}} P(i_2 t_2 | i_1 t_1) = 1 \; . \tag{12.5}$$

Equation (12.4) follows because at time $t_2 = t_1$ the probability is still fully concentrated at the state i_1. Equation (12.5) holds, since the system at any time t_2 will have to be *with certainty* in one of the states $\{i\}$ of the system.

From (12.3, 4, 5) it becomes clear that the conditional probability is nothing but a special probability distribution $P_{i_1 t_1}(i_2, t_2)$ namely that one which evolves from the initial distribution $P(i, t_1) = \delta_{i i_1}$.

A further important concept is the joint probability

$$P(i_n t_n, i_{n-1} t_{n-1}, \ldots, i_2 t_2, i_1 t_1) \; . \tag{12.6}$$

By definition the *n*-fold joint probability is the probability to find the system in state i_1 at t_1 and jointly in state i_2 at t_2 ... and jointly in state i_{n-1} at t_{n-1} as well as in state i_n at t_n. From this definition it follows that the lower joint probabilities can be obtained from the higher ones by the following reduction formula:

$$P(i_3 t_3, i_1 t_1) = \sum_{\{i_2\}} P(i_3 t_3, i_2 t_2, i_1 t_1) \tag{12.7}$$

or, in the general case, by

$$P(i_n t_n, \ldots, i_{k+1} t_{k+1}, i_{k-1} t_{k-1}, \ldots, i_1 t_1)$$

$$=\sum_{\{i_k\}} P(i_nt_n, \ldots, i_{k+1}t_{k+1}, i_kt_k, i_{k-1}t_{k-1}, \ldots, i_1t_1)\,. \tag{12.8}$$

Clearly, the summation in (12.7) of $P(i_3t_3, i_2t_2, i_1t_1)$ over all possible states at t_2 leads to the probability of being in state i_1 at t_1 and jointly in state i_3 at t_3 *irrespective* of the state at time t_2. That means it leads to $P(i_3t_3,\, i_1t_1)$.

If we make the Markov assumption, then all joint probabilities can be expressed by the simple probability (12.1) and the conditional probability (12.3). In particular, the two-fold joint probability has the form

$$P(i_2t_2, i_1t_1) = P(i_2t_2|i_1t_1)\,P(i_1t_1) \tag{12.9}$$

since the probability to find the system in state i_1 at t_1 *and* in state i_2 at t_2 is synonymous with the probability to find it in state i_1 at t_1 multiplied with the (conditional) probability to find it in state i_2 at t_2, *given that it was* in i_1 at t_1.

Generalizing this consideration and taking into account that the conditional probability does not depend on the previous history, if the Markov condition holds, we obtain

$$\begin{aligned} &P(i_nt_n, i_{n-1}t_{n-1}, \ldots, i_2t_2, i_1t_1) \\ &\quad = P(i_nt_n|i_{n-1}t_{n-1}), \ldots, P(i_2t_2|i_1t_1)\cdot P(i_1,t_1)\,. \end{aligned} \tag{12.10}$$

The *composition formulas* (12.9) and (12.10) may now be combined with the *reduction formula* (12.7) in order to derive the famous Chapman-Kolmogorov equation. Taking the sum over i_1 in (12,9) and using the reduction formula (12.8) yields

$$\begin{aligned} &\sum_{\{i_1\}} P(i_2t_2, i_1t_1) = P(i_2,t_2) \\ &\qquad\qquad = \sum_{\{i_1\}} P(i_2t_2|i_1t_1)\cdot P(i_1,t_1)\,. \end{aligned} \tag{12.11}$$

Equation (12.11) shows how the probability distribution $P(i,t)$ is propagated in the course of time by means of the conditional probability $P(i_2t_2|i_1t_1)$. Therefore, the latter is also referred to as *the propagator*. On inserting (12.9) and (12.10) in (12.7) there follows the relation

$$P(i_3t_3|i_1t_1)\,P(i_1,t_1) = \sum_{\{i_2\}} P(i_3t_3|i_2t_2)\,P(i_2t_2|i_1t_1)\,P(i_1,t_1) \tag{12.12a}$$

Since this equation must hold for an arbitrary initial distribution $P(i_1,t_1)$ one can conclude that also

$$P(i_3t_3|i_1t_1) = \sum_{\{i_2\}} P(i_3t_3|i_2t_2)\,P(i_2t_2|i_1t_1) \tag{12.12b}$$

holds. Equation (12.12b) is the well-known *Chapman-Kolmogorov equation.* It shows how the propagator from t_1 to t_3 can be decomposed into propagators from t_1 to t_2 and from t_2 to t_3.

12.2 The Master Equation

We have now seen that the propagator $P(i_2t_2|i_1t_1)$, that is the conditional probability, is the crucial quantity determining the evolution with time of any probability distribution $P(i,t)$. This follows from equation (12.11).

The master equation now is nothing but a differential equation in time for the propagator, or for the probability distribution. It can be derived by considering equation (12.11) for times $t_1=t$, $t_2=t+\tau$ where τ is an (infinitesimally) short time interval. Applying (12,11) in this way, we obtain the short-time evolution

$$P(i_2,t+\tau)=\sum_{\{i_1\}} P(i_2,t+\tau|i_1 t)P(i_1,t) \ . \qquad (12.13)$$

The short-term propagator is now being expanded in a Taylor series around t with respect to the variable $t_2=t+\tau$ yielding

$$P(i_2,t+\tau|i_1 t)=P(i_2 t|i_1 t)+\tau \left.\frac{\partial P(i_2 t_2|i_1 t)}{\partial t_2}\right|_{t_2=t} +O(\tau^2) \ . \qquad (12.14)$$

Making use of (12.4) and (12.5) in (12.14) one obtains

$$P(i_2 t\,|\,i_1 t)=\delta_{i_2 i_1}$$

$$\sum_{\{i\}} \left.\frac{\partial P(it_2|i_1 t)}{\partial t_2}\right|_{t_2=t} =0 \qquad (12.15)$$

where the sum extends over *all* states i of the system. Re-inserting of (12.15) in (12.14) yields

$$P(i_2,t+\tau|i_1 t)=\tau w_t(i_2|i_1)+O(\tau^2)$$

$$\text{for} \quad i_2 \neq i_1 \qquad (12.16a)$$

and

$$P(i_2,t+\tau|i_1 t)=1-\tau \sum_{i\neq i_1} w_t(i|i_1)+O(\tau^2)$$

$$\text{for} \quad i_2 = i_1 \qquad (12.16b)$$

where the probability transition rate

$$w_t(i_2|i_1)\equiv \left.\frac{\partial P(i_2 t_2|i_1 t)}{\partial t_2}\right|_{t_2=t} \qquad (12.17)$$

has been introduced for $i_1 \neq i_2$, and where the higher powers in τ can be neglected since they will play no role in the limit $\tau \to 0$.

What is the *interpretation* of (12.16)? Given that the system was in state i_1, the probability (12.16a) that it reaches state $i_2(\neq i_1)$ in the infinitesimally short time interval τ is proportional to that interval and to the transition rate from i_1 to i_2. On the other hand, the probability (12.16b) to remain in the same state during the interval τ is equal to one minus the probability transferred to all other states within that time interval.

The Master Equation is now established by inserting (12.16) into (12.13). Dividing through τ it follows after trivial rearrangements of terms by going to the limit $\tau \to 0$. *The Master Equation* reads

$$\frac{dP(i_2,t)}{dt} \equiv \lim_{\tau \to 0} \frac{P(i_2,t+\tau)-P(i_2,t)}{\tau}$$

$$= \sum_{i_1} w_t(i_2|i_1)P(i_1,t) - \sum_{i_1} w_t(i_1|i_2)P(i_2,t) \tag{12.18}$$

where the sums extend over all $i_1 \neq i_2$.

Since the master equation (12.18) is valid for *any* probability distribution $P(i,t)$, it holds in particular for the conditional probability $P(it|i_0t_0)$. Let us now give an illustrative interpretation of the master equation. The quantities $w_t(i_2|i_1)$ designate probability *transition rates per unit of time* in the following sense: $\tau \cdot w_t(i_2|i_1)P(i_1,t)$ is the probability transferred from state i_1 to the state i_2 in the time interval τ. The quantity $w_t(i_2|i_1)P(i_1,t)$ therefore is referred to as the *probability* flow (per unit of time) from i_1 to i_2. Then the master equation (12.18) can also be read as a *rate equation for probabilities*:

The change per unit of time of the probability of state i_2, that is the l.h.s. of (12.18) is the consequence of two counteractive terms: Firstly, there is the probability flow *from* all other states i_1 *into* state i_2: This is the first term of the r.h.s. of (12.18). Secondly, there is the probability flow *out of* state i_2 *into* all other states i_1. This is the second term of the r.h.s. of (12.18). Both terms have opposite sign and their net effect determines the increase or decrease per unit of time of the probability $P(i_2,t)$.

The master equation gives the most detailed knowledge about the evolution of a system under conditions of uncertainty and restricted information. The quantities representing this restricted information are the transition rates, namely, the transition probabilities per unit of time $w_t(i_2|i_1)$ to reach i_2 from i_1.

In our model the transition rates are represented as functions of dynamic utilities and mobilities (see Section 1.3).

It is now easy to convert the general master equation (12.18) into the migratory master equation (2.6). We only have to identify the states of the system and the special form of the probability transition rates. Evidently, the state of the system is defined by the population numbers in each of the L regions. This means we have to identify

$$i \Rightarrow \boldsymbol{n} = \{n_1, n_2, \ldots, n_L\} \ . \tag{12.19}$$

The probability transition rates $w_t(\boldsymbol{n}'|\boldsymbol{n})$ are given by (1.21a) and (1.21b):

$$w_t(\boldsymbol{n}^{(ji)}|\boldsymbol{n}) \equiv w_{ji}(\boldsymbol{n},t) = n_i p_{ji}(t)$$

$$\text{for} \quad \boldsymbol{n}' = \boldsymbol{n}^{(ji)} = \{n_1, \ldots, (n_i - 1), \ldots, (n_j + 1), \ldots, n_L\} \tag{12.20a}$$

and

$$w_t(\boldsymbol{n}'|\boldsymbol{n}) = 0$$

$$\text{for} \quad \boldsymbol{n}' \neq \boldsymbol{n}^{(ji)} . \tag{12.20b}$$

Therefore the sums on the right hand side of the master equation must only be extended over states $\boldsymbol{n}'$ with nonvanishing probability transition rates, thus arriving at the form

$$\frac{dP(\boldsymbol{n},t)}{dt} = \sum_{i,j}{}' w_{ji}(\boldsymbol{n}^{(ij)},t)\,P(\boldsymbol{n}^{(ij)},t) - \sum_{i,j}{}' w_{ji}(\boldsymbol{n},t)\,P(\boldsymbol{n},t) \tag{12.21}$$

which was also obtained in Section 2.

12.3 Individual and Configurational Probability Transition Rates

Finally the relations (12.20a, b) or equivalently (1.24a, b) will now be derived. For this purpose we have to establish the relationship between individual and configurational conditional probabilities (or propagators).

By definition, the *individual conditional probability* $P(j,t+\tau|i,t)$ is the probability to find the individual migrant in region j at time $t+\tau$, given that he was in region i at time t (see 1.9b)). On the other hand, the *configurational conditional probability* $P(\boldsymbol{n}',t+\tau|\boldsymbol{n},t)$ is the probability to find the configuration $\boldsymbol{n}'$ at time $t+\tau$, given that the configuration $\boldsymbol{n}$ was realized at time t (see (1.18)).

The relation between both kinds of conditional probabilities turns out to be a rather subtle combinatorial problem. Fortunately, however, this relation becomes simple for infinitesimally small τ, for which expansion terms of order $O(\tau^2)$ can be neglected.

Let us denote the conditional probability of an individual v (where $v=1,2,\ldots,N$) to migrate from region i_v to j_v in the time interval τ by $P(j_v,t+\tau|i_v,t)$. We assume, that initially (at time t) n_1 people are in region 1, n_i in region i, and n_L in region L. This evidently constitutes the initial migratory configuration $\boldsymbol{n}=\{n_1,\ldots,n_i,\ldots,n_j,\ldots,n_L\}$. Without restriction of generality the enumeration can be chosen such, that the first n_1 individuals $v=1,2,\ldots,n_1$ are in region $i_v=1$, the next n_2 individuals $v=n_1+1,\ldots,n_1+n_2$ in region $i_v=2\ldots$ and the last n_L individuals $v=N-n_L+1,\ldots,N$ in region $i_v=L$.

How probable is it now, to find at time $(t+\tau)$ simultaneously individual 1 in region j_1, individual v in j_v, and finally individual N in j_N? Since the individuals by assumption migrate in a statistically independent manner, this probability is given by the *product* of the individual conditional probabilities:

$$P(j_1,j_2,\ldots,j_N,t+\tau|i_1,i_2,\ldots,i_N,t)=\prod_{\nu=1}^{N}P(j_\nu,t+\tau|i_\nu,t)\ . \tag{12.22}$$

If the destination regions $j_1,j_2,\ldots,j_N$ of individuals 1, 2, ..., N contain n'_1 times region 1, n'_2 times region 2, ..., and n'_L times region L, the $j_1,\ldots,j_N$ build up the migratory configuration $\boldsymbol{n}'=\{n'_1,n'_2,\ldots,n'_i,\ldots,n'_j,\ldots,n'_L\}$.

Nevertheless $P(j_1,\ldots,j_N,t+\tau|i_1,\ldots,i_N,t)$ does not yet coincide with the configurational conditional probability $P(n'_1,\ldots,n'_L,t+\tau|n_1,\ldots,n_L,t)$ because many mutually exclusive sets of destination regions $\{j_1,j_2,\ldots,j_N\}$ can build up the same migratory configuration $\boldsymbol{n}'=\{n'_1,\ldots,n'_L\}$. Hence, in order to obtain $P(\boldsymbol{n}',t+\tau|\boldsymbol{n},t)$ from $P(j_1,j_2,\ldots,j_N,t+\tau|i_1,i_2,\ldots,i_N,t)$ one has to take the sum over all sets of destination regions $\{j_1,\ldots,j_N\}$ which build up the same configuration $\boldsymbol{n}'=\{n'_1,\ldots,n'_L\}$. That leads to the formal expression

$$\begin{aligned}P(\boldsymbol{n}',t+\tau|\boldsymbol{n},t)&=\sum_{\{j_1,\ldots,j_N\}}^{\boldsymbol{n}'}P(j_1,\ldots,j_N,t+\tau|i_1,\ldots,i_N,t)\\&=\sum_{\{j_1,\ldots,j_N\}}^{\boldsymbol{n}'}\prod_{\nu=1}^{N}P(j_\nu,t+\tau|i_\nu,t)\ .\end{aligned} \tag{12.23}$$

The explicit evaluation of (12.23) poses a difficult combinatorial problem even if the $P(j_\nu,t+\tau|i_\nu,t)$ are known. This problem however simplifies considerably, if we are only interested in the short term propagator up to terms linear in τ (see (1.21)):

$$P(\boldsymbol{n}',t+\tau|\boldsymbol{n},t)=\delta_{\boldsymbol{n}'\boldsymbol{n}}+\tau w_{\boldsymbol{n}'\boldsymbol{n}}(t)+O(\tau^2)\ . \tag{12.24}$$

In this case the right hand side of (12.23) must be expanded up to linear terms in τ, too, making use of (see (1.13))

$$P(j,t+\tau|i,t)=\delta_{ji}+\tau p_{ji}+O(\tau^2)\ . \tag{12.25}$$

It can now easily be seen for which configurations $\boldsymbol{n}'=\{n'_1,\ldots,n'_L\}$ the r.h.s. of (12.23) gives rise to nonvanishing terms linear in τ.

Let us first consider the end-configurations

$$\boldsymbol{n}'=\boldsymbol{n}^{(ji)}=\{n_1,\ldots,(n_i-1),\ldots,(n_j+1),\ldots,n_L\}.$$

The only terms on the r.h.s. of (12.23) giving contributions linear in τ in this case are those, where only one of the n_i individuals in region i migrates into region j, whereas all other people stay in their region. There exist n_i equal terms of this kind leading to the same configuration $\boldsymbol{n}'=\boldsymbol{n}^{(ji)}$, because each of the n_i people in region i might alternatively migrate into j. In this case the r.h.s. of (12.23) assumes the form:

$$\text{r.h.s.} = n_i P^{n_1}(1,t+\tau|1,t), \ldots, P^{(n_i-1)}(i,t+\tau|i,t)$$
$$\cdot P(j,t+\tau|i,t), \ldots, P^{n_j}(j,t+\tau|j,t), \ldots, P^{n_L}(L,t+\tau|L,t)$$
$$= n_i\tau \cdot p_{ji}(t) + O(\tau^2) \; . \tag{12.26}$$

Comparing (12.26) with (12.24) one obtains

$$w_{n'n}(t) \equiv w_t(\boldsymbol{n}'|\boldsymbol{n}) = n_i p_{ji}(t)$$
$$\text{for} \quad \boldsymbol{n}' = \boldsymbol{n}^{(ji)} \; . \tag{12.27}$$

All other end-configurations $\boldsymbol{n}' \neq \boldsymbol{n}$ differ from $\boldsymbol{n}$ in more than two integers. For such configurations the r.h.s. of (12.23) only contains terms of order $O(\tau^2)$ at least. As an example we consider the configuration

$$\boldsymbol{n}' = \{n_1, \ldots, (n_i-1), \ldots, (n_j+1), \ldots, (n_k-1), \ldots, (n_l+1), \ldots, n_L\}$$

which differs from $\boldsymbol{n}$ in four regions i, j, k, l. The transition from $\boldsymbol{n}$ to $\boldsymbol{n}'$ here implies at least two simultaneously migrating individuals: one from i to j and one from k to l or one from i to l and one from k to j. To lowest order in τ, the r.h.s. of (12.23) yields in this case

$$\text{r.h.s.} = \tau^2 n_i p_{ji} n_k p_{lk} + \tau^2 n_i p_{li} n_k p_{jk} + O(\tau^3) \; . \tag{12.28}$$

Generalizing this argumentation we can conclude that

$$w_t(\boldsymbol{n}'|\boldsymbol{n}) = 0$$
$$\text{for} \quad \boldsymbol{n}' \neq \boldsymbol{n} \quad \text{and} \quad \boldsymbol{n}' \neq \boldsymbol{n}^{(ji)} \; . \tag{12.29}$$

The results (12.27) and (12.29) end our proof of (1.24a, b).

13 Solutions of the Master Equation

Wolfgang Weidlich

Let us now consider the solutions of the general and in particular of the migratory master equation.

At first we may assume that the probability transition rates $w_t(j|i)$ do not depend on time t. Furthermore we assume that the transition rates $w(j|i)$ satisfy the condition of *detailed* balance. Fortunately this condition is fulfilled in the case of our migratory model as demonstrated in Section 13.1.

Whereas the stationary solution has a very complicated form in general, a simple representation can be given to it in the case of detailed balance. In particular the unique stationary solution of the migratory master equation is derived in Section 13.2.

Its importance derives from the fact, that all time dependent solutions of the master equation must finally evolve into the stationary solution. This remarkable property is shown in Section 13.3 by deriving the master equation version of the famous *H-Theorem* of L. Boltzmann. (This theorem was originally set up by Boltzmann in order to prove that closed thermodynamic systems approach an equilibrium state.)

Finally we go over to time dependent solutions of the master equation in Section 13.4. The general method of solution leads to an eigenvalue problem which in general can only be solved numerically. For the special case of the migratory master equation with probability transition rates *linear* in the population numbers, however, we shall find an easily interpretable exact time dependent solution of analytical form. This special case also serves for demonstrating the relation between solutions of the master equation and the meanvalue equations.

13.1 Detailed Balance

The stationary solution $P_{st}(i)$ of the general master equation (12.18) is defined by obeying the stationary master equation

$$\sum_j [w(i|j)\,P_{st}(j) - w(j|i)\,P_{st}(i)] = 0 \quad \text{for all} \quad i\,. \tag{13.1}$$

Here we have assumed that the probability transition rates do not depend on time t.

In general it is not easy to obtain a practicable form of $P_{st}(i)$ in terms of all $w(k|j)$ (We shall not discuss here the graph-theoretical solution according to the Kirchhoff theorem; see however [13.3].) In special cases, however, the transition rates fulfil the condition of *detailed balance*

$$\left.\begin{array}{l} w(i|j)\,P_{st}(j)=w(j|i)\,P_{st}(i) \\ \text{for all} \quad i \text{ and } j\ (i\neq j) \end{array}\right\} \tag{13.2}$$

Let us compare the meaning of (13.1) and (13.2):

The stationary master equation requires that under stationary conditions the *total* probability flow into each state i must be equal to the *total* flow out of this state.

The more specific condition of detailed balance requires that the stationary probability flow from i to j is equal to the stationary counterflow from j to i *separately for each pair* of states i and j. It is clear that (13.2) implies (13.1) since by (13.2) each bracket in the sum of (13.1) must vanish separately.

13.1.1 Stationary Solution of the Master Equation

Given that the condition of detailed balance holds, it is easy to construct the explicit form of the stationary solution (see [13.2]). For this aim we may take any chain $\mathscr{C}$ of states $i_0, i_1, \ldots, i_{n-1}$, until i_n from a reference state i_0 to an arbitrary state i_n so that all transition rates

$$w(i_1|i_0), w(i_0|i_1), w(i_2|i_1), w(i_1|i_2) \ldots w(i_n|i_{n-1}), w(i_{n-1}|i_n)$$

are nonvanishing. The repeated application of (13.2) then yields

$$P_{st}(i_n)=(\mathscr{C})\prod_{\nu=0}^{n-1}\frac{w(i_{\nu+1}|i_\nu)}{w(i_\nu|i_{\nu+1})}\,P_{st}(i_0) \tag{13.3}$$

where i_n can be *any state* of the system and $\mathscr{C}$ *any chain* of states i_ν with finite nonvanishing quotients $w(i_{\nu+1}|i_\nu)/w(i_\nu|i_{\nu+1})$ leading from the reference state i_0 to the terminal state i_n. Finally the value of $P_{st}(i_0)$ is determined by inserting (13.3) into the probability normalization condition

$$\sum_{i_n} P_{st}(i_n)=1 \ . \tag{13.4}$$

We now proceed to the question how for a given master equation it can be checked, whether detailed balance holds. The condition (13.2) is not directly appropriate for this test, because it implies the stationary solution which is not known beforehand and can only be written in the form (13.3), *if* detailed balance holds! Therefore we must first derive conditions for the transition rates which are *equivalent* to (13.2) but which *do not* contain the unknown stationary solution. In orther words: we must eliminate the stationary state in the condition of detailed balance.

At first a *set of necessary relations* for the transition rates can be derived from the assumption of detailed balance:

Since (13.2) implies formula (13.3) for *any* chain $\mathscr{C}$ and *any* terminal state i_n we can also apply (13.3) to chains $\mathscr{C}$ of states which are *closed loops* $\mathscr{L}$ and for which the terminal state i_n coincides with the initial reference state i_0. Evidently there follows for $i_n \hat{=} i_0$

$$P_{st}(i_0) = (\mathscr{L}) \prod_{\nu=0}^{n-1} \frac{w(i_{\nu+1}|i_\nu)}{w(i_\nu|i_{\nu+1})} P_{st}(i_0) \tag{13.5a}$$

or

$$\left.\begin{array}{l} (\mathscr{L}) \displaystyle\prod_{\nu=0}^{n-1} \frac{w(i_{\nu+1}|i_\nu)}{w(i_\nu|i_{\nu+1})} = 1 \\ \text{for every closed loop } \mathscr{L}\{i_0, i_1, \ldots i_n = i_0\} \\ \text{with finite nonvanishing quotients } \dfrac{w(i_{\nu+1}|i_\nu)}{w(i_\nu|i_{\nu+1})} \end{array}\right\} \tag{13.5b}$$

as a necessary consequence of detailed balance (13.2).

It will now be shown that the set of relations (13.5b), which do no more involve $P_{st}(i)$ is not only *necessary* but also *sufficient*, that means *equivalent* to the original condition (13.2) of detailed balance. In order to prove that (13.5b) is sufficient, too, for the validity of detailed balance (13.2) one has to *construct by definition* the stationary solution $P_{st}(i)$ presuming the relations (13.5b) only, and to show that this stationary solution indeed fulfills (13.2).

Let us construct the stationary solution $P_{st}(i_n)$ by the definition

$$P_{st}(i_n) = (\mathscr{C}) \prod_{\nu=0}^{n-1} \frac{w(i_{\nu+1}|i_\nu)}{w(i_\nu|i_{\nu+1})} P_{st}(i_0) \tag{13.6}$$

where i_0 is a reference state and $\mathscr{C}$ is a chain of states $\{i_0, i_1, \ldots, i_n\}$ with finite nonvanishing quotients $\dfrac{w(i_{\nu+1}|i_\nu)}{w(i_\nu|i_{\nu+1})}$.

This definition, however, is only consistent and unique, if it can be shown by using the presumed relations (13.5b) that (13.6) is *independent* of the choice of the chain from i_0 to i_n. In other words it must be shown that

$$P_{st}(j_{n'}) = (\mathscr{C}') \prod_{\nu=0}^{n'-1} \frac{w(j_{\nu+1}|j_\nu)}{w(j_\nu|j_{\nu+1})} P_{st}(j_0) \tag{13.7}$$

leads to the same result if $\mathscr{C}'$ is any other chain $\{j_0 = i_0, j_1, j_2, \ldots, j_{n'} = i_n\}$ with the same initial state $j_0 \hat{=} i_0$ and final state $j_{n'} \hat{=} i_n$.

Figure 13.1 illustrates this requirement for two chains $\mathscr{C}\{i_0, i_1, \ldots, i_6\}$ and $\mathscr{C}'\{j_0, j_1, \ldots, j_4\}$.

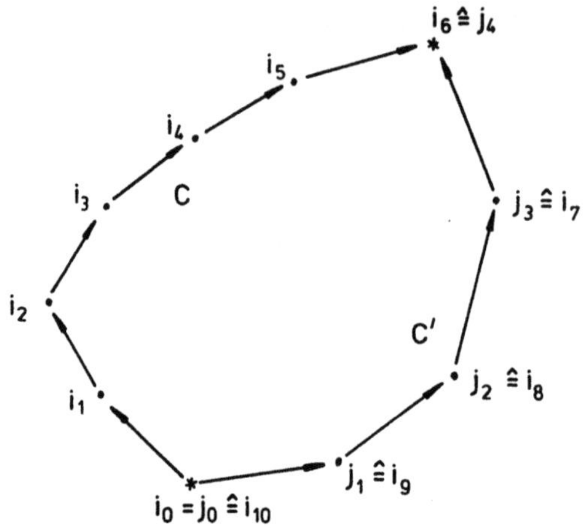

Fig. 13.1. Two equivalent chains $\mathscr{C}(i_0, \ldots, i_6)$ and $\mathscr{C}'(j_0, \ldots, j_4)$ of states for constructing the stationary solution $P_{st}(i_6) = P_{st}(j_4)$ starting from $P_{st}(i_0 = j_0)$

The requirement of the equivalence of (13.6) and (13.7) leads to the condition

$$(\mathscr{C}) \prod_{\nu=0}^{n-1} \frac{w(i_{\nu+1}|i_\nu)}{w(i_\nu|i_{\nu+1})} = (\mathscr{C}') \prod_{\nu=0}^{n'-1} \frac{w(j_{\nu+1}|j_\nu)}{w(j_\nu|j_{\nu+1})} \tag{13.8}$$

for two chains $\mathscr{C}$ and $\mathscr{C}'$ with $i_0 = j_0$ and $i_n = j_{n'}$ or to

$$\left.\begin{aligned} &(\mathscr{L}) \prod_{\nu=0}^{n+n'-1} \frac{w(i_{\nu+1}|i_\nu)}{w(i_\nu|i_{\nu+1})} = 1 \\ &\text{for the closed loop} \\ &\mathscr{L}\{i_0, i_1, \ldots, i_n \mathrel{\hat{=}} j_{n'}, i_{n+1} \mathrel{\hat{=}} j_{n'-1}, \ldots, i_{n+n'-1} \mathrel{\hat{=}} j_1, i_{n+n'} \mathrel{\hat{=}} j_0 \mathrel{\hat{=}} i_0\} \,. \end{aligned}\right\} \tag{13.9}$$

Since (13.9) is fulfilled by assumption (13.5b), we have proved that (13.6) is a consistent definition presuming that (13.5b) holds.

Detailed balance in the form (13.2) then follows, if we include the states i and j as the last but one ($i \mathrel{\hat{=}} i_{n-1}$) and last ($j \mathrel{\hat{=}} i_n$) state in a chain $\mathscr{C}\{i_0, i_1, \ldots, i_{n-1}, i_n\}$. Applying formula (13.6) we obtain:

$$\left.\begin{aligned} P_{st}(j) &= (\mathscr{C}) \prod_{\nu=0}^{n-1} \frac{w(i_{\nu+1}|i_\nu)}{w(i_\nu|i_{\nu+1})} P_{st}(i_0) \\ &= \frac{w(j|i)}{w(i|j)} (\mathscr{C}) \prod_{\nu=0}^{n-2} \frac{w(i_{\nu+1}|i_\nu)}{w(i_\nu|i_{\nu+1})} P_{st}(i_0) \\ &= \frac{w(j|i)}{w(i|j)} P_{st}(i) \,. \end{aligned}\right\} \tag{13.10}$$

The result (13.10) ends our proof of the equivalence of the two formulations (13.5b) and (13.2) of the condition of detailed balance.

13.1.2 Proof of Detailed Balance for the Migratory System

It will now be proved that in our model detailed balance is fulfilled.

As in Part I we use the notation, that

$$\boldsymbol{n}^{(ji)} = \{n_1, \ldots, (n_i - 1), \ldots, (n_j + 1), \ldots, n_L\}$$

is the configuration arising from $\boldsymbol{n}$ by one individual migrating from i to j. The configurations $\boldsymbol{n}^{(ji)}$ are called neighbouring to $\boldsymbol{n}$. Similarly

$$n^{(lk)(ji)} = \{n_1, \ldots, (n_i - 1), \ldots, (n_j + 1), \ldots, (n_k - 1), \ldots, (n_l + 1), \ldots, n_L\}$$

arises from $\boldsymbol{n}$ by one migration from i to j and one from k to l. We recall (see (12.20)) that nonvanishing transition rates $w(\boldsymbol{n}'|\boldsymbol{n})$ only exist if the final configuration $\boldsymbol{n}'$ is neighbouring to $\boldsymbol{n}$. That means $\boldsymbol{n}'$ must have the form $\boldsymbol{n}^{(ji)}$.

The explicit formula for the nonvanishing transition rates is (see (1.21))

$$\begin{aligned} w(\boldsymbol{n}^{(ji)}|\boldsymbol{n}) &= n_i p_{ji}(\boldsymbol{n}) \\ &= n_i v_{ji} \exp[u_j(n_j+1) - u_i(n_i)]. \end{aligned} \tag{13.11}$$

Here we have assumed that the utility u_i of region i depends on the population n_i of that region. Note, that the utility u_i of the *origin region* depends on the population n_i belonging to $\boldsymbol{n}$ *before* migration, whereas the utility u_j of the *destination region* depends on the population (n_j+1) belonging to $\boldsymbol{n}^{(ji)}$ *after* migration.

At first we prove the condition of detailed balance for the elementary closed loops connecting neighbouring configurations:

$$\begin{aligned} &\text{a)}\ \mathscr{L}\{\boldsymbol{n} \leftarrow \boldsymbol{n}^{(ki)} \leftarrow \boldsymbol{n}^{(ji)} \leftarrow \boldsymbol{n}\} \\ &\text{b)}\ \mathscr{L}\{\boldsymbol{n} \leftarrow \boldsymbol{n}^{(lk)} \leftarrow \boldsymbol{n}^{(lk)(ji)} \leftarrow \boldsymbol{n}^{(ji)} \leftarrow \boldsymbol{n}\} \\ &\text{c)}\ \mathscr{L}\{\boldsymbol{n} \leftarrow \boldsymbol{n}^{(li)} \leftarrow \boldsymbol{n}^{(lk)(ji)} \leftarrow \boldsymbol{n}^{(ji)} \leftarrow \boldsymbol{n}\} \\ &\text{d)}\ \mathscr{L}\{\boldsymbol{n} \leftarrow \boldsymbol{n}^{(jk)} \leftarrow \boldsymbol{n}^{(lk)(ji)} \leftarrow \boldsymbol{n}^{(ji)} \leftarrow \boldsymbol{n}\} \end{aligned} \tag{13.12}$$

In the formulas belonging to these loops (to be read from right to left) each arrow in (13.12) corresponds to one quotient of transition rates.

$$\left.\begin{aligned} &\text{a)}\ \frac{w(\boldsymbol{n}|\boldsymbol{n}^{(ki)})}{w(\boldsymbol{n}^{(ki)}|\boldsymbol{n})} \cdot \frac{w(\boldsymbol{n}^{(ki)}|\boldsymbol{n}^{(ji)})}{w(\boldsymbol{n}^{(ji)}|\boldsymbol{n}^{(ki)})} \cdot \frac{w(\boldsymbol{n}^{(ji)}|\boldsymbol{n})}{w(\boldsymbol{n}|\boldsymbol{n}^{(ji)})} = 1 \\ &\text{b)}\ \frac{w(\boldsymbol{n}|\boldsymbol{n}^{(lk)})}{w(\boldsymbol{n}^{(lk)}|\boldsymbol{n})} \cdot \frac{w(\boldsymbol{n}^{(lk)}|\boldsymbol{n}^{(lk)(ji)})}{w(\boldsymbol{n}^{(lk)(ji)}|\boldsymbol{n}^{(lk)})} \cdot \frac{w(\boldsymbol{n}^{(lk)(ji)}|\boldsymbol{n}^{(ji)})}{w(\boldsymbol{n}^{(ji)}|\boldsymbol{n}^{(lk)(ji)})} \cdot \frac{w(\boldsymbol{n}^{(ji)}|\boldsymbol{n})}{w(\boldsymbol{n}|\boldsymbol{n}^{(ji)})} = 1 \\ &\text{c)}\ \frac{w(\boldsymbol{n}|\boldsymbol{n}^{(li)})}{w(\boldsymbol{n}^{(li)}|\boldsymbol{n})} \cdot \frac{w(\boldsymbol{n}^{(li)}|\boldsymbol{n}^{(lk)(ji)})}{w(\boldsymbol{n}^{(lk)(ji)}|\boldsymbol{n}^{(li)})} \cdot \frac{w(\boldsymbol{n}^{(lk)(ji)}|\boldsymbol{n}^{(ji)})}{w(\boldsymbol{n}^{(ji)}|\boldsymbol{n}^{(lk)(ji)})} \cdot \frac{w(\boldsymbol{n}^{(ji)}|\boldsymbol{n})}{w(\boldsymbol{n}|\boldsymbol{n}^{(ji)})} = 1 \\ &\text{d)}\ \frac{w(\boldsymbol{n}|\boldsymbol{n}^{(jk)})}{w(\boldsymbol{n}^{(jk)}|\boldsymbol{n})} \cdot \frac{w(\boldsymbol{n}^{(jk)}|\boldsymbol{n}^{(lk)(ji)})}{w(\boldsymbol{n}^{(lk)(ji)}|\boldsymbol{n}^{(jk)})} \cdot \frac{w(\boldsymbol{n}^{(lk)(ji)}|\boldsymbol{n}^{(ji)})}{w(\boldsymbol{n}^{(ji)}|\boldsymbol{n}^{(lk)(ji)})} \cdot \frac{w(\boldsymbol{n}^{(ji)}|\boldsymbol{n})}{w(\boldsymbol{n}|\boldsymbol{n}^{(ji)})} = 1. \end{aligned}\right\} \tag{13.13}$$

It is left to the reader to insert (13.11) into (13.13) and to check explicitly that (13.13) is indeed fulfilled.

Secondly it will now be proved that (13.5b) is also fulfilled for *arbitrary loops* connecting neighbouring configurations. The procedure is illustrated in two Figures 13.2a and 13.2b showing how the general proof works.

In Figure 13.2a we choose $L=3$ and, to be explicit $N=n_1+n_2+n_3=g$. The possible configurations form a two dimensional lattice within a regular triangle. The figure shows an arbitrary closed loop of arrows between neighbouring configurations. Each arrow pointing from a $\boldsymbol{n}_1$ to a neighbouring $\boldsymbol{n}_2$ corresponds to the quotient $\dfrac{w(\boldsymbol{n}_2|\boldsymbol{n}_1)}{w(\boldsymbol{n}_1|\boldsymbol{n}_2)}$ (so that the inverse $\dfrac{w(\boldsymbol{n}_1|\boldsymbol{n}_2)}{w(\boldsymbol{n}_2|\boldsymbol{n}_1)}$ belongs to the inverted arrow.)

It must now be proved that the product of all quotients belonging to this closed loop of arrows is equal to one. This proof can be reduced to repeated use of (13.13a). For this aim we fill the closed loop with elementary triangles. The product of the arrows of each of them yields the factor 1 according to (13.13a); hence the product of all triangle arrows within the closed loop must be equal to one, too. On the other hand all contributions of the arrows of opposite direction at the boundary of two triangles cancel out, so that the only remaining contribution to the total product stems from the arrows along the closed loop. Hence eq. (13.5a) is fulfilled for this loop. q.e.d.

This argumentation – namely filling out the interior of a closed loop with elementary triangles or quadrangles and repeatedly applying (13.13) – can be generalized to $L>3$, where the lattice space of configurations is $(L-1)$-dimensional.

Figure 13.2b exemplifies the procedure of decomposing a closed loop into elementary quadrangles fulfilling (13.13b,c,d). Applying the above mentioned argumentation, it is easily seen, that detailed balance (13.5b) holds for the surrounding closed loop. Since all closed loops can be decomposed in this manner, this ends the proof of the validity of detailed balance for our migratory model.

13.2 The Stationary Solution for the Migratory Master Equation

Since detailed balance holds for the migratory master equation, we are able to construct its stationary solution using (13.6). Starting from an arbitrary but conveniently chosen reference configuration called reference state we must reach every state by a chain of states so that the transition rates between adjacent states are nonvanishing.

Starting from the reference state $\{N,0,0,\ldots,0\}$ we choose the following "standard chain" to the general state $\boldsymbol{n}=\{n_1,n_2,\ldots,n_L\}$:

$$\begin{aligned}\mathscr{C}=\{&(N,0,\ldots,0)\to(N-1,1,0,\ldots,0)\ldots\to(N-n_2,n_2,0,\ldots,0)\\ &\to(N-n_2-1,n_2,1,0,\ldots,0)\ldots\to(N-n_2-n_3,n_2,n_3,0,\ldots,0)\\ &\to(N-n_2-n_3-1,n_2,n_3,1,0,\ldots,0)\ldots\to(n_1,n_2,n_3,\ldots,n_L)\}\,.\end{aligned}\tag{13.14}$$

According to (13.6) and with the notation

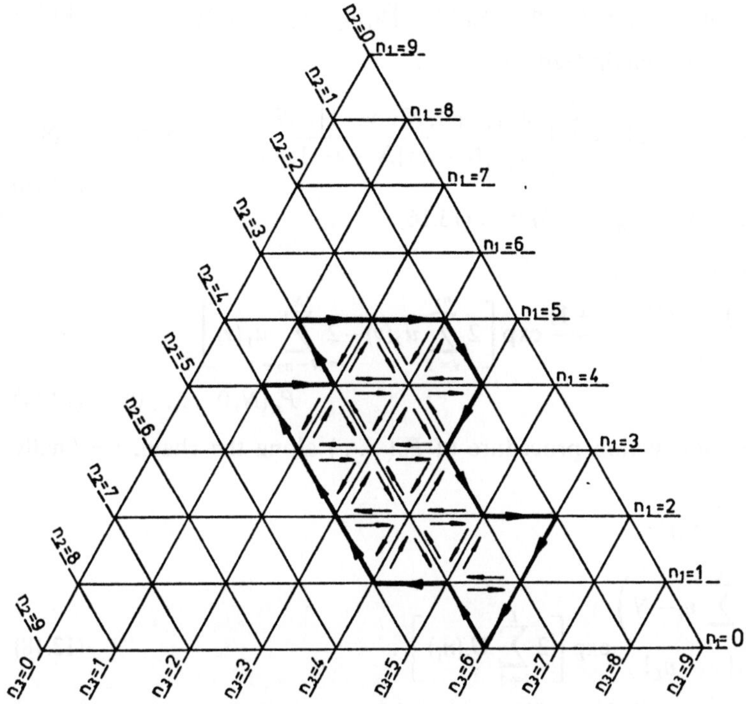

Fig. 13.2a. Elementary triangles filling the lattice space between a closed loop in the case $L=3$ and for $N=n_1+n_2+n_3=9$. The closed loop satisfies detailed balance according to equation (13.5b)

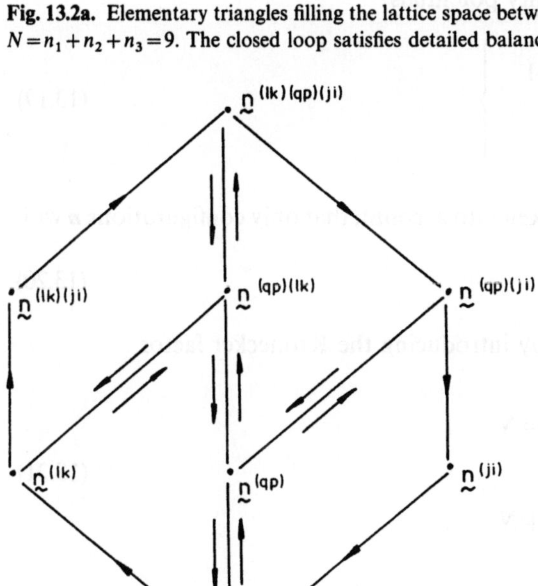

Fig. 13.2b. The closed loop $\mathscr{L}\{\boldsymbol{n}\rightarrow\boldsymbol{n}^{(lk)}\rightarrow\boldsymbol{n}^{(lk)(ji)}\rightarrow\boldsymbol{n}^{(lk)(qp)(ji)}\rightarrow\boldsymbol{n}^{(qp)(ji)}\rightarrow\boldsymbol{n}^{(ji)}\rightarrow\boldsymbol{n}\}$ fulfills detailed balance (13.5b) because it can be decomposed into elementary quadrangles, for which equation (13.13) holds

$$w(\boldsymbol{n}^{(ji)}|\boldsymbol{n}) \equiv w_{ji}(n_j+1, n_i-1|n_{ji}, n_i) = v_{ji} n_i \exp[u_j(n_j+1) - u_i(n_i)] \tag{13.15}$$

the first n_2 steps of the chain lead to

$$P_{st}(N-n_2, n_2, 0, \dots, 0) = \prod_{\nu=0}^{n_2-1} \frac{w_{21}(\nu+1, N-\nu-1|\nu, N-\nu)}{w_{12}(N-\nu, \nu|N-\nu-1, \nu+1)} \cdot P_{st}(N, 0, \dots, 0) \tag{13.16}$$

or explicitly, after inserting (13.15) into (13.16)

$$\begin{aligned} &P_{st}(N-n_2, n_2, 0, \dots, 0) \\ &= \frac{N(N-1)\dots(N-n_2+1)}{n_2!} \exp\left[2\sum_{\nu=1}^{n_2} u_2(\nu) - 2\sum_{\nu=N-n_2+1}^{N} u_1(\nu)\right] \\ &\qquad P_{st}(N, 0, \dots, 0) \,. \end{aligned} \tag{13.17}$$

Continuing the construction procedure of $P_{st}(\dots)$ along the chain, we finally obtain

$$\begin{aligned} &P_{st}(n_1, n_2, \dots, n_L) \\ &= \frac{Z^{-1}\delta\left(\sum_{i=1}^{L} n_i - N\right)}{n_1!\,n_2!\dots n_L!} \exp\left[2\sum_{i=1}^{L} U_i(n_i)\right] . \end{aligned} \tag{13.18}$$

Here we have introduced the "utility potentials"

$$\left.\begin{aligned} U_i(n_i) &= \sum_{\nu=1}^{n_i} u_i(\nu), \qquad \text{for} \quad n_i \geq 1 \\ U_i(0) &= 0 \,. \end{aligned}\right\} \tag{13.19}$$

Furthermore, we have explicitly taken into account, that only configurations $\boldsymbol{n}$ with

$$\sum_{i=1}^{L} n_i = N \tag{13.20}$$

have a nonvanishing probability by introducing the Kronecker factor

$$\delta\left(\sum_{i=1}^{L} n_i - N\right) = \begin{cases} 1 & \text{for} \quad \sum_{i=1}^{L} n_i = N \\ 0 & \text{for} \quad \sum_{i=1}^{L} n_i \neq N \,. \end{cases} \tag{13.21}$$

Finally we have put the reference probability equal to

$$P_{st}(N, 0, \dots, 0) = \frac{\exp[2U_1(N)]}{N!} \cdot \frac{1}{Z} \tag{13.22}$$

where the factor Z now follows from the normalization condition

$$\sum_{n} P_{st}(n_1, n_2, \ldots, n_L) = 1 \tag{13.23}$$

and is given by

$$Z = \sum_{n} \frac{\delta\left(\sum_{i=1}^{L} n_i - N\right) \exp\left[2 \sum_{i=1}^{L} U_i(n_i)\right]}{n_1! n_2! \ldots n_L!} . \tag{13.24}$$

The sum extends over all configurations $\boldsymbol{n}$.

If the utility functions $u_i(n_i)$ are explicitly known, the formula (13.18) may be further evaluated. Guided by the empiric results of Part 2 it can be suggested to put

$$u_i(n_i) = \delta_i + \kappa_i n_i + \sigma_i n_i^2 . \tag{13.25}$$

In this case we obtain

$$\left.\begin{aligned} U(n_i) &= \delta_i \sum_{\nu=1}^{n_i} 1 + \kappa_i \sum_{\nu=1}^{n_i} \nu + \sigma_i \sum_{\nu=1}^{n_i} \nu^2 \\ &= \delta_i n_i + \kappa_i \frac{n_i(n_i+1)}{2} + \sigma_i \frac{n_i(n_i+1)(2n_i+1)}{6} . \end{aligned}\right\} \tag{13.26}$$

For practical applications it is convenient to approximate the exact stationary solution (13.18) by using Stirlings formula for the factorials

$$n! \cong n^n e^{-n} \ \text{for}\ n \gg 1 \tag{13.27}$$

and by going over to the continuous approximation of the utility potentials:

$$U_i(n) = \int_0^n u_i(\nu)\, d\nu . \tag{13.28}$$

The approximate form of $P_{st}(\boldsymbol{n})$ reads

$$P_{st}(\boldsymbol{n}) = Z^{-1} \delta\left(\sum_{i=1}^{L} n_i - N\right) \exp\left[\sum_{i=1}^{L} \Phi_i(n_i)\right] \tag{13.29}$$

with

$$\Phi_i(n_i) = 2U_i(n_i) - [n_i \ln(n_i) - n_i] . \tag{13.30}$$

It is now easy to find the configuration – or the configurations – $\hat{\boldsymbol{n}} = \{\hat{n}_1, \hat{n}_2, \ldots, \hat{n}_L\}$ which lead to an extremum of $P_{st}(\boldsymbol{n})$. The variation of the exponent in (13.29) under the constraint (13.20) leads to:

$$\left.\begin{aligned} &\delta\left[\sum_{i=1}^{L} \Phi_i(n_i) - \lambda\left(\sum_{i=1}^{L} n_i - N\right)\right] \\ &= \sum_{i=1}^{L} \delta n_i \left[2 \frac{\partial U_i(n_i)}{\partial n_i} - \ln(n_i) - \lambda\right] = 0 . \end{aligned}\right\} \tag{13.31}$$

Solving for n_i leads to determining equations for the extremal configurations $\{\hat{n}_1, \ldots, \hat{n}_L\}$:

$$\left.\begin{aligned} \hat{n}_i &= e^{-\lambda} \exp[2u_i(\hat{n}_i)] \\ &= \frac{N \exp[2u_i(\hat{n}_i)]}{\sum_{j=1}^{L} \exp[2u_j(\hat{n}_j)]} \\ &\text{for} \quad i = 1, 2, \ldots, L \end{aligned}\right\} \tag{13.32}$$

where the Lagrangian factor λ is chosen in order to fulfil the constraint (13.20) (see also [13.3]).

It is interesting to note that equation (13.32) agrees with equation (2.41) for the stationary solutions of the meanvalue equations. Hence it turns out, that the maxima (minima) of the stationary probability distribution correspond to the stable (unstable) stationary states of the meanvalue equations.

If the utilities u_i do not depend on n_i, equation (13.32) already provides the explicit unique configuration $\{\hat{n}_1, \ldots, \hat{n}_L\}$ of maximal probability. If, however, the utilities are functions of n_i, equation (13.32) is a set of implicit transcendental equations for $\hat{n}$, which may have more than one solution. A model for this case describing the formation of metropolitan areas has been published in [13.4].

13.3 *H*-Theorem and Entropy

In this section we show, that the time dependent solutions $P(i,t)$ of the general master equation with constant transition rates

$$\frac{dP(i,t)}{dt} = \sum_j w(i|j) P(j,t) - \sum_j w(j|i) P(i,t) \tag{13.33}$$

have the remarkable property that they all approach the stationary state $P_{st}(i)$ for $t \to \infty$ independently of their initial state. In order to simplify the proof we also presume that detailed balance (13.2) is fulfilled.

The method of proof was first invented by L. Boltzmann who applied it to his equation in kinetic gas theory. With some modifications it can also be applied to the master equation. The idea is simple in principle and consists in finding a function $H(t)$ with the following properties:

1. $H(t)$ is a positive definite function of all $P(i,t)$
2. $\dfrac{dH(t)}{dt} \leq 0$ holds, if the $P(i,t)$ satisfy the master equation
3. The value $H = H_{\min}$ and $\dfrac{dH}{dt} = 0$ is attained, if and only if *all* $P(i,t)$ assume the value $P_{st}(i)$.

Evidently the existence of such a function $H(t)$ implies, that all $P(i,t)$ must approach $P_{st}(i)$: Since $H(t)$ has to remain positive definite and is on the other hand monotonously decreasing with time, it must approach a minimum value, where $\frac{dH}{dt}=0$. By 3. this can only happen *if all* $P(i,t)$ approach $P_{st}(i)$.

The function fulfilling the properties 1., 2. and 3. was found by Boltzmann and reads

$$H(t)=\sum_j P(j,t)\ln\left[\frac{P(j,t)}{P_{st}(j)}\right] . \tag{13.34}$$

It should be mentioned, that the function $S(t)=-H(t)$ is denoted as *entropy* and plays a central role in thermodynamics and statistical physics. According to property 2 and 3, the entropy $S(t)$ is a monotonously increasing function attaining its maximum in the stationary state. The entropy can be interpreted as a measure of the disorder in the system, so that its evolution with time describes the irreversible increase of disorder in a closed system.

Proof of Property 1:

With the abbreviation

$$r(i,t)=\frac{P(i,t)}{P_{st}(i)}\geq 0 \tag{13.35}$$

and taking into account the probability normalization

$$\sum_j P(j,t)=\sum_j P_{st}(j)=1 \tag{13.36}$$

the function $H(t)$ can be transformed into

$$\left.\begin{aligned} H(t) &=\sum_j [P(j,t)\ln r(j,t)-P(j,t)+P_{st}(j)] \\ &=\sum_j P_{st}(j)[r(j,t)\ln r(j,t)-r(j,t)+1] \, . \end{aligned}\right\} \tag{13.37}$$

Because of

$$r\ln r-r+1\equiv\int_1^r \ln(x)\,dx\geq 0 \tag{13.38}$$

for every $r\geq 0$ the expression (13.37) is positive semi-definite for *arbitrary choice* of the probabilities $P(i,t)$.

Proof of Property 2:

The time derivative of $H(t)$ reads

$$\frac{dH(t)}{dt} = \sum_j \frac{dP(j,t)}{dt} \ln r(j,t) + \sum_j \frac{dP(j,t)}{dt} \tag{13.39}$$

where the last term vanishes because of (13.36). Inserting the master equation (13.33) in (13.39) and simultaneously applying detailed balance one obtains

$$\left.\begin{aligned} \frac{dH(t)}{dt} &= \sum_{i,j} [w(j|i)P(i,t) - w(i|j)P(j,t)] \ln r(j,t) \\ &= \sum_{i,j} w(j|i)P_{st}(i)[r(i,t) - r(j,t)] \ln r(j,t) \ . \end{aligned}\right\} \tag{13.40}$$

After writing the r.h.s. once more, but interchanging the indices i and j the time derivative of $H(t)$ assumes the final form

$$2\frac{dH(t)}{dt} = -\sum_{i,j} w(j|i)P_{st}(i)\{[r(i,t) - r(j,t)][\ln r(i,t) - \ln r(j,t)]\} \ . \tag{13.41}$$

The bracket under the sum has the form

$$F(x,y) = (x-y)(\ln(x) - \ln(y)) \tag{13.42}$$

where the function $F(x,y)$ has the property

$$\left.\begin{aligned} &F(x,y) \geq 0 \quad \text{for all} \quad x>0 \quad \text{and} \quad y>0 \\ &F(x,y) = 0 \quad \text{only for} \quad x=y \ . \end{aligned}\right\} \tag{13.43}$$

Since the factor $w(j|i)P_{st}(i)$ is also positive semi-definite, every term of the sum is positive semi-definite, and we can conclude, that

$$\frac{dH(t)}{dt} \leq 0 \tag{13.44}$$

Proof of Property 3:

We have to show, that

$$\left.\begin{aligned} &\frac{dH(t)}{dt} = 0 \quad \text{if and only if } P(i,t) = P_{st}(i) \\ &\text{for all } i \ . \end{aligned}\right\} \tag{13.45}$$

The function H then assumes the lowest possible value $H=0$.

It is trivial that the condition $P(i,t) = P_{st}(i)$ or $r(i,t) = 1$ for all i is *sufficient* for the vanishing of the time derivative of H. But this condition is also *necessary*, what can be seen as follows:

The vanishing of $\frac{dH(t)}{dt}$ requires that at least all those terms $\{[r(i,t) - r(j,t)]$ $[\ln r(i,t) - \ln r(j,t)]\}$ vanish – so that $r(i,t) = r(j,t)$ must hold – for which $w(j|i)$ $P_{st}(i) > 0$. Now there must exist at least one chain of states $i, i_1, i_2, \ldots, j$ with nonvanishing transition rates between neighbour states from a reference state i to each state j of the system. (Otherwise the system would decompose into independent subsystems.) For this chain one obtains accordingly

$$r(i,t)=r(i_1,t)=\ \dots\ =r(j,t)=\rho\ ,$$

or (13.46)

$$P(i,t)=\rho P_{st}(i),\ \dots\ ,\ P(j,t)=\rho P_{st}(j)$$

with the same factor ρ for all states on the chain. The factor ρ must however be universal, since each state j can be reached on a chain from state i, and finally ρ must be equal to one because of the normalization (13.36). This ends the proof of property 3.

Although the now proved existence of the function $H(t)$ implies that every time-dependent solution $P(i,t)$ must approach $P_{st}(i)$, nothing can be said so far about the time interval in which this relaxation process takes place. A more detailed inspection of the evolution with time shows, that in the general case the evolution to $P_{st}(i)$ may consist of fast drift-dominated and extremely slow fluctuation-dominated processes.

13.4 Time Dependent Solutions

The general method of solving the time dependent master equation consists in decomposing $P(\boldsymbol{n},t)$ into eigensolutions of the form $P_j(\boldsymbol{n})\mathrm{e}^{-\lambda_j t}$. The eigenvalues λ_j and the eigenmodes $P_j(\boldsymbol{n})$ in general can only be found by numerical methods.

Instead of discussing this standard method it is the aim of the present section to derive a fully analytical solution of the time dependent master equation in the case of special transition rates and to demonstrate its relation to solutions of the meanvalue equations.

Once more we write down the full master equation describing both migration and birth-death processes (see (2.11)):

$$\frac{\mathrm{d}P(\boldsymbol{n},t)}{\mathrm{d}t}=\left(\frac{\partial P}{\partial t}\right)_M+\left(\frac{\partial P}{\partial t}\right)_{BD} \tag{13.47}$$

with the migration term

$$\left(\frac{\partial P}{\partial t}\right)_M=\sum_{i,j=1}^{L}[w_{ij}(\boldsymbol{n}^{(ji)})P(\boldsymbol{n}^{(ji)},t)-w_{ji}(\boldsymbol{n})P(\boldsymbol{n},t)] \tag{13.48}$$

and the birth-death-term

$$\begin{aligned}\left(\frac{\partial P}{\partial t}\right)_{BD}&=\sum_{i=1}^{L}[w_i^{\beta}(\boldsymbol{n}^{(i-)})P(\boldsymbol{n}^{(i-)},t)-w_i^{\beta}(\boldsymbol{n})P(\boldsymbol{n},t)]\\&\quad+\sum_{i=1}^{L}[w_i^{\mu}(\boldsymbol{n}^{(i+)})P(\boldsymbol{n}^{(i+)},t)-w_i^{\mu}(\boldsymbol{n})P(\boldsymbol{n},t)]\end{aligned} \tag{13.49}$$

where the migratory transition rates are given by

$$w_{ji,t}(\boldsymbol{n}) = n_i p_{ji}(t) = n_i \nu_{ji}(t) \exp[u_j(t) - u_i(t)] \ . \tag{13.50}$$

Here, however, we now assume, that the regional utilities u_j, u_i do not depend on n_j, n_i, so that $w_{ji,t}(\boldsymbol{n})$ is *linear* in the $\boldsymbol{n}$. The utilities u_j and mobilities $\nu_{ji} = \nu_{ij}$ may however be functions of time, so that also the transition rates can become time dependent.

The birth- and death-rate in region i are assumed to be proportional to n_i and given by

$$w_i^\beta(\boldsymbol{n}) = \beta_i n_i \tag{13.51}$$

and

$$w_i^\mu(\boldsymbol{n}) = \mu_i n_i \tag{13.52}$$

respectively.

Let us begin with the pure migratory process by first considering the pure migratory master equation

$$\frac{dP(\boldsymbol{n}, t)}{dt} = \left(\frac{\partial P}{\partial t}\right)_M . \tag{13.53}$$

The fact that the total number

$$\sum_{i=1}^{L} n_i = N \tag{13.54}$$

is conserved during purely migratory evolution, leads to the following lemma:

If $P(\boldsymbol{n}, t)$ is a solution of (13.53), then also

$$P_r(\boldsymbol{n}, t) = c\delta\left(\sum_{i=1}^{L} n_i - N\right) P(\boldsymbol{n}, t) \tag{13.55}$$

is a solution of (13.53). The index r stands for "restricted", since $P_r(\boldsymbol{n}, t)$ is restricted to the sheet of configurations fulfilling (13.54). The factor c re-adjusts the normalization of $P_r(\boldsymbol{n}, t)$.

The proof of the lemma follows because of

$$\delta\left(\sum_{l=1}^{L} n_l^{(ji)} - N\right) = \delta\left(\sum_{l=1}^{L} n_l - N\right) \tag{13.56}$$

so that the factor $\delta\left(\sum_{l=1}^{L} n_l - N\right)$ can be extracted out of the sum on the r.h.s. of (13.53) and cancels with the same factor on the l.h.s. of (13.53).

The lemma is useful if we have found a solution $P(\boldsymbol{n}, t)$ containing nonvanishing probabilities for different values of the sum $\sum_{l=1}^{L} n_l$. By (13.55) each sheet of that solution with fixed N is *also* a solution of the master equation (13.53), (see also [13.5]).

The form (13.18) of the stationary solution now suggests to try the following ansatz for a time dependent solution of (13.53)

$$P(\boldsymbol{n},t)=\frac{\pi(\boldsymbol{n},t)}{n_1!n_2!\dots n_L!} \tag{13.57}$$

from which there follows immediately

$$P(\boldsymbol{n}^{(ji)},t)=\frac{\pi(\boldsymbol{n}^{(ji)},t)}{n_1!n_2!\dots n_L!}\cdot\frac{n_i}{(n_j+1)} \ . \tag{13.58}$$

On inserting (13.57) in (13.53) and making use of (13.48) and (13.50) one obtains the equation of motion for $\pi(\boldsymbol{n},t)$:

$$\frac{d\pi(\boldsymbol{n},t)}{dt}=\sum_{i,j=1}^{L}\left[n_i p_{ij}(t)\pi(\boldsymbol{n}^{(j,i)},t)-n_i p_{ji}(t)\pi(\boldsymbol{n},t)\right] \ . \tag{13.59}$$

We try to solve (13.59) by the ansatz

$$\pi(\boldsymbol{n},t)=\prod_{l=1}^{L}\{[\nu_l(t)]^{n_l}e^{-\nu_l(t)}\} \tag{13.60}$$

where $\nu_l(t)$ has to be determined appropriately.

Before doing this, the meaning of (13.60) should be discussed. Inserting (13.60) into (13.57) one obtains

$$P(\boldsymbol{n},t)=\prod_{l=1}^{L}\left\{\frac{[\nu_l(t)]^{n_l}}{n_l!}e^{-\nu_l(t)}\right\} \tag{13.61}$$

so that the ansatz (13.57) and (13.60) amounts to a L-dimensional Poisson probability distribution, which is already correctly normalized because of

$$\sum_{\boldsymbol{n}}P(\boldsymbol{n},t)=\prod_{l=1}^{L}\left\{\sum_{n_l=0}^{\infty}\frac{[\nu_l(t)]^{n_l}}{n_l!}e^{-\nu_l(t)}\right\}=\prod_{l=1}^{L}1=1 \ . \tag{13.62}$$

It is well-known and can be seen immediately, that in this distribution the $\nu_i(t)$ have the meaning of the meanvalues $\bar{n}_i(t)$. Indeed there follows

$$\bar{n}_i(t)=\sum_{\boldsymbol{n}}n_i P(\boldsymbol{n},t)=\sum_{\boldsymbol{n}}\prod_{l=1}^{L}n_i\left\{\frac{[\nu_l(t)]^{n_l}}{n_l!}e^{-\nu_l(t)}\right\}$$

$$=\sum_{\boldsymbol{n}}\left[\frac{\partial P(\boldsymbol{n},t)}{\partial\nu_i(t)}+P(\boldsymbol{n},t)\right]\nu_i(t)=\nu_i(t) \ . \tag{13.63}$$

In the last line of (13.63), equation (13.62) and its consequence,

$$\frac{\partial}{\partial\nu_i}\sum_{\boldsymbol{n}}P(\boldsymbol{n},t)=0 \ , \tag{13.64}$$

has been used.

It must now be checked whether – and under which conditions for $v_i(t)$ – the ansatz (13.60) can solve equation (13.59).

First we note that according to (13.60) the relations hold:

$$\frac{d\pi(\boldsymbol{n},t)}{dt} \equiv \dot{\pi} = \sum_{i=1}^{L} \left(n_i \frac{\dot{v}_i}{v_i} - \dot{v}_i \right) \pi(\boldsymbol{n},t) \tag{13.65}$$

and

$$\pi(\boldsymbol{n}^{(ji)},t) = \frac{v_j}{v_i} \cdot \pi(\boldsymbol{n},t) \ . \tag{13.66}$$

Making use of them in (13.59), we obtain

$$\left.\begin{aligned} &\sum_{i=1}^{L} \left(n_i \frac{\dot{v}_i}{v_i} - \dot{v}_i \right) \pi(\boldsymbol{n},t) \\ &\quad = \sum_{i=1}^{L} n_i \pi(\boldsymbol{n},t) \cdot \sum_{j=1}^{L} \left[p_{ij}(t) \frac{v_j}{v_i} - p_{ji}(t) \right] \ . \end{aligned}\right\} \tag{13.67}$$

Comparing the coefficients of $n_i \pi(\boldsymbol{n},t)$ and of $\pi(\boldsymbol{n},t)$ under the sum $\sum_{i=1}^{L}$ in (13.67) we see that the following conditions for $v_j(t)$ must be fulfilled:

$$\dot{v}_i(t) = \sum_{j=1}^{L} [p_{ij}(t) v_j(t) - p_{ji}(t) v_i(t)] \qquad (i = 1, 2, \ldots, L) \tag{13.68}$$

and

$$\sum_{i=1}^{L} \dot{v}_i(t) = 0 \tag{13.69}$$

where evidently (13.69) is a consequence of (13.68). Of course (13.68) is nothing but the equation of motion for the meanvalues of the probability distribution already derived in Section 2 (see (2.29)).

In the case of time-independent rates p_{ji} the solution of the meanvalue equation (13.68) evolves into the unique stationary meanvalues

$$\bar{v}_i = \bar{n}_i = \frac{N \exp(2u_i)}{\sum_{j=1}^{L} \exp(2u_j)} \tag{13.70}$$

Correspondingly, the solution (13.61) of the master equation (13.53) evolves into the unique stationary probability distribution

$$\left.\begin{aligned} P_{st}(\boldsymbol{n}) &= \sum_{l=1}^{L}\left\{\frac{[\bar{n}_l]^{n_l}}{n_l!}e^{-\bar{n}_l}\right\} \\ &= \frac{\tilde{c}}{n_1!n_2!\dots n_L!}\exp\left[2\sum_{l=1}^{L}U_l(n_l)\right] \end{aligned}\right\} \tag{13.71}$$

where

$$U_l(n_l) = \sum_{\nu=1}^{n_l} u_l = u_l \cdot n_l \ . \tag{13.72}$$

Simultaneously, the reduced distribution

$$P_r(\boldsymbol{n}, t) = c\delta\left(\sum_{i=1}^{L} n_i - N\right)P(\boldsymbol{n}, t) \tag{13.73}$$

merges into the reduced stationary distribution

$$P_{r,st}(\boldsymbol{n}) = \frac{Z^{-1}\delta\left(\sum_{i=1}^{L} n_i - N\right)}{n_1!n_2!\dots n_L!}\exp\left[2\sum_{l=1}^{L}U_l(n_l)\right] \tag{13.74}$$

with $Z^{-1} = \tilde{c} \cdot c$, which agrees, taking the special utility potentials (13.72), with the general formula (13.18) for stationary distributions, as expected.

Summarizing the result, we have seen that in the case of transition rates linear in $\boldsymbol{n}$ the migratory master equation possesses an exact timedependent solution, namely the multi-dimensional Poisson distribution, whose meanvalues move according to the meanvalue equations (13.68). (It is also easy to see that the latter equations are exact and not approximate for transition rates *linear* in $\boldsymbol{n}$.) The Poisson distribution accompanying the meanvalues is of course the outcome of the fact, that the master equation automatically takes into account the statistical fluctuations around meanvalues in a probabilistic process.

For large population numbers $n_i \gg 1$ the Poisson distribution (13.61) can be approximated by a multi-dimensional Gauss distribution:

$$P(\boldsymbol{n}, t) \approx \frac{1}{(2\pi)^{L/2}\sigma_1(t)\dots\sigma_i(t)}\exp\left\{-\sum_{i=1}^{L}\frac{[n_i - \bar{n}_i(t)]^2}{2\sigma_i^2(t)}\right\} \tag{13.75}$$

with the variance characteristic of the Poisson distribution

$$\sigma_i^2(t) = \bar{n}_i(t) \ . \tag{13.76}$$

Finally let us prove that the moving multi-dimensional Poisson distribution – after a slight approximation – is also a solution of the full master equation (13.74) for migratory and birth-death processes. The approximation consists in making the substitution:

$$w_i^{(\beta)} = \beta_i n_i \rightarrow \beta_i \bar{n}_i \tag{13.77}$$

in the birth rate, which is certainly justified for not too small meanvalues $\bar{n}_i$.

On inserting the ansatz (13.57) into (13.47) one now obtains instead of (13.59)

$$\begin{aligned}\frac{d\pi(\boldsymbol{n},t)}{dt}=&\sum_{i,j}[n_i p_{ij}(t)\pi(\boldsymbol{n}^{(ji)},t)-n_i p_{ji}(t)\pi(\boldsymbol{n},t)]\\&+\sum_i[\beta_i(\bar{n}_i-1)n_i\pi(\boldsymbol{n}^{(i-)},t)-\beta_i\bar{n}_i\pi(\boldsymbol{n},t)]\\&+\sum_i[\mu_i\pi(\boldsymbol{n}^{(i+)},t)-\mu_i n_i\pi(\boldsymbol{n},t)]\ .\end{aligned}\tag{13.78}$$

Furthermore, we now obtain with (13.60) instead of (13.67)

$$\left.\begin{aligned}&\sum_{i=1}^{L}\left(n_i\frac{\dot{v}_i}{v_i}-\dot{v}_i\right)\pi(\boldsymbol{n},t)\\&\quad=\sum_{i=1}^{L}n_i\pi(\boldsymbol{n},t)\sum_{j=1}^{L}\left[p_{ij}(t)\frac{v_j}{v_i}-p_{ji}(t)\right]\\&\quad+\sum_{i=1}^{L}\left[\beta_i(\bar{n}_i-1)n_i\pi(\boldsymbol{n},t)\frac{1}{v_i}-\beta_i\bar{n}_i\pi(\boldsymbol{n},t)\right]\\&\quad+\sum_{i=1}^{L}[\mu_i\pi(\boldsymbol{n},t)\cdot v_i-\mu_i n_i\pi(\boldsymbol{n},t)]\ .\end{aligned}\right\}\tag{13.79}$$

Because of (13.63) we can identify $\bar{n}_i(t)$ with $v_i(t)$ in (13.79); for $\bar{n}_i \gg 1$ we can also put $(\bar{n}_i-1)\approx\bar{n}_i$. Comparing the coefficients of $n_i\pi(\boldsymbol{n},t)$ and $\pi(\boldsymbol{n},t)$ in (13.79), we now obtain the conditions for $v_i(t)$, generalizing (13.68) and (13.69)

$$\left.\begin{aligned}\dot{v}_i=&\sum_{j=1}^{L}[p_{ij}(t)v_j(t)-p_{ji}(t)v_i(t)]\\&+(\beta_i-\mu_i)v_i(t)\end{aligned}\right\}\tag{13.80}$$

and

$$\sum_{i=1}^{L}\dot{v}_i=\sum_{i=1}^{L}(\beta_i-\mu_i)v_i(t)\ .\tag{13.81}$$

The equation (13.81) turns out to be a consequence of (13.80). The latter is of course the equation of motion for meanvalues, if simultaneous migration and birth-death processes take place. Again, the meanvalues obeying (13.80) are accompanied by the moving multi-Poisson distribution being the solution of the full master equation.

14 Tests of Significance in the Ranking Regression Analysis

Günter Haag, Martin Munz and Rolf Reiner

In order to quantify how far our estimated trendparameters are able to "explain" the variation of the empirical migration data it is necessary to introduce statistical tests. A first impression of the quality of adjustment is already given by the graphical representation of theoretical versus empirical migration flows as shown in Figures 3.1 and 3.2. It is obvious that there is a high significance of the estimated trendparameters ($v_{ij}(t)$, $u_i(t)$) if $w_{ij}^{th}(t) \approx w_{ij}^{e}(t)$, or in other words if the points in Figures 3.1 and 3.2 are close to the 45°-line. In these figures deviations of single flows may optically lead to a wrong impression of the quality of the estimation. However, the relative frequency of these deviations has to be considered, too.

In Figure 14.1 the relative frequency of events is plotted versus the difference between the empirical and theoretical migration rates. Obviously, small deviations are more frequent than large ones. Because this behaviour is tacitly presumed as a necessary basis for all analyzing procedures it has to be verified in order to substantiate the functional form of the theoretical transition rates (1.40). Comparing the different assumptions for the mobility matrix (1.40a, 1.40b and 3.9) of course, it can be seen that the best result (Figure 14.1a) is obtained using the most general mobility matrix (1.40a). The number of parameters is considerably reduced by introducing the reduced matrix (1.40b). This is reflected in an enhancement of the width of the distribution function of residuals (Figure 14.1b). The most unsatisfactory result of this investigation is obtained by using the geographical distance in the deterrence function (Figure 14.1c). In this case the distribution function of the residuals is not symmetric, this means the deterrence function (3.9) introduces a bias in the residuals of the migration flows. Therefore, regarding both the number of parameters and the width of the distribution of residuals the reduced mobility matrix (1.40b) seems to be the most appropriate one (see Figure 14.2).

The statistical tests introduced below will be used to quantify the quality of such a graphical representation.

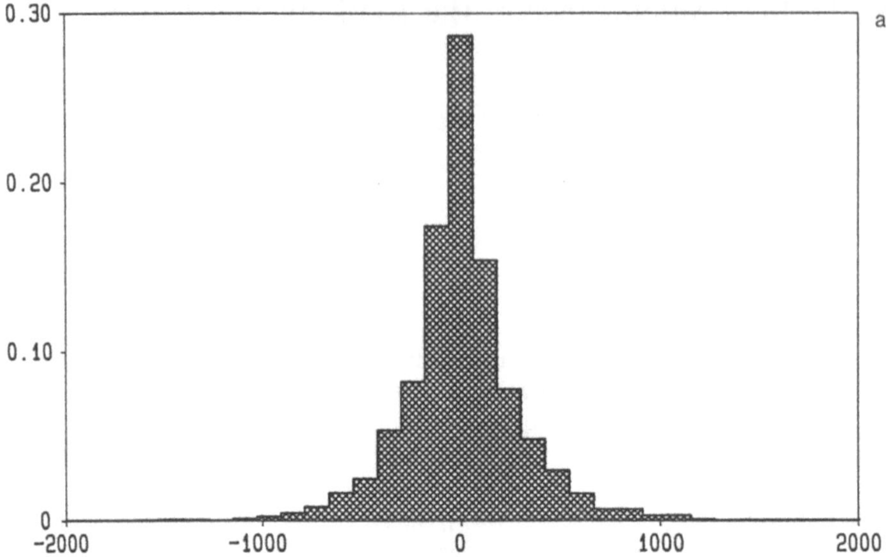

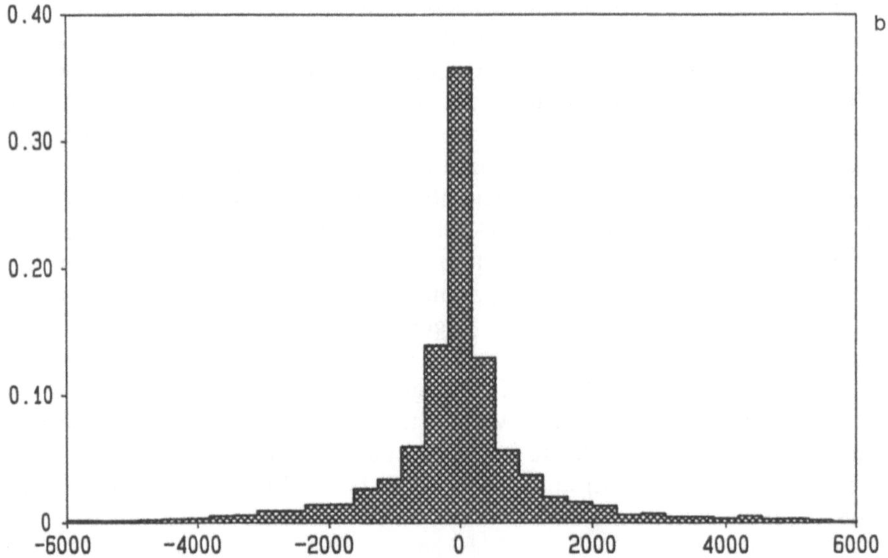

Fig. 14.1a–c. The relative frequency of events versus the difference between the empirical and theoretical migration rates of the Federal Republic of Germany, using (a): the total mobility matrix (1.40a), (b): the reduced mobility (1.40b), (c): the geographical distance in the deterrence function (3.9)

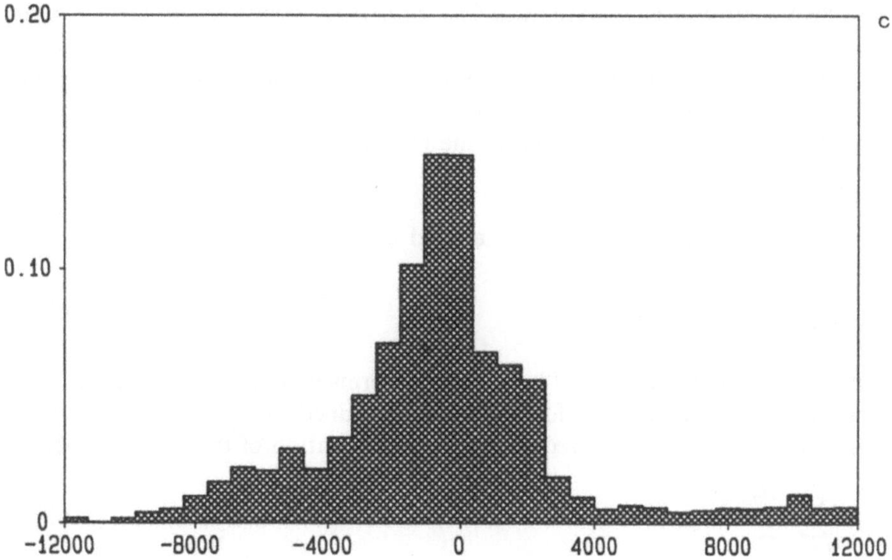

Fig. 14.1c. Legend see p. 346

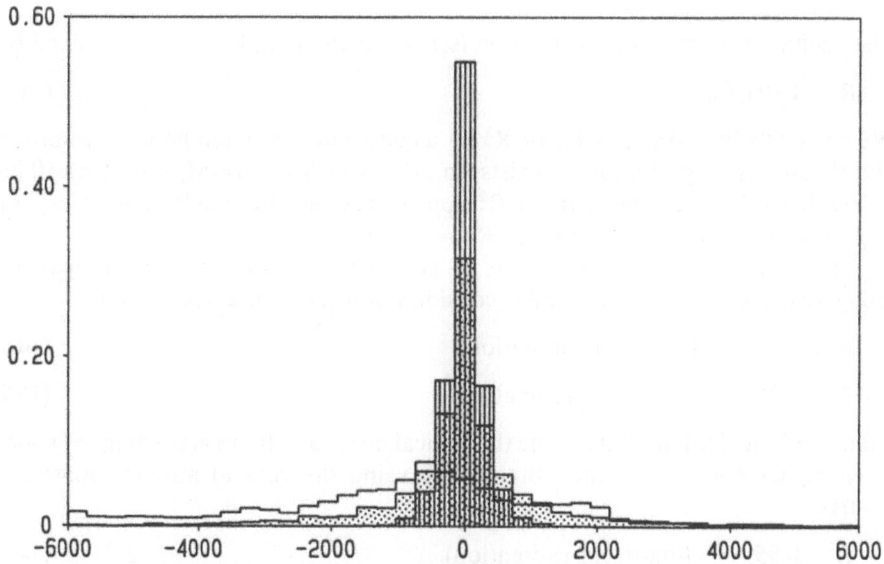

Fig. 14.2. Comparison of the relative frequency of events for the cases a, b and c of Figures 14.1

14.1 The Coefficient of Multiple Correlation (R^2)

The frequently used *coefficient of multiple correlation* R^2 measures the proportion of the variation in the empirical data which is described by the regression analysis. In order to define R^2 we first introduce the residuals

$$\varepsilon_{kl}(t)=w_{kl}^e(t)-w_{kl}^{th}(t) \tag{14.1}$$

and the sum of the squared residuals, denoted by

$$G=\sum_{t=1}^{T}\sum_{k,l=1}^{L}{}' \, [\varepsilon_{kl}(t)]^2 \tag{14.2}$$

The prime at the sum over k and l indicates that terms with $k=l$ are excluded. The expression (14.2) is a measure for the "non explained" part of the regression analysis, which must be compared with the total variation of the empirical data

$$G_0=\sum_{t=1}^{T}\sum_{k,l=1}^{L}{}' \, (w_{kl}^e(t)-\bar{w}^e)^2 \tag{14.3}$$

where

$$\bar{w}^e=\frac{1}{L(L-1)T}\sum_{t=1}^{T}\sum_{k,l=1}^{L}{}' \, w_{k.l}^e(t) \, . \tag{14.4}$$

The coefficient of multiple correlation (see for example [14.1]) is now defined by

$$R^2=1-G/G_0 \, . \tag{14.5}$$

We obviously have $0 \leq R^2 \leq 1$. For $R^2=1$ a complete agreement between empirical and theoretical migration flows exists. In other words, $\varepsilon_{kl}(t)=w_{kl}^e(t)-w_{kl}^{th}(t)=0$ for all residuals. On the other hand if R^2 approaches zero the model cannot explain anything. In Table 14.1 bounds for R are given.

The evaluation of the coefficient of multiple correlation in this comparitive study yields for all countries under consideration the satisfactory result

$$R^2>0.94 \qquad \text{(log-linear estimation)} \tag{14.6}$$

$$R^2>0.96 \qquad \text{(non-linear estimation)} \tag{14.7}$$

using the factorized mobility in the theoretical migration flows according to (1.40b). Even higher correlations are obtained by using the general mobility matrix in (1.40a).

$$R^2>0.959 \qquad \text{(log-linear estimation)} \tag{14.8}$$

$$R^2>0.994 \qquad \text{(non-linear estimation)} \tag{14.9}$$

More detailed results are presented in Part III (Chapters 10 and 11).

The coefficient of multiple correlation however has an unsatisfactory property: More independent fitting parameters used in the model to adjust the theoretical flows $w_{kl}^{th}(t)$ to the empirical $w_{kl}^e(t)$ will always lead to an increase of R^2. Hence, higher values of R^2 may only reflect the trivial fact, that more fitting parameters

Table 14.1. Bounds for R with f degrees of freedom. (Extracted from [14.2])

	Error probability				Error probability			
f	5	1	0.1	f	5	1	0.1	per cent
1	0.9969	0.9999	>0.9999	25	0.381	0.487	0.597	
2	0.9500	0.9900	0.9990	30	0.349	0.449	0.554	
3	0.8783	0.9587	0.9911	35	0.325	0.418	0.519	
4	0.811	0.917	0.974	40	0.304	0.393	0.490	
5	0.754	0.875	0.951	50	0.273	0.354	0.443	
6	0.707	0.834	0.925	60	0.250	0.325	0.408	
7	0.666	0.798	0.898	70	0.232	0.302	0.380	
8	0.632	0.765	0.872	80	0.217	0.283	0.357	
9	0.602	0.735	0.847	90	0.205	0.267	0.338	
10	0.576	0.708	0.823	100	0.195	0.254	0.321	
11	0.553	0.684	0.801	150	0.159	0.208	0.263	
12	0.532	0.661	0.780	200	0.138	0.181	0.230	
13	0.514	0.641	0.760	250	0.124	0.162	0.206	
14	0.497	0.623	0.742	300	0.113	0.148	0.188	
15	0.482	0.606	0.725	400	0.0978	0.128	0.164	
16	0.468	0.590	0.708	500	0.0875	0.115	0.146	
17	0.456	0.575	0.693	700	0.0740	0.0972	0.124	
18	0.444	0.561	0.679	1000	0.0619	0.0813	0.104	
19	0.433	0.549	0.665	1500	0.0505	0.0664	0.0847	
20	0.423	0.537	0.652	2000	0.0438	0.0575	0.0734	

lead to a better adjustment. But the significance of a theoretical model also derives from its attribute, that already as few as possible fitting parameters provide a good representation of empirical facts.

The introduction of the *corrected coefficient of multiple correlation* $\bar{R}^2$ eliminates the trivial dependence of the goodness of fit on the number of independent fitting parameters. It is defined by

$$\bar{R}^2 = 1 - s^2(\varepsilon)/s^2(w^e) \tag{14.10}$$

where

$$s^2(w^e) = G_0/(q-1) \tag{14.11}$$

is the variance of the empirical migration flows and

$$s^2(\varepsilon) = G/(q-k) \tag{14.12}$$

is the variance of the residuals. Here, q is the number of observations and k the number of independent explanatory variables. Thus $f=(q-k)$ denotes the degrees of freedom (compare (3.3)–(3.5)). The corrected coefficient of multiple correlation has the following properties:

a) $\bar{R}^2=R^2$ in the case $k=1$ or if $R^2=1$, and $\bar{R}^2<R^2$ in all other cases.

b) $\bar{R}^2$ may rise or fall, if new fitting variables are added to the model, whereas R^2 always increases with growing number of fitting variables.

Our migration model yields corrected coefficients of multiple correlation for the countries Federal Republic of Germany, Canada, France, Israel, Italy and Sweden

$$\bar{R}^2>0.93 \qquad \text{(log-linear estimation)} \tag{14.13}$$

$$\bar{R}^2>0.95 \qquad \text{(non-linear estimation)} \tag{14.14}$$

following (1.40b). The more general ansatz (1.40a) for the transition rates gives

$$\bar{R}^2>0.903 \qquad \text{(log-linear estimation)} \tag{14.15}$$

$$\bar{R}^2>0.985 \qquad \text{(non-linear estimation)}\,. \tag{14.16}$$

In the case of Germany, details are listed in Table 3.3, Section 3.

14.2 The F-Test

The F statistics can be used in multiple regression analysis to test the significance of the R^2 statistics. Strictly speaking, the F statistics with $(k-1)$ and $(q-k)$ degrees of freedom allows to test the null-hypothesis that none of the k explanatory variables such as utilities and mobilities are able to explain the variation of the migratory flows around their mean. It can be shown [14.3], that

$$F_{k-1,q-k}=\frac{R^2(q-k)}{(1-R^2)(k-1)} \tag{14.17}$$

is distributed as Fisher's F with $f_1=(k-1)$ and $f_2=(q-k)$ degrees of freedom. If the null-hypothesis is true, then we would expect R^2, and therefore F, to be close to zero. Thus we use a high value of the F statistics as a rationale for rejecting the null-hypothesis. The F-values for the different countries under consideration are listed in Table 14.2. It will be shown that in all cases the null-hypothesis can be rejected. In Table 14.3 bounds for F with error probability 0.1 per cent are given.

For the FRG (see Table 3.3), ($T=27$ years, $L=11$ regions) the number of observations is $q=T(L^2-L)=2970$. Following the non-linear estimation procedure, Section 3.1.2, we want to compare different assumptions of the mobility matrix $v_{ij}(t)$, namely (3.2) and (3.6). In the case $v_{ij}(t)=v_{ji}(t)$ without any further restriction, we have $k=1782$ trendparameters yielding the value $F=882$. With the more restrictive assumption (3.6) the number of trendparameters is reduced to

Table 14.2. *F*-values for the countries under consideration

	non - linear		log - linear	
country	(1.40a)	(1.40b)	(1.40a)	(1.40b)
FRG	882.4	355.7	248.1	331.0
Canada	328.1	138.1	53.4	118.9
France	207.9	476.5	51.9	245.4
Israel	115.6	248.2	17.1	112.9
Italy	305.8	663.8	73.1	518.4
Sweden	162.3	445.9	87.7	344.6

Table 14.3. Bounds for *F* with error probability 0.1 per cent. The degrees of freedom are $f_1=(k-1)$ in the numerator and $f_2=(q-k)$ in the denominator. (Extracted from [14.4])

$f_2 \setminus f_1$	5	10	20	30	50	100
5	29.8	26.9	25.4	24.9	24.4	24.1
10	10.5	8.75	7.80	7.47	7.19	6.98
30	5.53	4.24	3.49	3.22	2.98	2.79
50	4.90	3.67	2.95	2.68	2.44	2.24
100	4.48	3.30	2.59	2.32	2.07	1.87
200	4.29	3.12	2.42	2.15	1.90	1.68
500	4.18	3.02	2.33	2.05	1.80	1.57
∞	4.10	2.96	2.27	1.99	1.73	1.49

$k=379$ yielding the value $F=356$. A comparison of the *F*-values with the 0.1 per cent error bounds in Table 14.3 demonstrate that both assumptions reject the null-hypothesis with a probability of 0.999.

14.3 The *t*-Test

Another useful statistical test is the so-called *t*-test. For an expansion of a trendparameter $y_i(t)$ in terms of r relevant socio-economic variables $x_i^\alpha(t)$

$$y_i(t)=\sum_{\alpha=1}^{\alpha=r} a_\alpha x_i^\alpha(t) \tag{14.18}$$

the t-value of a coefficient a_α is defined as

$$t_\alpha = a_\alpha / s(a_\alpha) \tag{14.19}$$

where $s^2(a_\alpha)$ is the variance of the coefficient a_α. For a given confidence interval there exists a critical value of the t-distribution based on the number of degrees of freedom in the data and the desired level of significance. In Table 14.4 bounds for t with f degrees of freedom are given. For example the regression of the regional utilities for the FRG yields with $f=(TL-r)=(22*11\text{-}6)=236$, as a lower bound with 0.1 per cent error probability, for the significance of a socio-economic variable to be a key-factor, a t-value of $t_{min} \approx 3.3$ (for details see Chapter 4).

Table 14.4. Bounds for t with f degrees of freedom. (Extracted from [14.4])

f	5	2	1	0.1	per cent
1	12.71	31.28	63.66	636.6	
5	2.571	3.365	4.032	6.869	
10	2.228	2.764	3.169	4.587	
30	2.042	2.457	2.750	3.646	
50	2.009	2.403	2.678	3.495	
100	1.984	2.365	2.626	3.389	
200	1.972	2.345	2.601	3.339	
500	1.965	2.334	2.586	3.310	
∞	1.960	2.326	2.576	3.291	

14.4 The Durbin-Watson Test

In addition to the above introduced tests the Durbin-Watson test should be performed if a time series of a single variable is analyzed in terms of key-factors. We shall now consider this test for the hypothesis that no first order serial correlation is present ($\rho=0$) in the residuals ε_t, for $t=1, 2, \ldots, T$.

$$\varepsilon_t = \rho\, \varepsilon_{t-1} + \eta_t\,. \tag{14.20}$$

The test involves the calculation of a test statistics based on the residuals from the ordinary least-square procedure. Durbin and Watson [14.5] proposed the statistical measure

$$d = \frac{\sum_{t=2}^{T} (\varepsilon_t - \varepsilon_{t-1})^2}{\sum_{t=1}^{T} \varepsilon_t^2} \tag{14.21}$$

to quantify the first order serial autocorrelation of the residuals. From (14.21) it is obvious that if successive values of ε_t are close to each other, the statistics will be $d \approx 0$, indicating positive serial correlation. For $d = 2$ no first order serial correlation is present. In general we find $0 \leq d \leq 4$. If one is investigating the possibility of positive serial correlation, a value $d < d_l$ leads to the rejection of the null-hypothesis, whereas for $d > d_u$ the null-hypothesis (no first order serial autocorrelation) may be assumed. A value of $d_l \leq d \leq d_u$ means inconclusive results. For negative serial correlation the considerations have to be done with respect to the endpoint 4 instead of the endpoint 0. Then the null-hypothesis is acceptable if $d_u < d < 4 - d_u$,

Table 14.5. Five per cent significance points d_l and d_u for the Durbin-Watson test. (T = number of observations; r = number of explanatory variables, excluding the constant term). (Extracted from [14.3])

T	r = 1		r = 2		r = 3		r = 4		r = 5	
	d_l	d_u	d_l	d_u	d_l	d_u	d_l	d_u	d_l	d_u
15	1.08	1.36	0.95	1.54	0.82	1.75	0.69	1.97	0.56	2.21
16	1.10	1.37	0.98	1.54	0.86	1.73	0.74	1.93	0.62	2.15
17	1.13	1.38	1.02	1.54	0.90	1.71	0.78	1.90	0.67	2.10
18	1.16	1.39	1.05	1.53	0.93	1.69	0.82	1.87	0.71	2.06
19	1.18	1.40	1.08	1.53	0.97	1.68	0.86	1.85	0.75	2.02
20	1.20	1.41	1.10	1.54	1.00	1.68	0.90	1.83	0.79	1.99
21	1.22	1.42	1.13	1.54	1.03	1.67	0.93	1.81	0.83	1.96
22	1.24	1.43	1.15	1.54	1.05	1.66	0.96	1.80	0.86	1.94
23	1.26	1.44	1.17	1.54	1.08	1.66	0.99	1.79	0.90	1.92
24	1.27	1.45	1.19	1.55	1.10	1.66	1.01	1.78	0.93	1.90
25	1.29	1.45	1.21	1.55	1.12	1.66	1.04	1.77	0.95	1.89
26	1.30	1.46	1.22	1.55	1.14	1.65	1.06	1.76	0.98	1.88
27	1.32	1.47	1.24	1.56	1.16	1.65	1.08	1.76	1.01	1.86
28	1.33	1.48	1.26	1.56	1.18	1.65	1.10	1.75	1.03	1.85
29	1.34	1.48	1.27	1.56	1.20	1.65	1.12	1.74	1.05	1.84
30	1.35	1.49	1.28	1.57	1.21	1.65	1.14	1.74	1.07	1.83
40	1.44	1.54	1.39	1.60	1.34	1.66	1.29	1.72	1.23	1.79
50	1.50	1.59	1.46	1.63	1.42	1.67	1.38	1.72	1.34	1.77
100	1.65	1.69	1.63	1.72	1.61	1.74	1.59	1.76	1.57	1.78

otherwise the test is inconclusive. In Table 14.5, the 5 per cent significance points of d_l and d_u for the Durbin-Watson test are given following [14.3]. We used this test in this book in the regression analysis of the global mobility described in Part II and Chapter 15. One aim of the regression was to extract "systematic" effects from the mobility. This can only be done, if the null-hypothesis of the Durbin-Watson test is acceptable. In the case of Italy we have $T=18$ observations (from 1965 to 1982) and $r=3$ explanatory variables yielding the bounds $d_l=0.93$ and $d_u=1.69$ according to Table 14.5. In Chapter 8 we obtained the value $d=2.21$. Because of $1.69<d<2.31$ the null-hypothesis is acceptable here.

15. Ranking Regression Analysis of the Global Mobility

Martin Munz and Rolf Reiner

The global mobility $v_0(t)$ introduced in Chapter 1 is an average global measure for the frequency of changing location between any two regions of the considered country. This quantity can be determined from empirical migration data using our estimation procedure (compare Chapter 3). The global mobility is time-dependent, in general showing a long term trend superimposed by 3–5 years short term variations. This behaviour suggests to look for correlations between the mobility and appropriate socio-economic variables denoted as key-factors. Such an analysis can be performed in analogy to Section 3.2, where the utilities are analyzed in terms of socio-economic variables. In this regression analysis of the regional utilities global time trends of the variables are eliminated by normalizing the utilities and the socio-economic variables (see (3.69)). But this procedure cannot be adapted to the analysis of the global mobility. Instead we use a slightly modified procedure which enables us to introduce separate trends in time if necessary. Such separate trends – having no deeper explanatory value of their own – should only be taken into account if there exist no explanatory variables automatically implying the trend of the global mobility. Thus, we present a procedure allowing for both, pure trend variables and socio-economic variables implying trends.

The global mobility $v_0(t)$ of a population is the spatial average of the total mobility matrix (1.29). Therefore, temporal changes of $v_0(t)$ only can depend on spatially aggregated socio-economic variables $\Omega^\alpha(t)$ of the country.

15.1 The General Procedure

Let us assume, that a number A of such socio-economic variables are available. Because of the possibility of time lags between the economic variables and the global mobility it is desirable to have the economic variables for a time period beginning earlier than the period $[t=1, 2, \ldots, T]$ of the mobility:

$$\Omega^\alpha(t), \quad t=\ldots,-2,-1,0,1,2,\ldots,T \qquad \alpha=1,2,\ldots,A\,. \tag{15.1}$$

In this case the whole period $[t=1, 2, \ldots, T]$ of the estimated global mobility can be used for the regression analysis, even if there should appear a time delay in the functional dependence of the global mobility on key-factors.

In order to test whether a detrending of the mobility and the socio-economic variables is necessary or not we formally extend the space of "test-variables" $\Omega^\alpha(t)$ by adding the pure trend variables

$$\Omega^\beta(t)=t^\gamma, \qquad \text{with} \quad \beta=A+\gamma+1 \tag{15.2}$$

for $\gamma=0, 1, 2, \ldots, \Gamma$. Obviously in equation (15.2) we take into account a constant for $\gamma=0$, the linear trend for $\gamma=1$, and nonlinear trends for $\gamma\geq 2$. The t-test described in Chapter 14 enables us to check the significance of such possible additional linear or nonlinear trends in the global mobility.

It is assumed that the global mobility depends on a set of r key-factors. The expansion of the global mobility in terms of economic key-factors yields

$$\nu_0(t)=\sum_{\alpha=1}^{r} a_\alpha \Omega^\alpha(t-T_\alpha)\,. \tag{15.3}$$

In order to estimate the coefficients a_α as well as the time-lags T_α the conventional ordinary least square procedure could be used. This means, that the square sum of the residuals

$$\varepsilon_t=\nu_0(t)-\sum_{\alpha=1}^{r} a_\alpha \Omega^\alpha(t-T_\alpha) \tag{15.4}$$

has to be minimized by an optimal choice of a_α and T_α

$$\sum_{t=1}^{T} \varepsilon_t^2=\text{min!} \tag{15.5}$$

This condition yields the well known set of linear equations for the coefficients a_α for given T_α. The optimal time lags T_α are those leading to the minimal value of the sum (15.5). Usually this method is unpracticable because of the high number of possible combinations which should be tested.

As already mentioned we apply a modification of the ordinary least square method consisting of the stepwise regression procedure presented in Chapter 3. This is now adapted to the analysis of the global mobility which is a function of time only.

Let us assume that $(k-1)$ variables $\Omega^1, \ldots, \Omega^{k-1}$ with certain time-lags $T_1, \ldots, T_{k-1}$ have already been selected out of the total set $\{\Omega^\alpha\}$. In order to select the next (the k-th) key-factor out of the remaining set of variables, we have to look for the variable describing best the not yet explained part of the global mobility. Therefore, we first go over to the orthogonalized variables $\hat{\Omega}^\alpha(t-\tau_\alpha)$

$$\hat{\Omega}^\alpha(t-\tau_\alpha)=\Omega^\alpha(t-\tau_\alpha)-\sum_{\beta=1}^{k-1} \frac{\langle\hat{\Omega}^\beta(T_\beta)|\Omega^\alpha(\tau_\alpha)\rangle}{\langle\hat{\Omega}^\beta(T_\beta)|\hat{\Omega}^\beta(T_\beta)\rangle}\,\hat{\Omega}(t-T_\beta) \tag{15.6}$$

where $\quad \alpha=k, k+1, \ldots, A+\Gamma$, and time lags $\tau_\alpha=0, 1, 2, \ldots$

in which the part of time-dependence already contained in the previous key-factors $\Omega^\beta(t-T_\beta)$, $\beta=1, 2, \ldots, (k-1)$, has been subtracted. The scalar products $\langle\Omega^\beta(\tau_\beta)|\Omega^\alpha(\tau_\alpha)\rangle$ appearing in (15.6) are defined as follows:

$$\langle \Omega^{\beta}(\tau_\beta)|\Omega^{\alpha}(\tau_\alpha)\rangle = \sum_{t=1}^{T} \Omega^{\beta}(t-\tau_\beta)\,\Omega^{\alpha}(t-\tau_\alpha)\,. \tag{15.7}$$

Similarly, the residual part of the global mobility not yet represented in terms of the previous key-factors $\hat{\Omega}^{\beta}(t-T_\beta)$, $\beta=1, 2, \ldots, (k-1)$ is given by:

$$\hat{v}_0(t) = v_0(t) - \sum_{\beta=1}^{k-1} \frac{\langle \hat{\Omega}^{\beta}(T_\beta)|v_0\rangle}{\langle \hat{\Omega}^{\beta}(T_\beta)|\hat{\Omega}^{\beta}(T_\beta)\rangle}\,\hat{\Omega}^{\beta}(t-T_\beta)\,. \tag{15.8}$$

Now let us consider the question which of the variables $\hat{\Omega}^{\alpha}(t-\tau_\alpha)$, $\alpha=k, k+1, \ldots$ are appropriate to represent optimally the residual mobility $\hat{v}_0(t)$ in terms of

$$\hat{v}_0(t) = b_\alpha \hat{\Omega}^{\alpha}(t-\tau_\alpha) + \hat{\hat{v}}_0(t) \tag{15.9}$$

where $\hat{\hat{v}}_0(t)$ is the nonexplained rest of $\hat{v}_0(t)$. For this purpose we introduce the mean square of $\hat{v}_0(t)$

$$S_{\hat{v}}^2 = \frac{1}{T-1} \sum_{t=1}^{T} \hat{v}_0^2(t)\,. \tag{15.10}$$

Since we have used – without anticipating the detrending procedure – the constant $\Omega^{A+1}(t)=1$, $t=1, 2, \ldots, T$ as the first "explanatory" variable, the meanvalue of $\hat{v}_0(t)$ is vanishing and the mean square (15.10) of $\hat{v}_0(t)$ is equivalent to its variance. The mean square (variance) of the nonexplained rest $\hat{\hat{v}}_0(t)$ of $\hat{v}_0(t)$

$$S_{\text{res}}^2 = \frac{1}{T-1} \sum_{t=1}^{T} (\hat{v}_0(t) - b_\alpha \hat{\Omega}^{\alpha}(t-\tau_\alpha))^2 = \frac{1}{T-1} \sum_{t=1}^{T} \hat{\hat{v}}_0^2(t) \tag{15.11}$$

can now be minimized for each $\hat{\Omega}^{\alpha}(t-\tau_\alpha)$ by chosing the coefficient

$$b_\alpha = \frac{\langle \hat{\Omega}^{\alpha}(T_\alpha)|\hat{v}_0\rangle}{\langle \hat{\Omega}^{\alpha}(T_\alpha)|\hat{\Omega}(T_\alpha)\rangle} = \frac{\langle \hat{\Omega}^{\alpha}(T_\alpha)|v_0\rangle}{\langle \hat{\Omega}^{\alpha}(T_\alpha)|\hat{\Omega}(T_\alpha)\rangle} \tag{15.12}$$

and the optimal time lag $\tau_\alpha = T_\alpha$. In order to check the significance of each variable $\Omega^{\alpha}(t-\tau_\alpha)$ (to be a key-factor) one can make use of the corresponding t-test (see equation (14.19) and Table 14.4) and the correlation coefficient R^2 (see equation (14.5) and Table 14.1). The value of the t-test is obtained by dividing the coefficient b_α by the variance of the residual mobility S_{res}.

$$t_\alpha = \frac{\langle \hat{\Omega}^{\alpha}(T_\alpha)|\hat{v}_0\rangle}{\langle \hat{\Omega}_{\alpha}(T_\alpha)|\hat{\Omega}^{\alpha}(T_\alpha)\rangle^{1/2} S_{\text{res}}} \tag{15.13}$$

and the correlation coefficient follows from (14.5)

$$R^2 = 1 - S_{\text{res}}^2 / S_{\hat{v}_0}^2\,. \tag{15.14}$$

The comparison of the values of t and/or R^2 of the different socio-economic variables $\hat{\Omega}^{\alpha}(t-\tau_\alpha)$ with different reasonable time lags τ_α leads to the selection of the optimal $\Omega^{\alpha}(t-T_\alpha)$ to be the k-th key-factor $\Omega^{k}(t-T_k)$. Of course, this has to be done under consideration of the error bounds in Tables 14.1 and 14.4.

If the remaining term $\hat{\hat{v}}_0(t)$ can be neglected, the formula (15.9) with $\alpha=k$ can be inserted in (15.8) yielding the representation of $v_0(t)$ in terms of k key-factors

$$v_0(t)=\sum_{\beta=1}^{k}\frac{\langle\hat{\Omega}^\beta(T_\beta)|v_0\rangle}{\langle\hat{\Omega}^\beta(T_\beta)|\hat{\Omega}^\beta(T_\beta)\rangle}\hat{\Omega}(t-T_\beta)\,. \tag{15.15}$$

Then the selection process of key-factors ends here and we may identify k with the final number r of key-factors.

If, on the other hand, the rest term $\hat{\hat{v}}_0(t)$ cannot be neglected one obtains by inserting (15.9) with (15.12) into (15.8):

$$\hat{\hat{v}}_0(t)=v_0(t)-\sum_{\beta=1}^{k}\frac{\langle\hat{\Omega}^\beta(T_\beta)|v_0\rangle}{\langle\hat{\Omega}^\beta(T_\beta)|\hat{\Omega}^\beta(T_\beta)\rangle}\hat{\Omega}^\beta(t-T_\beta)\,. \tag{15.16}$$

This formula has the same form as (15.8) with k instead of $(k-1)$ and $\hat{\hat{v}}_0(t)$ instead of $\hat{v}_0(t)$ now playing the role of the residual part of the global mobility. The procedure of finding an optimal representation of $\hat{\hat{v}}_0(t)$ in terms of the remaining variables $\hat{\Omega}^\alpha(t-\tau_\alpha)$, $\alpha=k+1, k+2, \ldots$ (orthogonal to $\hat{\Omega}^1(t-T_1), \hat{\Omega}^2(t-T_2), \ldots, \hat{\Omega}^k(t-T_k)$) must now be continued on the same lines.

If the process of selecting key-factors has ended with $k=r$, the global mobility $v_0(t)$ can be represented as the linear combination (15.15) of the r orthogonalized key-factors $\hat{\Omega}^\beta(t-T_\beta)$, $\beta=1, 2, \ldots, r$. It is however more suggesting and intuitively appealing to rewrite $v_0(t)$ as a linear combination of the *original variables* $\Omega^\beta(t-T_\beta)$, $\beta=1, 2, \ldots, r$, which are not orthogonal in general:

$$v_0(t)=\sum_{\beta=1}^{r}a_\beta\Omega^\beta(t-T_\beta)\,. \tag{15.17}$$

Comparing (15.17) with (15.15) after expressing $\Omega^\beta(t-T_\beta)$ by the $\hat{\Omega}^\gamma(t-T_\gamma)$ with the help of (15.6) one obtains the relations

$$a_\alpha=\frac{\langle\hat{\Omega}^\alpha(T_\alpha)|v_0\rangle}{\langle\hat{\Omega}^\alpha(T_\alpha)|\hat{\Omega}^\alpha(T_\alpha)\rangle}-\sum_{\beta=\alpha+1}^{r}\frac{\langle\hat{\Omega}^\alpha(T_\alpha)|\Omega^\beta(\tau_\beta)\rangle}{\langle\hat{\Omega}^\alpha(T_\alpha)|\hat{\Omega}^\alpha(T_\alpha)\rangle}a_\beta \tag{15.18}$$

for $\alpha=r, r-1, \ldots, 1$

which allow for a recursive determination of the coefficients a_β. Finally, the test-values characterizing the quality of our analysis are determined. The t-values of the coefficients a_α of the key-factors become

$$t_\alpha=\frac{\langle\hat{\Omega}^\alpha(T_\alpha)|v_0\rangle}{\langle\hat{\Omega}_\alpha(T_\alpha)|\Omega^\alpha(T_\alpha)\rangle^{1/2}\,S_{\mathrm{res}}} \tag{15.19}$$

and the corrected coefficient of multiple correlation reads

$$\bar{R}^2=1-S_{\mathrm{res}}^2/S_v^2 \tag{15.20}$$

where

$$S_{\mathrm{res}}^2=\frac{1}{T-r}\sum_{t=1}^{T}\left(v_0(t)-\sum_{\alpha=1}^{r}a_\alpha\Omega^\alpha(t-T_\alpha)\right)^2 \tag{15.21}$$

and

$$S_v^2 = \frac{1}{T-1} \sum_{t=1}^{T} v_{\text{det}}^2(t) \tag{15.22}$$

where v_{det} is the detrended part of the global mobility. Of course, the quality of the explanatory content cannot depend on pure trend variables. This requires the use of (15.22) instead of the simple variance of the global mobility.

Since the trend variables are treated in the same way as the socio-economic variables, the procedure allows to decide whether it is necessary or not to use pure trend variables beyond the socio-economic variables in the representation of $v_0(t)$. If a trend variable turns out to be relevant, it is indicated to start the selection procedure with the significant trends.

$$\Omega^{\beta}(t-T_{\beta}) = t^{\beta-1} \tag{15.23}$$

with $\beta = 1, 2, \ldots, g+1$. The following socio-economic key-factors $\hat{\Omega}^{\beta}(t-T_{\beta})$, $\beta = g+2, g+3, \ldots$ then are *automatically* detrended by formula (15.6). The same holds for the residual detrended mobility

$$\hat{v}_{\text{det}}(t) = v_0(t) - \sum_{\beta=1}^{g+1} \frac{\langle \hat{\Omega}^{\beta}(T_{\beta}) | v_0 \rangle}{\langle \hat{\Omega}^{\beta}(T_{\beta}) | \hat{\Omega}^{\beta}(T_{\beta}) \rangle} \hat{\Omega}^{\beta}(t-T_{\beta}) \,. \tag{15.24}$$

The above described ranking procedure seems to determine an *unique sequence* of key-factors out of the set of all socio-economic variables. In practice, however, the situation is not always so ideal. Starting from two different socio-economic variables of comparably high correlation to the mobility (so that each of them can be considered as the first-ranking key-factor) one may be lead to two sequences containing *different* key-factors. Nevertheless both final representations of the global mobility may be good fits of comparable significance according to all tests including the Durbin-Watson test. The ambivalence of the ensuing interpretations is evident. Such a non-uniqueness of representations can be seen as a hint to hidden cross-correlations between variables of different direct interpretations.

15.2 An Explicit Example

In order to illustrate this kind of problems in the key-factor analysis of the global mobility, we shall now explicitly treat a characteristic example.

The time evolution of the global mobility of Sweden (Figure 15.1) yields in the first step of the analysis the test results for the socio-economic and trend variables depicted in extracts in Table 15.1.

Obviously, t-test and the correlation coefficient R^2 yield in this case the same ranking of the variables. Without any manipulation of the procedure the computer program itself would select the variable E with a time-lag of 2 years in the first step. (Negative time-lags are only considered to characterize possible maxima in the test values for time-lag zero. Of course we cannot chose a negative time-lag because the

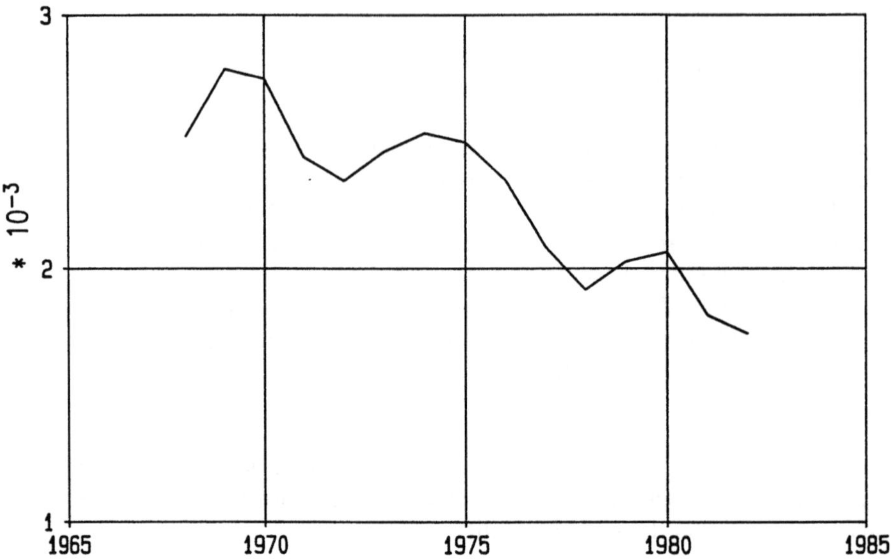

Fig. 15.1. The global mobility of Sweden

Table 15.1. "Test-Table" of the first step of the regression analysis of the global mobility of Sweden

Timelag (in years)	-1		0		1		2		3		4	
Variable*	t	R^2	t	R^2	t	R^2	t	R^2	t	R^2	t	R^2
L	-5.8	0.72	-7.3	0.79	-10.	0.89	-9.5	0.88	-6.7	0.80	-5.0	0.72
I	-6.6	0.77	-9.2	0.86	-11.	0.89	-8.5	0.86	-6.3	0.78	-5.4	0.74
E	-5.0	0.65	-6.0	0.72	-9.6	0.88	-12.	0.92	-7.3	0.83	-4.7	0.69
VR	2.3	0.30	3.8	0.51	1.1	0.09	-0.7	0.04	0.1	0.00	1.3	0.15
CPC	4.7	0.63	6.3	0.74	7.4	0.81	7.8	0.83	7.2	0.83	5.9	0.77
UR	-1.8	0.19	-2.0	0.23	0.3	0.01	2.2	0.28	1.4	0.15	0.5	0.02
LT	-6.8	0.78	-8.1	0.82	-9.1	0.86	-7.7	0.83	-6.4	0.79	-6.3	0.80

*abbreviations of the variables:

L: total labour force, I: total income, E: employment, VR: rate of vacancies with respect to L, CPC: new constructions per capita, UR: unemployment rate, LT: linear trend

Table 15.2. "Test-Table" of the second step of the regression analysis of the global mobility of Sweden using CPC with a time-lag of 2 years as the first key-factor

Timelag (in years)	-1		0		1		2		3		4	
Variable	t	R^2	t	R^2	t	R^2	t	R^2	t	R^2	t	R^2
L	0.4	0.01	-0.4	0.01	-1.8	0.21	-2.4	0.32	-0.3	0.01	0.4	0.01
I	-0.2	0.00	-1.7	0.19	-2.3	0.31	-1.7	0.19	-0.3	0.00	-0.7	0.04
E	0.8	0.05	0.9	0.05	-1.5	0.16	-3.8	0.55	-0.7	0.03	0.5	0.02
VR	1.0	0.09	13.	0.93	1.0	0.07	-4.2	0.60	-1.7	0.20	1.1	0.10
UR	-1.5	0.17	-5.7	0.73	-0.6	0.03	2.1	0.27	0.8	0.05	-0.5	0.03
LT	-0.8	0.05	-1.2	0.12	-1.2	0.12	-1.2	0.12	-0.1	0.00	-0.5	0.02

socio-economic variables are causative input-variables inducing a later effect on the mobility but not vice versa.) It would not be very instructive to present all our efforts to find the best fit. But it should be mentioned that the best result of the regression analysis using in the first step the variable E with the time-lag of 2 years gives $R^2=0.97$ and $F=147$. This result is however only moderately good. Selecting instead the variable CPC with a time-lag of 2 years, as first variable, the "test" Table 15.2 is obtained in the second step.

Here the computed test values give evidence that – with a great distance to all other variables – VR with a zero time-lag is the most important second variable. All other variables which were significant in the first step beside CPC have lost their relevance in the second step. This can be explained by a high cross correlation of these variables with CPC. To illustrate the results obtained in both steps of the regression analysis Figures 15.2 and 15.3 show the fit in each case.

Figure 15.2 shows that CPC describes the long term evolution of the mobility only. Therefore, the completion of new dwellings can be taken as an indicator of the declining "trend" in the time behaviour of the global mobility of Sweden. This is satisfactory, because it is desirable to find an explanatory socio-economic variable instead of fitting this decline by a simple linear or quadratic trend. The short term cycles of the mobility are also reflected in the fit, if VR without any time-lag is added (see Figure 15.3). Measuring CPC in units of newly finished dwellings per year and thousand people and VR in per cent of the total labour force, the results presented in Table 15.3 are obtained.

Because all test criteria were fulfilled to a satisfactory degree (compare the tables presented in Chapter 14), the analysis already could be finished after the second step. In general, more variables must be taken into account, in particular if it turns out that the test results for additional variables are of about the same quality. Further results of the regression analysis of the different countries are presented in detail and discussed in Part II.

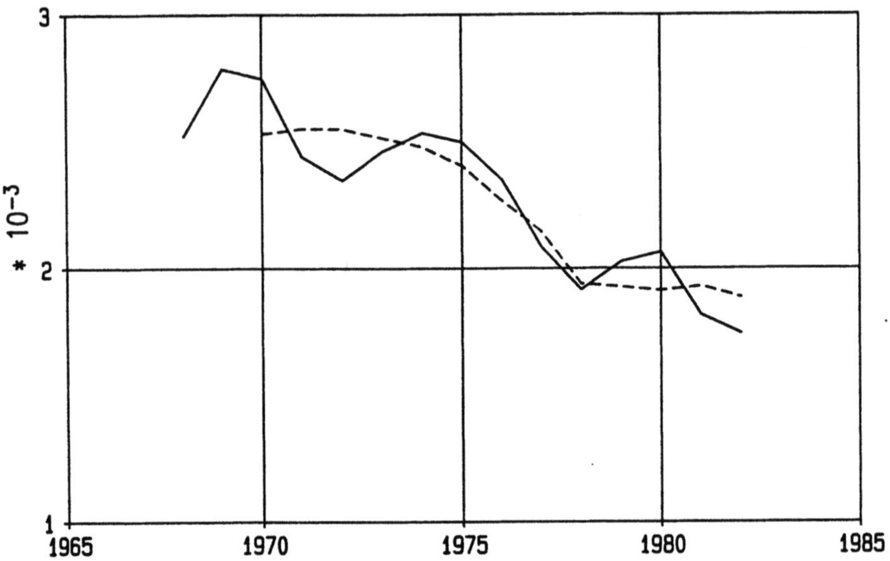

Fig. 15.2. Global mobility of Sweden (solid line) and the fit using only CPC with a time-lag of 2 years (dashed line)

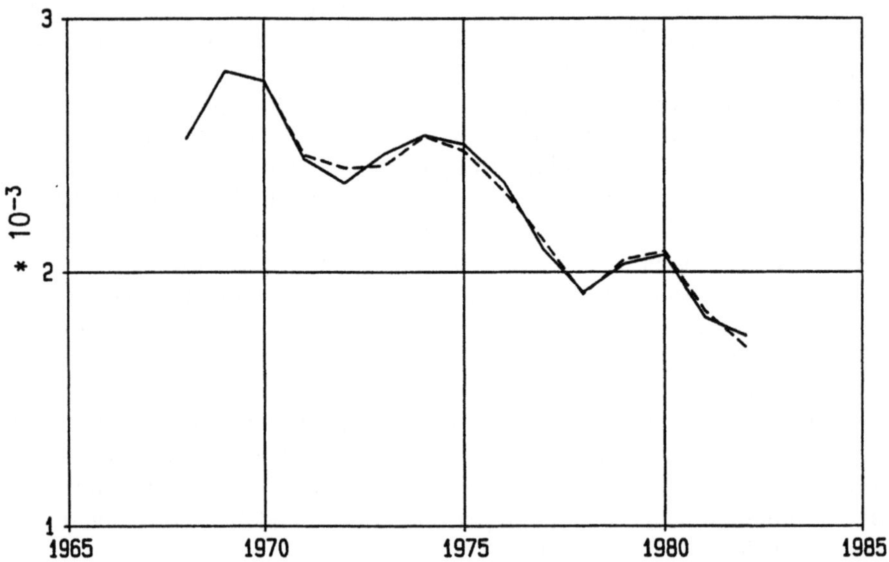

Fig. 15.3. Global mobility of Sweden (solid line) and the fitted curve using CPC and VR with time-lag zero (dashed line)

Table 15.3. Results of the regression analysis of the global mobility of Sweden

variable	timelag	coefficient	t - value
CPC	2 years	$7.48\ 10^{-5}$	20.6
VR	0 years	$4.47\ 10^{-2}$	10.7
constant	-	$1.04\ 10^{-3}$	-

correlation R^2 : 0.989
corrected $\bar{R}^2$: 0.987
F - test value : 451.6
Durbin-Watson test : 2.29

16 A Computer Program for the Estimation of Utilities and Mobilities

Martin Munz and Rolf Reiner

In this section a computer program written in Fortran 77 for the estimation of the utilities and mobilities is listed. For simplicity the log-linear estimation procedure is used. A program for the non-linear estimation procedure is obtained from the listed program by substituting the subroutine ESTIMA by a more complicated one for solving the non-linear equations arising in this case.

The program enables also to perform a decomposition of the total mobility matrix to a time-dependent global mobility function and a time-independent deterrence matrix (a function of the "Migration Distance"). Statistical tests including the correlation coefficient, the adjusted correlation and Fishers F-value, are computed using the total mobility and the reduced mobility.

The program was developed on a VAX 11/750 computer system using the standard Fortran F77-compiler. To adapt this program to another computer system the two PARAMETER statements defining the logical unit numbers for terminal input and output have to be changed.

```
PARAMETER LUNINP= ...        ! for Terminal input
PARAMETER LUNOUT= ...        ! for Terminal output
```

The listed program is designed for the evaluation of the data of Federal Republic of Germany. For using this program to analyze the migration data of another country the three PARAMETER statements defining the number of regions and the number of years for the available data set must be changed.

```
PARAMETER         L= ...     ! number of regions
PARAMETER YEAR1  = ...       ! first year of migration data
PARAMETER YEAR2  = ...       ! last year of migration data
```

In all cases the input procedure for the empirical migration datas has to be set up by the user. This means there has to be included a subroutine DATAIN, in which the population numbers as well as the transition rates are read, for example from a data file. $N(I, Y)$ is the population number of region I in year Y and $W(I, J, Y)$ is the transition rate from region J to region I in year Y. The statement

```
PARAMETER FILE1= ...
```

can be used to define the file name of the data file.

The subroutine DATAIN must be of the following structure

```
      SUBROUTINE DATAIN (N, W, L, YEAR, FILE1)
      INTEGER YEAR
      DIMENSION W(L, L, YEAR), N(L, YEAR)
      REAL N
      CHARACTER*(*) FILE1
C
      OPEN(UNIT=10, FILE=FILE1, STATUS='OLD')
C
C     Here include appropriate READ statements
C     for the input of N(I, K) and W(I, J, K)
C     where I, J=1, 2, . . . , L and K=1, 2, . . . , YEAR
C
      CLOSE(UNIT=10)
      RETURN
      END
```

<u>The estimation program :</u>

```
      PROGRAM ESTIMATION
C
C  Estimation of the utilities and mobilities using
C  the log - linear procedure
C  several statistical tests are also computed
C
C
      PARAMETER LAND = 'Federal Republic of Germany'
      PARAMETER LUNINP = 5                    ! logical unit number
                                              ! for Terminal Input
      PARAMETER LUNOUT = 6                    ! logical unit number
                                              ! for Terminal Output
      PARAMETER L = 11                        ! Number of regions
      PARAMETER YEAR1 = 1982                  ! first year
      PARAMETER YEAR2 = 1983                  ! last year
      PARAMETER YEAR = 1 + YEAR2 - YEAR1      ! Number of years
```

```
      PARAMETER FILE1 - '[MUNZ.MIGRA]BRDN.DAT'! Input filename
      PARAMETER FILE2 - 'BRDOUT.DAT'          ! Output Filename
C
      DIMENSION W(L,L,YEAR)      ! Migration - matrix
      DIMENSION WS(L,L,YEAR)     ! smoothed migration - matrix
      DIMENSION N(L,YEAR)        ! population in region L
      DIMENSION U(L,YEAR)        ! Utility of region L
      DIMENSION NUE(L,L,YEAR)    ! mobility in region L
      DIMENSION NUEO(YEAR)       ! spatial - averaged mobility
      DIMENSION NUEM(L,L)        ! spatial deterrence matrix
      DIMENSION R(YEAR)          ! correlation coefficient
      DIMENSION RA(YEAR)         ! adjusted correlation coefficient
      DIMENSION F(YEAR)          ! Fisher's F
      DIMENSION S1(YEAR)         ! variance of the utilities
      DIMENSION S2(YEAR)         ! variance of migration - matrices
      DIMENSION S3(YEAR)         ! square sum of residuals
      DIMENSION RNUE(L,L,YEAR)   ! working matrix
      REAL N,NUE,NUEO,NUEM
      CHARACTER*120 STRICH
      CHARACTER*13  FORM
      LOGICAL  ZERO
      COMMON/GEKA/ U,NUE
C
C
C   Input of migration data W(I,J,Y) and N(I,Y)
C
C   W(I,J,Y) is the empirical transition rate from region J
C   to region I in year Y
C
C   N(I,Y) is the population number in region I at year Y
C
      CALL DATAIN(N,W,L,YEAR,FILE1)   ! Note : This SUBROUTINE has to
C                                     ! be supplied by the user
C
      ZERO - .FALSE.
      DO 1 K - 1,YEAR
        DO 1 I - 1,L
          DO 1 J - I+1,L
          IF(W(I,J,K).EQ.0. .OR. W(J,I,K).EQ.0.) ZERO - .TRUE.
1     CONTINUE
C
```

```
C
      WRITE(LUNOUT,*)' Estimation of utilities and mobilities'
      WRITE(LUNOUT,*)' for ',LAND
      WRITE(LUNOUT,*)' '
      IF( .NOT.ZERO )  THEN
        WRITE(LUNOUT,*)' smoothing the migration - matrix ?   0/1'
        WRITE(LUNOUT,*)' 0   :   no'
        WRITE(LUNOUT,*)' 1   :   yes'
        READ(LUNINP,*) IS
      ELSE
        WRITE(LUNOUT,*)' smoothing is absolutely necessary'
        WRITE(LUNOUT,*)' there are zeros in the migration matrices'
        IS  =  1
      ENDIF
      IF(IS .EQ. 1) THEN
              WRITE(LUNOUT,*)' width of smoothing ?'
              WRITE(LUNOUT,*)' proposal : 50 - 100'
              READ(LUNINP,*) WO
              IF( WO.EQ.0. )  WO = 50.
              CALL SMOOTH(W,WS,L,YEAR,WO)
              WRITE(LUNOUT,*)'smoothing is finished'
      ELSE
              DO 10 I=1,L
                DO 10 J=1,L
                  DO 10 K=1,YEAR
                  WS(I,J,K) = W(I,J,K)
10            CONTINUE
      ENDIF
C
C
C
      CALL ESTIMA(WS,L,YEAR,N,U,NUE)

C
C   coefficient of multiple correlation
C
      CALL VARO(W,N,U,NUE,L,YEAR,S1,S2,S3)
C
      DATEN = FLOAT(L*(L - 1))
      PARM = FLOAT(L*(L + 1))/2. + 1.
      FAKTOR = (DATEN - 1.)/(DATEN - PARM)
```

```
      FAKTOR2 = (DATEN - PARM)/(PARM - 1.)
      DO 15 K = 1,YEAR
        R(K) = 1. - S3(K)/S2(K)
        RA(K) = 1. - S3(K)/S2(K)*FAKTOR
        F(K)  = R(K)/(1. - R(K))*FAKTOR2
15    CONTINUE
C
      CALL MOBRED(NUE,NUEM,NUEO,L,YEAR)
C
C
      CALL VAR1(W,N,U,NUE,L,YEAR,S20,S30)
C
      RM = 1. - S30/S20                          ! total squared correlation
      F1 = FLOAT(YEAR*L*(L-1))
      F2 = FLOAT(YEAR*(L*(L+1)/2)) + 1.
      RAM = 1. - (F1 - 1.)*S30/(S20*(F1 - F2))  ! adjusted total correlation
      FM = RM/(1. - RM)*(F1 - F2)/(F2 - 1.)     ! total Fisher' s F
C
      DO 20 K = 1,YEAR
        DO 20 I = 1,L
          DO 20 J = 1,L
            RNUE(I,J,K) = NUEO(K)*NUEM(I,J)
20    CONTINUE
C
C  with the separated mobility
C
      CALL VAR1(W,N,U,RNUE,L,YEAR,S20,S30)
C
      F2 = FLOAT(L*(L-1)/2 + (L+1)*YEAR) + 1.
      RM1 = 1. - S30/S20                          ! total squared correlation
      RAM1 = 1. - (F1 - 1.)*S30/(S20*(F1 - F2)) ! adjusted total correlation
      FM1 = RM1/(1. - RM1)*(F1 - F2)/(F2 - 1.)  ! total Fisher' s F
C
C
C   Output of the computed utilities, mobilities and number of
C   population in each region and for each year
C
      LUNF = LUNINP + LUNOUT
      OPEN(UNIT = LUNF,FILE = FILE2, STATUS = 'NEW')
C
      WRITE(LUNF,*)' '
```

```
      WRITE(LUNF,*)'Migration - Analysis of ',LAND
      WRITE(LUNF,*)' '
      DO 30 I = 1,80
        STRICH(I:I) = '*'
30    CONTINUE
      WRITE(LUNF,*)STRICH
      WRITE(LUNF,*)' '
      WRITE(LUNF,*)'The Migration datas has been analyzed using'
      WRITE(LUNF,*)'the log - linear estimation procedure'
      IF( IS.EQ.0 ) THEN
              WRITE(LUNF,*)'No smoothing of the migration matrix'
      ELSE
              WRITE(LUNF,'(1X,A,I3)')'Width of smoothing            :',INT(W0)
      ENDIF
      WRITE(LUNF,*)' '
      WRITE(LUNF,*)'Number of Regions             :',L
      WRITE(LUNF,*)'Number of Years               :',YEAR
      WRITE(LUNF,*)' '
      WRITE(LUNF,*)'Tests using the total Mobility Matrix'
      WRITE(LUNF,*)'Squared Correlation           :',RM
      WRITE(LUNF,*)'Adjusted Squared Correlation :',RAM
      WRITE(LUNF,*)'Fishers F - Value             :',FM
      WRITE(LUNF,*)' '
      WRITE(LUNF,*)'Tests using the reduced Mobilities'
      WRITE(LUNF,*)'Squared Correlation           :',RM1
      WRITE(LUNF,*)'Adjusted Squared Correlation :',RAM1
      WRITE(LUNF,*)'Fishers F - Value             :',FM1
C
      WRITE(LUNF,*)' '
      WRITE(LUNF,*) STRICH
      WRITE(LUNF,*)' '
C
      DO 40 I = 1,80
        STRICH(I:I) = '-'
40    CONTINUE
C
      FORM(1:1) = '('
      LANG = INT(ALOG10(REAL(L))) + 1
      IF( LANG.EQ.1 )  THEN
        WRITE(FORM(2:2),'(I1)')  L
      ELSE
```

```
          WRITE(FORM(2:3),'(I2)')  L
        ENDIF
        FORM(2+LANG:13) = '(1X,F6.3))'
C
        WRITE(LUNF,*) ' '
        WRITE(LUNF,*) ' '
        WRITE(LUNF,*)' spatial deterrence function - time averaged mobility '
        WRITE(LUNF,*)' '
        WRITE(LUNF,*) STRICH
        WRITE(LUNF,*)' '
        DO 45 I = 1,L
          WRITE(LUNF,FORM) ( NUEM(I,J),J=1,L)
45      CONTINUE
C
        DO 50 K = 1,YEAR
          WRITE(LUNF,*)' '
          WRITE(LUNF,*)' '
          WRITE(LUNF,*) STRICH
          WRITE(LUNF,*)' '
          WRITE(LUNF,*)' '
          WRITE(LUNF,'(1X,A,2X,I4)')'Year under consideration          '
     1  ,YEAR1 + K - 1
          WRITE(LUNF,*)'Spatial averaged mobility       :',NUEO(K)
          WRITE(LUNF,*)'Variance of the utilities       :',S1(K)
          WRITE(LUNF,*)'Fishers F - value               :',F(K)
          WRITE(LUNF,*)' '
          WRITE(LUNF,*) ' '
          WRITE(LUNF,*)' Region            Utility             Population '
          WRITE(LUNF,*)' '
          WRITE(LUNF,*)STRICH
          WRITE(LUNF,*)' '
          DO 60 I =1,L
            WRITE(LUNF,'(4X,I2,12X,F9.5,12X,E10.4)')I,U(I,K),N(I,K)
60        CONTINUE
50      CONTINUE
        CLOSE(UNIT = LUNF)

C
        WRITE(LUNOUT,*)'The results are in the file'
        WRITE(LUNOUT,*) FILE2
C
```

```
      END
C
C
C
      SUBROUTINE ESTIMA(W,L,YEAR,N,U,NUE)
      INTEGER YEAR
      DIMENSION W(L,L,YEAR),U(L,YEAR),NUE(L,L,YEAR),N(L,YEAR)
      REAL NUE,N
C
C   Estimation of the utilities
C
      DO 10 K=1,YEAR
        DO 10 I=1,L
        SUM = 0.
          DO 20 J=1,L
          IF(J.NE.I)SUM = SUM + ALOG(N(I,K)*W(I,J,K)/(N(J,K)*W(J,I,K)))
20        CONTINUE
        U(I,K) = SUM/FLOAT(2*L)
10    CONTINUE
C
C   Estimation of the mobilities
C
      DO 30 K=1,YEAR
        DO 30 I=1,L
          DO 30 J=I+1,L
          NUE(I,J,K) = SQRT(W(I,J,K)*W(J,I,K)/(N(I,K)*N(J,K)))
30    CONTINUE
C
      DO 40 K=1,YEAR
        DO 40 I=1,L
          DO 40 J=I+1,L
          NUE(J,I,K) = NUE(I,J,K)
40    CONTINUE
      RETURN
      END
C
C
C
      SUBROUTINE SMOOTH(W,WS,L,YEAR,WO)
C
C   smoothing of the empirical migration - matrix
```

```
C   W0 : width of the smoothing
C
        INTEGER YEAR
        DIMENSION W(L,L,YEAR),WS(L,L,YEAR)
C
        DO 10 I=1,L
          DO 10 J=1,L
          WM = 0.
            DO 20 K1=1,YEAR
            WM = WM + W(I,J,K1)
20          CONTINUE
            WM = WM/FLOAT(YEAR)
              DO 30 K=1,YEAR
              SUM = 0.
                DO 40 K2=1,YEAR
                CB = 0.
                  DO 50 K3=1-K2,YEAR-K2
                  CB = CB + EXP(-WM*ABS(FLOAT(K3))/W0)
50                CONTINUE
                  CB = 1./CB
                  SUM = SUM + CB*W(I,J,K2)*EXP(-WM*ABS(FLOAT(K-K2))/W0)
40              CONTINUE
                WS(I,J,K) = SUM
30            CONTINUE
10      CONTINUE
        RETURN
        END
C
C
C
        SUBROUTINE VARO(W,N,U,NUE,L,YEAR,S1,S2,S3)
C
C     Subroutine VARO computes the spatial variance of utilities
C     and the correlation between theoretical and empirical migration
C     matrices. The square sum of the residuals is computed for
C     each year too.
C
C
        INTEGER YEAR
        DIMENSION W(L,L,YEAR)    ! Migration - matrix
        DIMENSION N(L,YEAR)      ! population
```

```
      DIMENSION U(L,YEAR)      ! utilities
      DIMENSION NUE(L,L,YEAR) ! mobilities
      DIMENSION S1(YEAR)       ! variance of the utilities
      DIMENSION S2(YEAR)       ! variance of migration - matrixes
      DIMENSION S3(YEAR)       ! variance of residuals
      REAL NUE,N
C
C
      DO 10 K=1,YEAR
      UM = 0.
        DO 20 I=1,L
          UM = UM + U(I,K)
20      CONTINUE
        UM = UM/FLOAT(L)
        SUM = 0.
        DO 30 I=1,L
          SUM = SUM + (U(I,K)-UM)**2
30      CONTINUE
      S1(K) = SUM/FLOAT(L-1)
10    CONTINUE
C
C
C
      DO 40 K=1,YEAR
      WME = 0.
      WMT = 0.
        DO 50 I=1,L
          DO 50 J=1,L
           IF(I.NE.J) THEN
             WME = WME + W(I,J,K)
             WMT = WMT + NUE(I,J,K)*N(J,K)*EXP(U(I,K) - U(J,K))
           ENDIF
50      CONTINUE
      WME = WME/FLOAT(L*(L-1))
      WMT = WMT/FLOAT(L*(L-1))
      S2(K) = 0.
      S3(K) = 0.
      RM = WME - WMT
        DO 60 I=1,L
          DO 60 J=1,L
           IF(I.NE.J) THEN
```

```
            WT = NUE(I,J,K)*N(J,K)*EXP(U(I,K)-U(J,K))
            S3(K) = S3(K) + (W(I,J,K) - WT - RM)**2
            S2(K) = S2(K) + (W(I,J,K) - WME)**2
          ENDIF
60      CONTINUE
40    CONTINUE
C
      RETURN
      END
C
C
C
      SUBROUTINE VAR1(W,N,U,NUE,L,YEAR,S20,S30)
C
C   Subroutine VAR1 computes the total correlation between
C   theoretical and empirical migration matrices.
C
      INTEGER YEAR
C
      DIMENSION W(L,L,YEAR)       ! migration matrix
      DIMENSION N(L,YEAR)         ! population numbers
      DIMENSION U(L,YEAR)         ! utilities
      DIMENSION NUE(L,L,YEAR)     ! total mobilities
      REAL N,NUE
      WME = 0.
      WMT = 0.
      DO 10 K=1,YEAR
        DO 10 I=1,L
          DO 10 J=1,L
           IF(I.NE.J) THEN
             WME = WME + W(I,J,K)
             WMT = WMT + NUE(I,J,K)*N(J,K)*EXP(U(I,K) - U(J,K))
           ENDIF
10    CONTINUE
      WME = WME/(FLOAT(L*(L-1))*FLOAT(YEAR))
      WMT = WMT/(FLOAT(L*(L-1))*FLOAT(YEAR))
      RM = WME - WMT
      S20 = 0.
      S30 = 0.
      DO 20 K = 1, YEAR
        DO 20 I=1,L
```

```
          DO 20 J-1,L
           IF(I.NE.J) THEN
            WT - NUE(I,J,K)*N(J,K)*EXP(U(I,K)-U(J,K))
            S30 - S30 + (W(I,J,K) - WT - RM)**2
            S20 - S20 + (W(I,J,K) - WME)**2
           ENDIF
20      CONTINUE
C
        RETURN
        END
C
C
C
        SUBROUTINE MOBRED(NUE,NUEM,NUEO,L,YEAR)
C
C  Decomposition of the total mobility into a time - dependent
C  global mobility factor and a time - independent deterrence matrix
C
        INTEGER YEAR
C
        DIMENSION NUE(L,L,YEAR)     ! total mobility
        DIMENSION NUEO(YEAR)        ! spatial averaged mobility
        DIMENSION NUEM(L,L)         ! time averaged mobility
        REAL NUE,NUEO,NUEM
        DO 10 K-1,YEAR
          NUEO(K) - 0.
          DO 20 I-1,L
            DO 20 J-I+1,L
            NUEO(K) - NUEO(K) + NUE(I,J,K)
20        CONTINUE
          NUEO(K) - NUEO(K)*2./FLOAT(L*(L-1))
10      CONTINUE
        SUM - 0.
        DO 30 I-1,L
          DO 30 J-I+1,L
          NUEM(I,J) - 0.
            DO 40 K1-1,YEAR
              NUEM(I,J) - NUEM(I,J) + NUE(I,J,K1)/NUEO(K1)
40          CONTINUE
          NUEM(I,J)-NUEM(I,J)/FLOAT(YEAR)
30      CONTINUE
```

```
C
      DO 50 I=1,L
        DO 50 J=1,I-1
          NUEM(I,J) = NUEM(J,I)
50    CONTINUE
      DO 60 I=1,L
        NUEM(I,I) = 0.
60    CONTINUE
C
      RETURN
      END
```

Testing of the Program

In order to demonstrate the working of the program we present the migration data of the Federal Republic of Germany for the years 1982 and 1983. In the following the population numbers (in thousands) as well as the migration matrices of the eleven federal states of the FRG are listed (from [4.1]). The regions are designed by the same numbers as in Chapter 4. An element $w(i, j)$ of the migration matrices denotes the number of migrants from region j to region i. For example in 1982 the number of migrants from region 1 to region 6 is 2807.

Table 16.1. Regional population numbers of the FRG (in thousands)

region	1982	1983
1	2620	2617
2	1631	1617
3	7263	7252
4	689	682
5	17008	16900
6	5607	5584
7	3639	3634
8	9281	9257
9	10963	10965
10	1060	1054
11	1879	1861

Table 16.2a. Migration matrix for 1982

	1	2	3	4	5	6	7	8	9	10	11
1	0	16227	10657	1315	7679	2606	1526	3579	3437	264	2026
2	20290	0	11975	826	3612	1762	698	2026	2306	116	880
3	10712	9150	0	13693	31029	10694	3434	9133	8798	524	5404
4	1462	817	15832	0	2128	860	401	1023	938	62	471
5	10120	4260	35163	2362	0	23153	23817	23257	23335	1970	7236
6	2807	1537	10028	814	17707	0	14229	15788	15035	1280	2813
7	1551	636	3305	413	20082	14735	0	15074	6844	4712	1516
8	3198	1736	7512	857	17208	16346	14864	0	36544	2683	3739
9	2992	1663	6659	721	15459	12928	5640	32801	0	1105	4121
10	312	130	656	76	2820	1851	5698	3214	1665	0	401
11	2932	1162	7011	632	7185	4196	1656	4573	7049	397	0

Table 16.2b. Migration matrix for 1983

	1	2	3	4	5	6	7	8	9	10	11
1	0	16040	10513	962	6605	2089	1491	3059	3379	269	1906
2	20547	0	11161	600	2509	1252	654	1579	1969	92	975
3	10862	8895	0	11581	24665	7869	3164	7201	8550	541	5472
4	1490	705	14773	0	1624	646	343	847	906	65	575
5	9649	3632	32908	1720	0	16081	21398	17799	21675	1963	6362
6	2822	1453	9709	654	14081	0	14772	13570	15050	1387	3143
7	1545	547	3122	308	15526	11792	0	12329	6719	5243	1563
8	3053	1485	6791	633	13011	12798	14193	0	34633	2442	3693
9	2782	1335	6497	584	12730	10288	5303	27455	0	1077	4217
10	344	139	646	47	1918	1360	5665	2695	1869	0	409
11	2261	834	6381	450	4193	2538	1411	3033	5704	342	0

Listing of the Output

In order to give an impression how this program works the listing of the output is added. The following results represent the regional migratory system of the Federal Republic of Germany for the years 1982 and 1983 using the above listed migration data of the FRG.

<u>Output of the estimation program</u>

```
Migration - Analysis of Federal Republic of Germany

****************************************************************************

The Migration datas has been analyzed using
the log - linear estimation procedure
No smoothing of the migration matrix

Number of Regions              :          11
Number of Years                :           2

Tests using the total Mobility Matrix
Squared Correlation            :  0.9985312
Adjusted Squared Correlation   :  0.9963025
Fishers F - Value              :   448.0570

Tests using the reduced Mobilities
Squared Correlation            :  0.9956403
Adjusted Squared Correlation   :  0.9920436
Fishers F - Value              :   276.8188

****************************************************************************
```

spatial deterrence function - time averaged mobility

0.000	6.806	1.895	0.743	0.972	0.517	0.383	0.504	0.452	0.138	0.785
6.806	0.000	2.298	0.536	0.508	0.382	0.201	0.338	0.326	0.069	0.423
1.895	2.298	0.000	4.804	2.136	1.153	0.489	0.717	0.654	0.164	1.265
0.743	0.536	4.804	0.000	0.440	0.291	0.178	0.254	0.218	0.056	0.359
0.972	0.508	2.136	0.440	0.000	1.390	1.965	1.078	1.006	0.390	0.842
0.517	0.382	1.153	0.291	1.390	0.000	2.366	1.561	1.298	0.462	0.744
0.383	0.201	0.489	0.178	1.965	2.366	0.000	1.871	0.745	2.099	0.454
0.504	0.338	0.717	0.254	1.078	1.561	1.871	0.000	2.502	0.677	0.691
0.452	0.326	0.654	0.218	1.006	1.298	0.745	2.502	0.000	0.316	0.877
0.138	0.069	0.164	0.056	0.390	0.462	2.099	0.677	0.316	0.000	0.212
0.785	0.423	1.265	0.359	0.842	0.744	0.454	0.691	0.877	0.212	0.000

Year under consideration : 1982
Spatial averaged mobility : 1.3649918E-03
Variance of the utilities : 0.3045665
Fishers F - value : 492.9634

Region	Utility	Population
1	-0.12009	0.2620E+07
2	-0.43182	0.1631E+07
3	0.36499	0.7263E+07
4	-0.83750	0.6890E+06
5	0.70347	0.1701E+08
6	0.26717	0.5607E+07
7	0.01260	0.3639E+07
8	0.52173	0.9281E+07
9	0.67884	0.1096E+08
10	-0.71864	0.1060E+07
11	-0.44076	0.1879E+07

Year under consideration : 1983

Spatial averaged mobility : 1.2289982E-03

Variance of the utilities : 0.2926963

Fishers F - value : 389.2261

Region	Utility	Population
1	-0.08803	0.2617E+07
2	-0.41149	0.1617E+07
3	0.38828	0.7252E+07
4	-0.93465	0.6820E+06
5	0.64064	0.1690E+08
6	0.15043	0.5584E+07
7	0.03230	0.3634E+07
8	0.48275	0.9257E+07
9	0.71367	0.1097E+08
10	-0.68041	0.1054E+07
11	-0.29348	0.1861E+07

References

Chapter 1

1.1 Weidlich, W., Haag, G. (1983) Concepts and Models of a Quantitative Sociology. Springer, Berlin Heidelberg New York

1.2 Haken, H. (1978) Synergetics. An Introduction, 2nd. ed., Springer, Berlin Heidelberg New York

1.3 Haag, G., Weidlich, W. (1984) A Stochastic Theory of Interregional Migration, Geographical Analysis, vol. 16, no. 4, p. 331

1.4 Weidlich, W., Haag, G. (1980) Migration Behaviour of Mixed Population in a Town, Collective Phenomena, vol. 3, p. 89

1.5 Reiner, R., Munz, M., Haag, G., Weidlich, W. (1986) Chaotic Evolution of Migratory Systems, Sistemi Urbani, 2/3, pp. 285–308

1.6 Sonis, M. (1980) Locational Push-Pull Analysis of Migration Streams, Geographical Analysis, vol. 12, no. 1, p. 80

Chapter 2

2.1 Weidlich, W., Haag, G. (1987) A Dynamic Phase Transition Model for Spatial Agglomeration Processes, Journal of Regional Science, vol. 27, *4*, pp. 529–569

2.2 Haag, G. (1986) A Stochastic Theory on Sudden Urban Growth, in: Konzepte SFB 230, Heft 25, p. 255

2.3 Kanaroglou, P., Liaw, K. L., Papageorgiou, Y. (1985) An Operational Framework for the Analysis of Migratory Systems, QSEP Research Report, no. 136, McMaster University

Chapter 3

3.1 Sen, A., Pruthi, R. K. (1983) Least Squares Calibration of Gravity Model Parameters when Intrazonal Flows are Unknown, Environment and Planning A, 15, pp. 1545–1550

3.2 Sen, A. (1986) Estimation of Parameters for the Poisson Gravity Model, Proceedings of "The Seventh European Advanced Summer Institute in Regional Science", Umea, June 9–27, 1986

3.3 Haag, G., Weidlich, W. (1986) A Dynamic Migration Theory and its Evaluation for Concrete Systems, Regional Science and Urban Economics 16, p. 57

Chapter 4

4.1 Statistische Jahrbücher 1955–1984, Wiesbaden

4.2 INFAS: Räumliche Mobilität-Präferenzen, Motive, Tendenzen, Bonn-Bad Godesberg, Sept. 1972

4.3 Birg, H. (1975) Analyse und Prognose der Bevölkerungsentwicklung in der Bundesrepublik Deutschland und in ihren Regionen bis zum Jahr 1990. Deutsches Institut für Wirtschaftsforschung, Heft 35

4.4 Birg, H., Huinink, J., Koch, H., Vorholt, H. (1984) Kohortenanalytische Darstellung der Geburtenentwicklung in der Bundesrepublik Deutschland, IBS-Materialien Nr. 10, Universität Bielefeld

4.5 Birg, H., Filip, D., Hilge, K. (1983) Verflechtungsanalyse der Bevölkerungsmobilität zwischen den Bundesländern von 1950 bis 1980. IBS-Materialien Nr. 8, Universität Bielefeld

4.6 Herberger, L. (1971) Die Bevölkerung des Bundesgebietes nach der Volkszählung am 27.5.1970. Wirtschaft und Statistik, Heft 12

4.7 Koch, R., Gatzweiler, H.-P. (1980) Migration and Settlement: 9. Federal Republic of Germany. IIASA, RR-80-37, Laxenburg

4.8 Weidlich, W., Haag, G. (1983) Chapter 5 in "Concepts and Models of a Quantitative Sociology", Springer Series in Synergetics

4.9 Bevölkerungsstruktur und Wirtschaftskraft der Bundesländer 1984, Statistisches Bundesamt Wiesbaden, Mainz, 1985

4.10 Bevölkerung und Kultur, Fachserie A, Reihe 6. Statistisches Bundesamt Wiesbaden, Mainz, 1967

4.11 Der Außenhandel der Bundesrepublik Deutschland, Reihe 1. Statistisches Bundesamt Wiesbaden, Mainz

4.12 Finanzen und Steuern, Reihe 5. Sonderbeiträge zur Finanzstatistik, Statistisches Bundesamt Wiesbaden, Mainz

4.13 Statistische Berichte des Statistischen Amtes des Saarlandes, Saarbrücken

4.14 Volkswirtschaftliche Gesamtrechnungen für Berlin, 1984

4.15 Amtliche Nachrichten der Bundesanstalt für Arbeit, Nürnberg

4.16 Beschäftigtenstatistik der Bundesanstalt für Arbeit. Nürnberg

4.17 Zahlen zur wirtschaflichen Entwicklung der Bundesrepublik Deutschland, Ausgabe 1984. Institut der deutschen Wirtschaft, Köln, 1984

4.18 Mensch, G., Kaasch, K., Kleinknecht, A., Schnopp, R. (1980) Innovation Trends, and Switching between Full- and Under-Employment Equilibria, 1950–1978, International Institute of Management, Berlin

4.19 Gerstenberger, W. (1977) Zuviel Rationalisierungsinvestitionen? Ifo Institut München

Chapter 6

6.1 INSEE (1984) Recensement général de la population de 1982. Les Collections de L'INSEE, série D, no 97

6.2 Courgeau, D. (1975) Migrations et découpage du territoire. Population, no 3, pp. 511–537

6.3 Tugault, Y. (1973) La mesure de la mobilité; Cinq études sur les migrations internes. Paris, PUF, 180 p., INED, Travaux et Documents, Cahier no 67

6.4 Courgeau, D., Pumain, D. (1984) Baisse de la mobilité résidentielle. Population et Sociétés, no 179

6.5 Fabre, C., Taffin, C. (1981) Qui a déménagé entre 1973 et 1978, et pourquoi? Economie et Statistique, pp. 33–43

6.6 Pumain, D. (1983) Déconcentration urbaine. Population et Sociétés, no 166

6.7 Laurent, L. (1981) Les migrations interrégionales de population active en France: mesure des effets de quelques facteurs explicatifs. Lille, Actes du Colloque Migrations internes et externes en Europe occidentale, pp. 352–360

6.8 Puig, J. P. (1981) Le rôle des villes dans les migrations d'actifs. Economie et Statistique, no 133, pp. 57–75

6.9 Giard, V. (1974) Emploi et espace, Paris, DATAR, 381 p., Travaux de Recherche et de Prospective, no 49

6.10 Poudoul, J., Paur, J. P. (1986) Trente ans de migrations intérieures, Espace, Population, Société, no 2, pp. 292–302
6.11 Courgeau, D. (1978) Les migrations internes en France de 1954 à 1975, Vue d'ensemble. Population, no 3, pp. 525–545
6.12 Courgeau, D., Lefebvre, M. (1982) Migrations et urbanisation, Population, no 2, pp. 341–370
6.13 Gravier, J. F. (1947) Paris et le désert français. Paris, Le Porfulan, 315 p.
6.14 Vining, D. R., Pallone, R. (1982) Migration between peripheral and core regions: a description and tentative explanation of the patterns in 22 countries. Geoforum, vol. 13, no 4, pp. 339–410
6.15 Pumain, D. (1986) Les migrations interrégionales de 1954 à 1982: Directions préférentielles et effets de barrière. Population, no 2, pp. 378–389
6.16 Poulain, M. (1981) Contribution à l'analyse spatiale d'une matrice de migration interne. Louvain-la-Neuve, Cabay, 225 p., Recherches démographiques, no 3

Chapter 7

7.1 Bachi, R. The Population of Israel, CICRED (World Population Year), Jerusalem
7.2 Central Bureau of Statistics of Israel, *Series of Special Publications*; – Censuses of Population and Housing, 1961, 1972, Jerusalem, 1964, 1976. – Internal migration of Jews in Israel, no. 262, 342, 499, 721. – Supplements to the Monthly Bulletin of Statistics, vol. XXVIII, no. 3, vol. XXX. no. 8, 1979
7.3 Lipshitz, G. (1986) Internal migration and regional development – Theoretical framework and trends in Israel, *Geografiska Annaler*
7.4 Lowry, I. S. (1966) *Migration and Metropolitan Growth*, Chandler, San Francisco
7.5 Greenwood, M. J. (1981) *Migration and Economic Growth in the USA*; *national, regional metropolitan perspectives*, Academic Press, New York
7.6 Gonen, A., Hasson, S. (1975) A centripetal pattern of intra-urban mobility in Israel medium-sized towns, *Geografiska Annaler*, vol. 57, pp. 55–62
7.7 Shachar, A. (1986) Urban ecology in the Israeli metropolitan areas, In: *Studies in Population of Israel, Scripta Hierosolymitana*, vol. XXX, pp. 381–414
7.8 Sonis, M. (1980) Push-Pull analysis of migration streams, *Geographical Analysis*, vol. 12, no. 1, pp. 80–97
7.9 Tobler, W. (1981) A model of geographical movement, *Geographical Analysis*, vol. 13, pp. 1–20

Chapter 8

8.1 Vitali, O. (1978) La crisi italiana: il problema della popolazione. Franco Angeli, Milano
8.2 Campisi, D., La Bella, A., Rabino, G. (1982) Migration and settlement: 17. Italy, RR-82-33, I.I.A.S.A., Laxenburg
8.3 Rogers, A. (1975) Introduction to multiregional mathematical demography, Wiley, New York
8.4 Rabino, G. (1984) L'evoluzione demografica del Piemonte: scenari al 1989, Rapporto di Ricerca I.R.E.S., Torino
8.5 Golini, A. (1987) La popolazione dell'Italia e del Piemonte alla soglie del XXI secolo, Rapporto di Ricerca I.R.E.S; Torino
8.6 Rogers, A., Raquillet, R., Castro, L. (1977) Model migration schedules and their application, RR-77-57, I.I.A.S.A., Laxenburg
8.7 Rabino, G. (1987) Migrazioni ed evoluzione della struttura socio-economica e territoriale: parte 1. I movimenti migratori in Italia, Rapporto di Ricerca I.R.E.S., Torino
8.8 I.S.T.A.T., Popolazione e movimento anagrafico dei comuni, ISTAT, Roma, several years
8.9 I.S.T.A.T., Compendio statistico italiano, ISTAT, several years
8.10 I.S.T.A.T., Annuario di statistiche del lavoro, ISTAT, several years
8.11 I.S.T.A.T., Annuario di contabilità nazionale, ISTAT, several years

8.12 Lombardo, S. T., Rabino, G. A. (1986) A compartmental analysis of residential mobility in Turin, Sistemi Urbani, vol. VIII, n. 2/3, p. 263
8.13 Haag, G., (1984) working paper, University of Stuttgart
8.14 Bagnasco, A. (1977) Tre Italie: la problematica territoriale dello sviluppo italiano, Il mulino, Bologna
8.15 C. E. E., Sistema europeo dei conti economici integrati – SEC, Istituto statistico delle Communità, Europee Lussemburgo, 1981

Chapter 9

9.1 Dickinson, R. E, (1964) City and region. A geographical interpretation, Routledge & Kegan, London
9.2 Richardson, H. W. (1969) Regional Economics, Weidenfeld & Nicolson, London
9.3 Persson-Tanimura, I. (1980) Studier kring arbetsmarknad och information, Lund Econ. Studies, no 19
9.4 Holmlund, B. (1975) "The Behaviour of Unemployment and Vacancies". Oxford Bul. of Econ. and Stat.
9.5 Andersson, Å. E. and Mantsinen, J. (1980) "Mobility of Resources: Accessibility of Knowledge and Economic Growth". Behavioural Science, Vol 25
9.6 SOU 1980: 52 Langtidsutredningen 1980, Stockholm
9.7 Sääski, N. (1980) "Förändrigar i arbetsmarknadsens struktur och funktionsätt" NU A 1980:5
9.8 Graves, P. E. and Linneman, P. D. (1979) "Household Migration: Theoretical and Empirical Results". Journal of Urban Economics, 6
9.9 Kmenta, J. (1971) Elements of Econometrics, Macmillan, New York
9.10 Holmlund, B. (1984) Labour Mobility, IUI, Stockholm

Chapter 10

10.1 Schultz, P. (1973) A preliminary survey of economic analyses of fertility, *American Economic Review*, vol. 63
10.2 Andersson, Å. E. Lundquist, L. (1976) Regional analysis of consumption patterns, *Papers of the Regional Science Association*, vol. 36
10.3 Nikaido, H. (1970) Introduction to Sets and Mappings in Modern Economics, North Holland
10.4 Rogers, A. (Ed.), (1980) *Essays in Multistate Mathematical Demography*, IIASA RR-80–10
10.5 Hotelling, H. (1978) A Mathematical Theory of Migration, republished in *Environment and Planning*, vol. A10
10.6 Puu, T. (1984) A Simplified Model of Spatio-Temporal Population Dynamics, *Umea Economic Studies*, no. 139
10.7 Andersson, Å., E., Ferraro, G. (1983) Accesibility and density distributions in metropolitan areas: Theory and empirical studies, *Papers of the Regional Science Association*, vol. 52
10.8 Andersson, Å., E., Karlquist, A. (1976) Population and Capital in Geographical Space, Los, J., Los, M., (Eds.), *Computing Equilibria: How and Why*, North Holland
10.9 Beckmann, M. J. (1976) Spatial equilibrium in the dispersed city, *Mathematical Land Use Theory*, Papageorgiou, G., J., (Ed.), Lexington Books
10.10 Leonardi, G., Casti, J. (1986) Agglomerative tendencies in the distribution of populations, *Regional Science and Urban Economics*, vol. 16
10.11 McFadden, D. (1978) Modeling the choice of residential location, *Spatial Interaction Theory and Planning Models*, Karlquist et. al. (Eds.), North Holland
10.12 Malinvand, E. (1977) The Theory of Unemployment Reconsidered, Blackwell
10.13 Keyfitz, N. (1968) Introduction to the Mathematics of Population, Addison Wesley
10.14 Haag, G., Weidlich, W. (1986) A dynamic migration theory and its evaluation of concrete systems, *Regional Science and Urban Economics*, vol. 16

10.15 Andersson, Å., E., Philipov, D. (1982) Economic models of migration, *Regional Development Modeling: Theory and Practice*, Albegov, M., Andersson, Å., E., Snickars, F., (Eds.), North Holland

Chapter 11

11.1 H. Birg (1986) Regionale Demographie und Regionalwissenschaftliche Analyse, IBS-Materialien Nr. 21

Chapter 12

12.1 Stratonovich, R. L. (1963 and 1967) Topics in the Theory of Random Noise. Vols. 1 and 2, Gordon and Breach, New York
12.2 Bharucha-Reid, A. T. (1960) Elements of the Theory of Markov Processes and Their Applications, McGraw-Hill, New York
12.3 Wax, N. (ed.) (1954) Selected Papers on Noise and Stochastic Processes. Dover, New York
12.4 Gardiner, C. W. (1983) Handbook of Stochastic Methods for Physics, Chemistry and the Natural Sciences, Springer-Series in Synergetics, Vol. 13. Springer, Berlin Heidelberg New York
12.5 Van Kampen (1981) Stochastic Processes in Physics and Chemistry, North-Holland, Amsterdam

Chapter 13

13.1 Weidlich, W. (1978) On the Structure of Exact Solutions of Discrete Master Equations. Zeitschrift für Physik *B30*, 345
13.2 Haken, H. (1977) Synergetics, An Introduction, 2nd ed., Springer, Series in Synergetics, Vol. 1. Springer, Berlin Heidelberg New York
13.3 Haag, G., Weidlich, W. (1984) A Stochastic Theory of Interregional Migration. Geographical Analysis *16*, 331
13.4 Weidlich, W., Haag, G. (1987) A Dynamic Phase Transition Model for Spatial Agglomeration Processes. Journal of Regional Science
13.5 Haag, G. (1978) Transition Factor Method for Discrete Master Equations; and Application to Chemical Reactions. Zeitschrift für Physik *29*, 153

Chapter 14

14.1 Granger, C. W. J., Newbold, P. (1977) Forecasting Economic Time Series. Academic Press
14.2 Sachs, L. (1974) Angewandte Statistik. Springer, Berlin Heidelberg New York
14.3 Pindyck, R. S., Rubinfeld, D., L. (1985) Econometric Models and Economic Forecasts, McGraw-Hill
14.4 van der Waerden, B. L. (1969) Mathematical Statistics, Die Grundlehren der mathematischen Wissenschaften in Einzeldarstellungen Band 156, Springer, Berlin Heidelberg New York
14.5 Durbin, J., Watson, G. S. (1951) Biometrika, vol. 38, 159

MIX
Papier aus verantwortungsvollen Quellen
Paper from responsible sources
FSC® C105338

If you have any concerns about our products,
you can contact us on
ProductSafety@springernature.com

In case Publisher is established outside the EU,
the EU authorized representative is:
Springer Nature Customer Service Center GmbH
Europaplatz 3, 69115 Heidelberg, Germany

Printed by Libri Plureos GmbH
in Hamburg, Germany